Biodiversity

BIODIVERSITY
A Biology of Numbers and Difference

EDITED BY KEVIN J. GASTON

Royal Society University Research Fellow
Department of Animal and Plant Sciences
University of Sheffield

Blackwell Science

© 1996 by
Blackwell Science Ltd
Editorial Offices:
Osney Mead, Oxford OX2 0EL
25 John Street, London WCIN 2BL
23 Ainslie Place, Edinburgh EH3 6AJ
238 Main Street, Cambridge
  Massachusetts 02142, USA
54 University Street, Carlton
  Victoria 3053, Australia

Other Editorial Offices:
Arnette Blackwell SA
  224, Boulevard Saint Germain
  75007 Paris, France

Blackwell Wissenschafts-Verlag GmbH
  Kurfürstendamm 57
  10707 Berlin, Germany

  Zehetnergasse 6, A-1140 Wien
  Austria

First published 1996

Set by Excel Typesetters, Hong Kong
Printed and bound in Great Britain
at the University Press, Cambridge

The Blackwell Science logo is a
trade mark of Blackwell Science Ltd,
registered at the United Kingdom
Trade Marks Registry

DISTRIBUTORS

  Marston Book Services Ltd
  PO Box 269
  Abingdon
  Oxon OX14 4YN
  (*Orders*: Tel: 01235 465500
            Fax: 01235 465555)

USA
  Blackwell Science, Inc.
  238 Main Street
  Cambridge, MA 02142
  (*Orders*: Tel: 800 215-1000
            617 876-7000
            Fax: 617 492-5263)

Canada
  Copp Clark, Ltd
  2775 Matheson Blvd East
  Mississauga, Ontario
  Canada, L4W 4P7
  (*Orders*: Tel: 800 263-4374
            905 238-6074)

Australia
  Blackwell Science Pty Ltd
  54 University Street
  Carlton, Victoria 3053
  (*Orders*: Tel: 03 9347 0300
            Fax: 03 9349 3016)

A catalogue record for this title
is available from the British Library

ISBN 0–86542–804–2

Library of Congress
Cataloging-in-Publication Data

Biodiversity: a biology of numbers and
difference / edited by Kevin J. Gaston.
      p.   cm.
  Includes bibliographical references
  and index.
  ISBN 0–86542–804–2
  1. Biological diversity.   2. Biological
  diversity conservation
  I. Gaston, Kevin J.
  QH75.B523   1996
  333.95'11–dc20
                        95-20924
                        CIP

Contents

8708344

Colour plates 2.1, 2.2, 8.1, 8.2, 14.1 and 14.2 appear between pp. 214 and 215

List of contributors

TIMOTHY G. BARRACLOUGH *Department of Zoology, University of Oxford, South Parks Road, Oxford OX1 3PS, UK*

BRUNO BAUR *Conservation Biology Research Group (NLU), University of Basel, St Johanns-Vorstadt 10, CH-4056 Basel, Switzerland*

TRACY L. BENNING *Department of Biological Sciences, Stanford University, Stanford, CA 94305, USA and ESPM, 108 Hilgard Hall, University of California, Berkeley, CA 94720, USA*

SCOTT L. COLLINS *Ecological Studies Program. Rm 635, National Science Foundation, Arlington, VA 22230, USA and Department of Zoology, University of Maryland, College Park, MD 20742, USA*

JAMES A. DRAKE *Complex Systems Group, Department of Ecology and Evolutionary Biology, University of Tennessee, Knoxville, TN 37996, USA*

MADHAV GADGIL *Centre for Ecological Sciences, Indian Institute of Science, Bangalore, 560012, India, and Biodiversity Unit, Jawaharlal Nehru Centre for Advanced Scientific Research, Jakkur, Bangalore 560064, India*

KEVIN J. GASTON *Department of Animal and Plant Sciences, University of Sheffield, Sheffield S10 2TN, UK*

PAUL H. HARVEY *Department of Zoology, University of Oxford, South Parks Road, Oxford OX1 3PS, UK*

CHAD L. HEWITT *Complex Systems Group, Department of Ecology and Evolutionary Biology, University of Tennessee, Knoxville, TN 37996, USA*

CHRIS J. HUMPHRIES *Biogeography and Conservation Laboratory, The Natural History Museum, Cromwell Road, London SW7 5BD, UK*

List of contributors

GARY R. HUXEL *Complex Systems Group, Department of Ecology and Evolutionary Biology, University of Tennessee, Knoxville, TN 37996, USA*

JUREK KOLASA *Department of Biology, McMaster University, Hamilton, Ontario L8S 4K1, Canada*

WILLIAM E. KUNIN *NERC Centre for Population Biology, Imperial College at Silwood Park, Ascot, Berkshire SL5 7PY, UK and Department of Biology, University of Leeds, Leeds LS2 9JT, UK*

JOHN H. LAWTON *NERC Centre for Population Biology, Imperial College at Silwood Park, Ascot, Berkshire SL5 7PY, UK*

RIK LEEMANS *Department of Terrestrial Ecology and Global Change, National Institute of Public Health and Environmental Protection, RIVM, PO Box 1, 3720 BA Bilthoven, The Netherlands*

JAMES MALLET *Galton Laboratory, Department of Biology, 4 Stephenson Way, London NW1 2HE, UK*

NEO D. MARTINEZ *Bodega Marine Laboratory, University of California at Davis, Bodega Bay, CA 94923–0247, USA*

SEAN NEE *Department of Zoology, University of Oxford, South Parks Road, Oxford OX1 3PS, UK*

BERNHARD SCHMID *Institut für Umweltwissenschaften, Universität Zürich, Winterthurerstrasse 190, CH-8057 Zürich, Switzerland*

R.I. VANE-WRIGHT *Biogeography and Conservation Laboratory, The Natural History Museum, Cromwell Road, London SW7 5BD, UK*

PAUL H. WILLIAMS *Biogeography and Conservation Laboratory, The Natural History Museum, Cromwell Road, London SW7 5BD, UK*

Preface

As a topic of scientific investigation, biodiversity is a point at which many traditional fields of study blurr into and inform one another. As a rallying cry for conservation, it is a point at which species-, habitat-, and ecosystem-based concerns meet, and at which biological, economic and socially driven motivations merge. As a measure of the variety of the natural world, it is a point at which humankind commits itself to an on-going programme of destruction, or an enlightened programme of preservation.

The principal objective of this volume is to explore aspects of the first of these intersections. Each chapter takes as its theme a separate area of the investigation of biodiversity and reviews some of the main conceptual issues surrounding it and our present understanding of those issues. There has been no attempt to make the coverage comprehensive. It is, nonetheless, wide-ranging.

There are a number of ways in which a volume of this kind could be organized. None seems entirely satisfactory. Following an opening chapter (Chapter 1) which explores what is meant by 'biodiversity', the approach taken here has been to divide the topic into three broad areas. The first of these is concerned with the measurement of biodiversity, with chapters addressing this topic from the perspectives of genetic diversity (Chapter 2), character diversity (Chapter 3), species diversity (Chapter 4), functional diversity (Chapter 5) and higher levels of organization (Chapter 6). The second part of the book is concerned with spatial and temporal patterns in biodiversity. The principal themes of the chapters echo those of the previous section. Thus, there are chapters addressing patterns in genetic diversity (Chapter 7), taxonomic diversity (Chapters 8 and 9), and functional diversity (Chapter 10). Conservation provides the core topic for the third part of the book, it being by far the most significant applied dimension to the study of biodiversity. The chapters of this section tackle four major linked issues: the case for conservation (Chapter 11), how to prioritize for conservation action (Chapter 12), how to manage for biodiversity (Chapter 13), and the effects on biodiversity of global environmental change (Chapter 14).

Edited volumes are by definition collaborative affairs, and I would like to express my gratitude to the contributors for their enthusiastic support. Likewise, I thank the numerous reviewers who shared insights and viewpoints, as well as sometimes obscure pieces of knowledge. It has, as always, been a pleasure to work with Simon Rallison and Susan Sternberg of Blackwell Science.

Kevin J. Gaston

1: What is biodiversity?

KEVIN J. GASTON

1.1 Introduction

Bandwagon, buzz-word, growth industry, global resource, issue and phenomenon. Responses to, and perceptions of, biodiversity (a contraction of 'biological diversity' or 'biotic diversity') are anything but uniform. At the heart of why this is so lies the issue of what biodiversity actually is. More authors express bewilderment or proffer weak descriptions than seek to provide robust and closely reasoned definitions. No doubt this is, at least in part, a response to two things. First, the fact that 'biodiversity' is a pseudocognate term in that many users assume that everyone shares the same intuitive definition (Williams, 1993). Second, the variety of viewpoints about biodiversity which have developed. These viewpoints broadly can be grouped under three headings: those which regard biodiversity as a concept; those which regard it as a measurable entity; and those which regard it as predominantly a social or political construct.

1.2 Biodiversity as a concept

The most prevalent usage of the term 'biodiversity' is as a synonym for the 'variety of life' (for discussions of the origins of the term see Wilson & Peter, 1988; Shetler, 1991; Harper & Hawksworth, 1994; Norse, 1994). Indeed, many published definitions are simply expressions of, embellishments to, or expansions on this basic theme. They often emphasize the multiple dimensions and levels at which this variety, diversity or heterogeneity can be observed (Table 1.1; McAllister, 1991; Gaston, 1994). The definition given by the US Congress Office of Technology Assessment (OTA, 1987) is perhaps the most widely cited, and opens thus: 'Biological diversity refers to the variety and variability among living organisms and the ecological complexes in which they occur. . . .' (See Table 1.1 for a fuller statement.) As such, biodiversity is essentially an abstract concept, albeit one of which most of us would say we had some intuitive understanding. However, expressed as the 'variety of life', the breadth of the concept of biodiversity is not simply wide, but is so wide as in fact to be exceedingly difficult to comprehend. It amounts to the irreducible complexity of the totality of life (Williams *et al.*, 1994). One

1

Table 1.1 Definitions of 'biological diversity' and 'biodiversity'.

'Biological diversity refers to the variety and variability among living organisms and the ecological complexes in which they occur. Diversity can be defined as the number of different items and their relative frequency. For biological diversity, these items are organized at many levels, ranging from complete ecosystems to the chemical structures that are the molecular basis of heredity. Thus, the term encompasses different ecosystems, species, genes, and their relative abundance.' (OTA, 1987)

'Biodiversity is the variety of the world's organisms, including their genetic diversity and the assemblages they form. It is the blanket term for the natural biological wealth that undergirds human life and well-being. The breadth of the concept reflects the interrelatedness of genes, species and ecosystems.' (Reid & Miller, 1989)

'"Biological diversity" encompasses all species of plants, animals, and microorganisms and the ecosystems and ecological processes of which they are parts. It is an umbrella term for the degree of nature's variety, including both the number and frequency of ecosystems, species, or genes in a given assemblage.' (McNeely *et al.*, 1990)

'Biodiversity is the genetic, taxonomic and ecosystem variety in living organisms of a given area, environment, ecosystem or the whole planet.' (McAllister, 1991)

'Biodiversity is the total variety of life on earth. It includes all genes, species and ecosystems and the ecological processes of which they are part.' (ICBP, 1992)

'Biological diversity (=Biodiversity). Full range of variety and variability within and among living organisms, their associations, and habitat-oriented ecological complexes. Term encompasses ecosystem, species, and landscape as well as intraspecific (genetic) levels of diversity.' (Fiedler & Jain, 1992)

'[biodiversity] The variety of organisms considered at all levels, from genetic variants belonging to the same species through arrays of species to arrays of genera, families, and still higher taxonomic levels; includes the variety of ecosystems, which comprise both the communities of organisms within particular habitats and the physical conditions under which they live.' (Wilson, 1992)

'"Biological diversity" means the variability among living organisms from all sources including, *inter alia*, terrestrial, marine and other aquatic ecosystems and the ecological complexes of which they are part; this includes diversity within species, between species and of ecosystems.' (The Convention on Biological Diversity; Johnson, 1993)

'. . . biodiversity – the structural and functional variety of life forms at genetic, population, species, community, and ecosystem levels . . .' (Sandlund *et al.*, 1993)

suspects that one consequence of this breadth has been the expression of dissatisfaction with the concept of biodiversity on the grounds that it is both imprecise, and that it runs the danger of being defined so broadly that it equates to the whole of biology.

A number of schemes have been suggested by which the major features of biodiversity can be distinguished and some sense made of what constitutes the 'variety of life'. Three, increasingly conventional, divisions are between genetic diversity, species or taxonomic diversity and ecosystem diversity (e.g. McAllister, 1991; Solbrig, 1991; Stuart & Adams, 1991; Groombridge, 1992; Heywood, 1994; Norse, 1994), for which Harper and Hawksworth (1994) favour the terms genetic, organismal and ecological diversity. In the context of conservation strategies, Soulé (1991) distinguishes five divisions: genes; populations; species; assemblages (associations and communities); and whole systems at the landscape or ecosystem level. Another classification distinguishes three interdependent sets of attributes, compositional levels (the identity and variety of elements), structural levels (the physical organization or pattern of elements) and functional levels (ecological and evolutionary processes) (Noss, 1990).

Whatever the scheme, emphasis is usually given to a hierarchical perspective of biological phenomena, either by basing the classification on different levels of a proposed hierarchy (e.g. genes, species, ecosystems), or by recognizing that such levels exist within each division of the classification (e.g. Noss, 1990). Although schemes for recognizing different components of biodiversity may perhaps be helpful in facilitating a grasp of its breadth, it must be remembered that often some of the divisions employed by such schemes have no or only limited reality, or have somewhat blurred edges. Predominantly, they serve as convenient human constructs.

Some debate exists over the limits to the basic concept of biodiversity. The primary of these concerns the extent to which the concept embraces processes (e.g. energy and nutrient flow) rather than simply entities (e.g. individual organisms, taxa, habitat types). Most definitions of biodiversity lay explicit emphasis primarily on entities, and appear to pay little direct heed to process. It may perhaps be argued, however, that they implicitly embrace process in two ways: first, by recognizing that one dimension of biodiversity is the variety of functions some entities (e.g. species) perform; second, through the inclusion of ecosystems as a component of biodiversity. Nonetheless, it bears emphasis that in many discussions (as opposed to definitions) biodiversity is taken more explicitly to include the support systems on which life-forms depend (Western, 1992).

The notion of biodiversity as a wide-ranging abstract concept has led to the strongly interdisciplinary flavour to its study. Work on biodiversity impinges on, amongst others, the fields of biogeography, botany, conservation biology, ecology, genetics, palaeontology, systematics, and zoology. This interdisciplinary character has particularly served to highlight questions which have tended to remain peripheral to the research agendas of more traditional individual fields of study (e.g. how many species are there?, in what way do taxonomic diversity and ecosystem processes interact?, which biological entities are functionally most important?).

1.3 Biodiversity as a measurable entity

Haila and Kouki (1994) document a veritable explosion in the numbers of publications applying the term 'biodiversity' over the period 1984–1992 (see also Harper & Hawksworth, 1994). In major part this is a consequence of the belief that biodiversity is not simply an abstract concept but a measurable entity. That is, the concept can in some sense be made operational.

The desirability of measuring biodiversity has been stated repeatedly, principally on the grounds that it enables broad questions raised by the concept of biodiversity to be subjected to rigorous empirical enquiry (Solbrig, 1991; Harper & Hawksworth, 1994). Here, it is important to make a fundamental distinction between two ideas. First, that biodiversity can be quantified, and second that different facets or dimensions of biodiversity can be quantified. Many studies are written in such a way as to imply that their findings, although based only on single variables, concern biodiversity *per se*. However, the abstract concept of biodiversity as the 'variety of life', expressed across a range of hierarchical scales, cannot be encapsulated in a single variable. The complexity is in this sense irreducible, and the search for the all-embracing measure of biodiversity, however desirable it might seem, will be a fruitless one. As Norton (1994) writes: 'it appears that scientists can offer a very large number of possible "diversity measures", but that these measures cannot be aggregated into a unique measure of *the* diversity of the system'.

The choice and derivation of a measure of biodiversity will depend fundamentally on the use to which it will be put. As Williams and Humphries (Chapter 3) discuss, in choosing the particular aspect of biodiversity to be measured, one is placing a value on that aspect for the purposes of the exercise in hand. Measures based on concepts of 'option value' for conservation may not be appropriate for consideration of the ecology of interactions between biodiversity and nutrient cycling.

Most explicit attempts to quantify some dimension of biodiversity have concerned primarily genetic and specific levels of the hierarchy of biological phenomena. Indeed, a large number of papers open with the recognition that species richness is only one measure of biodiversity but proceed to treat it as if it were *the* measure of biodiversity. This is typified by the emphasis placed on species extinction in discussion of biodiversity loss and the biodiversity crisis, often at the expense of habitat loss and particularly local population loss and decline (Ehrlich & Daily, 1993; Ehrlich, 1994). Other levels included in definitions of biodiversity, such as ecosystems and landscapes, have received rather little attention. This perhaps reflects a lack of clarity as to what biodiversity is at these levels, and debate as to the definition of the entities themselves (for disscussion of the definition of 'ecosystem', for example, see O'Neill *et al.*, 1986; Lidicker, 1988; Cousins, 1990; Golley, 1991, 1993; Shrader-Frechette & McCoy, 1993). Where these 'higher' levels have purportedly been the objects of study, biodiversity has in fact usually been quantified in terms of the genetic or specific diversity which individual higher level units contain (e.g. species within ecosystems or landscapes).

The measurement of diversity is complex. Broadly, there are two kinds of measures, those which simply count entities and those which additionally attempt to incorporate some elements of their difference [Hurlbert's (1971) proposal, in the context of species diversity, that only the latter be termed diversity measures has largely gone unheeded]. There are numerous ways particularly of incorporating difference, these themselves differ in the relative weighting given to the two components and the manner in which this is done. Much of what we know about diversity measures in biology was garnered in the 1960s and early 1970s in the exploration of ecological measures of diversity, derived originally from information theory, which combined number of species and the evenness or equability of their abundances (see Grassle *et al.*, 1979; Magurran, 1988 for overviews). These indices have essentially fallen out of favour, usually being dependent on sample size and sampling pattern, and often difficult to interpret. Whether developing these approaches or entirely novel ones, fresh ways of measuring diversity need to be considered in the context of biodiversity. Some, typically associated with taxa, have been discussed (Chapter 3, *this volume*; Cousins, 1991; Vane-Wright *et al.*, 1991; Wayne & Bazzaz, 1991; Faith, 1994; French, 1994; Williams *et al.*, 1994).

The current high level of interest in the measurement of biodiversity serves to highlight the comparative youth of the subject. A science can be viewed as passing through three stages as it matures, the What? stage, the How? stage and the Why? stage (Wiegert, 1988). Biodiversity remains emphatically at the What? stage, with much discussion of issues of measurement and pattern, and comparatively little of issues of mechanism.

1.4 Biodiversity as a social/political construct

Use of the term 'biodiversity' arose in the context of, and has remained firmly wedded to, concerns over the loss of the natural environment and its contents. The importance of this connection cannot be understated. As a result, the term has gained a wide currency. Few other words can have entered the accepted vocabularies of science, the media, and the public arena (in parts at least) on such a large geographic scale with such rapidity (although amongst the general public, awareness of the term 'biodiversity' and understanding of its meaning is far less than has occasionally been claimed). There is general acceptance in many communities that biodiversity is *per se* a good thing, that its loss is bad, and hence that something should be done to maintain it. Indeed, in some sense these connections have been carried to the point where the term 'biodiversity' itself is seen to embody concepts not only of the variety of life, but additionally of the importance of that variety, of the crisis represented by its loss, and of the need for conservation action. Bowman's (1993) description of biodiversity as a synonym for 'nature conservation' represents one, perhaps rather extreme, manifestation of this situation. In this wider arena, biodiversity is not a neutral scientific concept, biodiversity is perceived as a value (often an option value), or as having a value.

The connection made between biodiversity and conservation raises a number of issues, of which I want to touch on three. First, there is the magnitude and the nature of the 'biodiversity crisis'. Doubtless, it is complex to answer the question of what proportion of the natural world has and will be destroyed or degraded, where, on what time scale, and with what consequences. Nonetheless, what we do know of past, current and projected rates of habitat change, of specific and subspecific (e.g. local population) extinction, and of human population growth and development, provides the basis for some profoundly disturbing conclusions (e.g. Simberloff, 1986; Ehrlich & Wilson, 1991; McNeely, 1992; Myers, 1993a, 1993b; Bongaarts, 1994; Ehrlich, 1994; Houghton, 1994; Gaston, 1995). Whilst existing evidence and the severity of the prognosis should of themselves be sufficient to engender action, much more information is required to optimize responses to the situation. This provides the climate in which biodiversity studies are performed, regardless of the motives underlying their instigation.

Second, if we return to the view that biodiversity is an abstract concept, only facets of which can be made operational and measured, how can the goals of, and success at, conserving biodiversity be quantified? Much discussion has concerned identification of the measure, expression or consequences of diversity (e.g. species numbers, hotspots, evolutionary potential, ecosystem health) around which conservation action should be centred and to what end (e.g. Walker, 1992; Franklin, 1993; Angermeier & Karr, 1994; Norton, 1994; Vogler & DeSalle, 1994; Williams *et al.*, 1994). Some consensus seems to be crystallizing on the overall significance of maintaining ecosystem integrity and function. This gives rather greater emphasis to process than is immediately evident from many definitions of biodiversity (Table 1.1), and means that different conservation strategies may be appropriate than would be the case were the goal primarily to preserve entities (e.g. species) (Western, 1992).

The final issue follows from the second, and concerns the connection between the biodiversity of an area and the significance placed on its conservation. It is important to distinguish carefully between quantity and priority. Areas high in biodiversity (e.g. species numbers, phylogenetic disparity) need not be of the highest priority for conservation action, because priorities must embrace other considerations, such as level of threat and contribution to a broad conservation goal. For example, the potential value of an area to overall plant biodiversity conservation may depend on which plant species that area contains, not how many. Likewise, distinction has to be made between levels of biodiversity which can be regarded as native (naturally evolved) and as artificial (human generated – as a result of changes by such actions as gene transfer, introduction of exotic species, and landscape fragmentation) and between levels which include and exclude vagrants and tourists, and the possible limitations of conservation policies directed to maximizing biodiversity must be recognized (see also Angermeier, 1994; Norton, 1994; Hambler & Speight, 1995).

1.5 Conclusion

'Biodiversity' is one of many terms which, whilst applied widely in biology, remain inconsistently or at times inadequately defined. However, the problem is perhaps amplified in this case, because the term has usage in other spheres (legal, media, political, etc.). It will be difficult, perhaps impossible, to put an end to the element of confusion which results. Nonetheless, whether viewed as an abstract biological concept which can or cannot be operationalized, or as something scientifically less neutral, it has served to draw attention to and provide fresh perspectives on a number of scientific concepts. These provide the backdrop to the rest of this book.

Acknowledgments

I am grateful to Paul Williams for much stimulating discussion. Hefin Jones, Natasha Loder and Paul Williams kindly commented on the manuscript.

References

Angermeier P.L. (1994) Does biodiversity include artificial diversity? *Conserv. Biol.* **8**, 600–602. [1.4]

Angermeier P.L. & Karr J.R. (1994) Biological integrity versus biological diversity as policy directives. *BioScience* **44**, 690–697. [1.4]

Bongaarts J. (1994) Population policy in the developing world. *Science* **263**, 771–776. [1.4]

Bowman D.M.J.S. (1993) Biodiversity: much more than biological inventory. *Biodiv. Lett.* **1**, 163. [1.4]

Cousins S.H. (1990) Countable ecosystems deriving from a new food web entity. *Oikos* **57**, 270–275. [1.3]

Cousins S.H. (1991) Species diversity measurement: choosing the right index. *Trends Ecol. Evol.* **6**, 190–192. [1.3]

Ehrlich P.R. (1994) Energy use and biodiversity loss. *Phil. Trans. Roy. Soc., Lond. B* **344**, 99–104. [1.3] [1.4]

Ehrlich P.R. & Daily G.C. (1993) Population extinction and saving biodiversity. *Ambio* **22**, 64–68. [1.3]

Ehrlich P.R. & Wilson E.O. (1991) Biodiversity studies: science and policy. *Science* **253**, 758–762. [1.4]

Faith D.P. (1994) Phylogenetic pattern and the quantification of organismal biodiversity. *Phil. Trans. Roy. Soc., Lond. B* **345**, 45–58. [1.3]

Fiedler P.L. & Jain S.K. (eds) (1992) *Conservation Biology: the theory and practice of nature conservation, preservation and management.* Chapman and Hall, New York. [1.2]

Franklin J.F. (1993) Preserving biodiversity: species, ecosystems, or landscapes? *Ecol. Appl.* **3**, 202–205. [1.4]

French D.D. (1994) Hierarchical richness index (HRI): a simple procedure for scoring 'richness', for use with grouped data. *Biol. Conserv.* **69**, 207–212. [1.3]

Gaston K.J. (1994) Biodiversity — measurement. *Prog. Phys. Geog.* **18**, 565–574. [1.2]

Gaston K.J. (1995) Biodiversity—loss. *Prog. Phys. Geog.*, **19**, 255–264. [1.4]

Golley F.B. (1991) The ecosystem concept: a search for order. *Ecol. Res.* **6**, 129–138. [1.3]

Golley F.B. (1993) *A History of the Ecosystem Concept in Ecology: more than the sum of the parts.* Yale University Press, New Haven. [1.3]

Grassle J.F., Patil G.P., Smith W. & Taillie C. (eds) (1979) *Ecological Diversity in Theory and Practice.* International Co-operative Publishing House, Fairland, Maryland. [1.3]

Groombridge B. (ed.) (1992) *Global Biodiversity: status of the Earth's living resources.* Chapman & Hall, London. [1.2]

Haila Y. & Kouki J. (1994) The phenomenon of biodiversity in conservation biology. *Ann. Zool. Fennici* **31**, 5–18. [1.3]

Hambler C. & Speight M.R. (1995) Biodiversity conservation in Britain: science replacing tradition. *British Wildlife* **6**, 137–147. [1.4]

Harper J.L. & Hawksworth D.L. (1994) Biodiversity: measurement and estimation. Preface. *Phil. Trans. Roy. Soc., Lond. B* **345**, 5–12. [1.2]

Heywood V.H. (1994) The measurement of biodiversity and the politics of implementation. In: *Systematics and Conservation Evaluation* (eds P.L. Forey, C.J. Humphries & R.I. Vane-Wright), pp. 15–22. Oxford University Press. [1.2]

Houghton R.A. (1994) The worldwide extent of land-use change. *BioScience* **44**, 305–313. [1.4]

Hurlbert S.H. (1971) The nonconcept of species diversity: a critique and alternative parameters. *Ecology* **52**, 577–586. [1.3]

ICBP (1992) *Putting Biodiversity on the Map: priority areas for global conservation.* ICBP (BirdLife International), Cambridge. [1.2]

Johnson S.P. (1993) *The Earth Summit: the United Nations Conference on Environment and Development (UNCED).* Graham and Trotman, London. [1.2]

Lidicker Jr. W.Z. (1988) The synergistic effects of reductionist and holistic approaches in animal ecology. *Oikos* **53**, 278–281. [1.3]

McAllister D.E. (1991) What is biodiversity? *Can. Biodiv.* **1**, 4–6. [1.2]

McNeely J.A. (1992) The sinking ark: pollution and the worldwide loss of biodiversity. *Biodiv. Conserv.* **1**, 2–18. [1.4]

McNeely J.A., Miller K.R, Reid W.V., Mittermeier R.A. & Werner T.B. (1990) *Conserving the World's Biodiversity.* IUCN, WRI, CI, WWF and World Bank, Washington DC. [1.2]

Magurran A.E. (1988) *Ecological Diversity and its Measurement.* Croom Helm, London. [1.3]

Myers N. (1993a) Biodiversity and the precautionary principle. *Ambio* **22**, 74–79. [1.4]

Myers N. (1993b) Population, environment, and development. *Environ. Conserv.* **20**, 205–216. [1.4]

Norse E.A. (ed.) (1994) *Global Marine Biological Diversity: a strategy for building conservation into decision making.* Island Press, Washington DC. [1.2]

Norton B.G. (1994) On what we should save: the role of culture in determining conservation targets. In: *Systematics and Conservation Evaluation* (eds P.L. Forey, C.J. Humphries & R.I. Vane-Wright), pp. 23–29. Oxford University Press. [1.3] [1.4]

Noss R.F. (1990) Indicators for monitoring biodiversity: a hierarchical approach. *Conserv. Biol.* **4**, 355–364. [1.2]

O'Neill R.V., DeAngelis D.L., Waide J.B. & Allen T.F.H. (1986) *A Hierarchical Concept of Ecosystems.* Princeton University Press, New Jersey. [1.3]

OTA (US Congress Office of Technology Assessment) (1987) *Technologies to Maintain Biological Diversity.* US Government Printing Office, Washington DC. [1.2]

Reid W.V. & Miller K.R. (1989) *Keeping Options Alive: the scientific basis for conserving biodiversity.* World Research Institute, Washington. [1.2]

Sandlund O.T., Hindar K. & Brown A.H.D. (eds) (1992) *Conservation of Biodiversity for Sustainable Development.* Scandinavian University Press, Oslo. [1.2]

Shetler S.G. (1991) Biological diversity: are we asking the right questions? In: *The Unity of Evolutionary Biology: Proceedings of the Fourth International Congress of Systematic and Evolutionary Biology* (2 vols) (ed. E.C. Dudley). Dioscorides Press, Portland. [1.2]

Shrader-Frechette K.S. & McCoy E.D. (1993) *Method in Ecology: strategies for conservation.* Cambridge University Press. [1.3]

Simberloff D.S. (1986) Are we on the verge of a mass extinction in tropical rainforests? In: *Dynamics of Extinction* (ed. D.K. Elliot), pp. 165–180. Wiley, New York. [1.4]

Solbrig O.T. (ed.) (1991) *From Genes to Ecosystems: a research agenda for biodiversity.* IUBS, Cambridge, Massachusetts. [1.2]

Soulé M.E. (1991) Conservation tactics for a constant crisis. *Science* **253**, 744–749. [1.2]

Stuart S.N. & Adams R.J. (1991) *Biodiversity in sub-Saharan Africa and its Islands.* World Conservation Union, Gland, Switzerland. [1.2]

Vane-Wright R.I., Humphries C.J. & Williams P.H. (1991) What to protect? Systematics and the agony of choice. *Biol. Conserv.* **55**, 235–254. [1.3]

Vogler A.P. & DeSalle R. (1994) Diagnosing units of conservation management. *Conserv. Biol.* **8**, 354–363. [1.4]

Walker B.H. (1992) Biodiversity and ecological redundancy. *Conserv. Biol.* **6**, 18–23. [1.4]

Wayne P.M. & Bazzaz F.A. (1991) Assessing diversity in plant communities: the importance of within-species variation. *Trends Ecol. Evol.* **6**, 400–404. [1.3]

Western D. (1992) The biodiversity crisis: a challenge for biology. *Oikos* **63**, 29–38. [1.2] [1.4]

Wiegert R.G. (1988) Holism and reductionism in ecology: hypotheses, scale and systems. *Oikos* **53**, 267–269. [1.3]

Williams P.H. (1993) Choosing conservation areas: using taxonomy to measure more of biodiversity. In: *Manus. Col. ISBC KEI* (ed. T.-Y. Moon), pp. 194–227. Korean Entomological Institute, Seoul. [1.1]

Williams P.H. Gaston K.J. & Humphries C.J. (1994) Do conservationists and molecular biologists value differences between organisms in the same way? *Biodiv. Lett.* **2**, 67–78. [1.2] [1.4]

Wilson E.O. (1992) *The Diversity of Life.* Allen Lane/The Penguin Press, London. [1.2]

Wilson E.O. & Peter F.M. (eds) (1988) *Biodiversity.* National Academy Press, Washington DC. [1.2]

PART 1: MEASURING BIODIVERSITY

2: The genetics of biological diversity: from varieties to species

JAMES MALLET

2.1 Measurement of genetic diversity

2.1.1 What is biodiversity?

Biodiversity consists of the variety of morphology, behaviour, physiology, and biochemistry in living things. Underlying this phenotypic diversity is a diversity of genetic blueprints, nucleic acids that specify phenotypes and direct their development. Biodiversity is often used as a synonym for species diversity, but I argue in this chapter that the species level is not clearly demarcated in the diversity either of blueprints or of phenotypes. Within species, we can find ecological interactions expected between closely related species, such as competition, parasitism (selfish behaviour), and even predation (cannibalism). Communication, cooperation and sex, normally thought of as within-species phenomena, also occur to some extent between species. Morphological divergence between populations and species is continuous, and differs quantitatively rather than qualitatively. Indeed Darwin's (1859) realization that varieties formed a continuum with species led him to his revolutionary solution to the species problem: evolution by natural selection. Because the diversity of life is fundamentally genetic, rather than a matter of species counts, we can use a variety of genetic methods to investigate diversity both within and between species. As I shall show, Darwin's insight about species is little altered by modern genetic studies; the diversity of blueprints, rather than their arrangement into taxonomic units, is perhaps the most important level of consideration for biodiversity studies. On the other hand, genetic discontinuities between individual pairs of species undoubtedly exist, and genetics provides superior tools for revealing them.

2.1.2 What do we measure?

There is a functional hierarchy from blueprint to phenotype. Genes consist of coding DNA, rarely RNA, and are interconnected on long chains, or chromosomes. Genes specify proteins, and proteins form the basis of physiology, development, appearance, and behaviour of organisms. Genetic variation can be examined at any of these functional levels. Cutting across this functional hierarchy there is a classification hierarchy, of classes of structural entities within which genetic material interacts: genes, chromosomes, cells, organisms, local populations, races, species, and higher taxa. Baur and Schmid (Chapter 7) call this a hierarchy of 'structural levels of organization'. Although some argue that the gene is the only true unit of selection (Dawkins, 1976, 1982) and that the species is the only 'real' taxon (Ghiselin, 1975), all levels of classification are important in evolution. Genetic diversity is measurable at any functional level from blueprint to phenotype; and we may be interested in this diversity at any classification level from populations to higher taxa.

In this chapter, I discuss how functional genetic traits – morphology, chromosomes, enzyme loci and DNA – can be sampled to measure diversity and how this diversity might evolve. For each genetic level, I explore the classification levels – local populations, metapopulations, races and species – at which genetic diversity can be measured. Finally, I give examples of the applications of genetic diversity with which I am familiar within local populations, between populations, and in the definition of closely related species.

2.2 The classification hierarchy of genetic diversity

Groups of individuals can be classified into local populations, meta-populations, races, and species. By *local populations* are usually meant groups containing individuals that have a high probability of mixing or recombining with each other (i.e. are *sympatric*), and have genotypes in approximate Hardy–Weinberg equilibrium (i.e. are *panmictic*; see Box 2.1a for measures of deviation from Hardy–Weinberg). Local populations can be defined by Wright's (1943, 1969) 'neighbourhood population size', $N_b = 4\pi\sigma_x^2 d$, consisting of individuals at density d within a circle of radius $2\sigma_x$, twice the standard deviation of the distance along one axis, x, between the birthplaces of parent and offspring (see also Chapter 7). Assuming dispersal is Gaussian, one neighbourhood will contain 86.5% of the progeny of a parent giving birth in the centre. The use of a standard deviation as opposed to the mean radial parent–offspring distance, $\bar{r}$, comes from diffusion theory (see also Chapter 7); σ_x can be estimated from measurements of $\bar{r}$ because, assuming a Gaussian distribution,

$$\bar{r} = \sqrt{\frac{\pi}{2}}\sigma_x \quad \text{and} \quad \overline{r^2} = 2\sigma x^2$$

This *gene flow* distance σ_x is proportional to the minimum patch width over which adaptation is likely, and to the width of clines maintained by selection (Endler, 1977; Barton & Gale, 1993), as well as the spatial scale of random

---

Box 2.1 Measures of genetic diversity.

(a) Genetic diversity within populations
Numbers of alleles at a single locus in a sample or population (richness):

n_a

Frequency of the ith allele in a population (evenness):

$\pi_i = f_i/2N$ (N is the total population size; f_i is the number of the ith allele in the population)

Heterozygosity or allelic diversity at a single locus (richness and evenness):

$H_{dc} = \dfrac{1}{N} \displaystyle\sum_{i<j\leq n_a} f_{ij}$ (frequency of heterozygotes, based on direct counts, f_{ij} of heterozygotes with alleles i,j within a population)

$H_e = 1 - \displaystyle\sum_{i=1}^{n_a} \pi_i^2$ (i.e. expected heterozygosity on basis of Hardy–Weinberg)

Deviation from Hardy–Weinberg, *I*nbreeding within a *S*ubpopulation (*is*):

$F_{is} = \dfrac{H_e - H_{dc}}{H_e}$

Deviation from two-locus (*A, B*) equilibrium, gametic or linkage disequilibrium:

$D_{AB} = \Pi_{AB} - \pi_A\pi_B$ (deviation between actual frequency of AB gametes, Π_{AB}, and that expected on basis of random assortment, $\pi_A\pi_B$)

$R_{AB} = \dfrac{D_{AB}}{\sqrt{\pi_A\pi_B(1-\pi_A)(1-\pi_B)}}$ (correlation coefficient of disequilibrium between alleles at two loci A and B)

(b) Genetic diversity between subpopulations in a metapopulation
Deviation from Hardy–Weinberg, inbreeding due to *S*ubdivision relative to *T*otal population (*st*), standardized variance of allele frequencies, or proportion of total allelic diversity found between populations:

$F_{st} = \dfrac{H_{e,m} - \overline{H}_e}{H_{e,m}}$

($H_{e,m}$ is the H_e based on overall allele frequencies in the entire metapopulation, and $\overline{H}_e$ is the average of H_e over subpopulations)

---

gene frequency fluctuations or *genetic drift* (Wright, 1951; Slatkin & Barton, 1989).

Populations of populations, or *metapopulations*, consist of groups of local populations loosely connected via occasional gene flow over distances $\gg\sigma_x$ so that extinct populations are not immediately recolonized. *Geographic races* consist of genetically similar geographic metapopulations or genotypic

15

clusters that differ sharply from other races at one or more genetic elements (human 'races' are not geographic races in this sense; see Marks, 1994). Races may be geographically isolated (*allopatric*) or connected at their boundaries (*parapatric*), and are often named as *subspecies*. In areas where races abut, narrow *hybrid zones* are formed between them (Hewitt, 1988; Harrison, 1993). Finally, *species* are similar to and form a continuum with races, though genetic differences are usually somewhat greater. If two species overlap, intermediates are rare or absent, and do not approach the number or variety expected under random mating in the absence of selection. See Section 2.4.3 for a genotypic definition.

2.3 The functional hierarchy of genetic diversity

In this section, I summarize the main types of variation, their evolutionary dynamics, and their uses in biodiversity studies. I structure this discussion from phenotype to blueprint, rather than building up from the fundamental level of DNA, both for historical reasons, and because detectable diversity typically increases in this direction.

2.3.1 Morphological variation

Morphological variation is usually a good clue to genetic variation, but the two do not correlate perfectly. Much morphological variation, though genetically programmed, depends on seasonal or maturity changes that affect many individuals in a population regardless of genotype. A glance at any bird book will give good examples of winter and breeding plumage, particularly in males. More extreme are larvae, pupae and adults of insects and other invertebrates; stage-specific expression differs strongly to produce radically different morphologies and ecologies within single genotypes. Finally, there are 'plastic' phenotypes in which growth form depends on the environment, especially in plants (Stearns, 1989; Schmid, 1992). The usually adaptive set of phenotypes produced by a genotype in different environments are collectively known as the norm of reaction of that genotype (Gabriel & Lynch, 1992). Phenotypic plasticity is also common in many animals; for example, eyespots in the satyrid butterfly genus *Bicyclus* are smaller in the dry season, when predation is more intense (Brakefield & Larsen, 1984); age and size at maturity in guppies and many other fish depends strongly on nutrition (Reznick, 1990).

Genetically based morphological diversity can be continuous or discrete; and may be determined by many or few genes together with some input from the environment. Genes affecting morphology are among the most interesting because morphology is often under strong selection. Such genes strongly affect an organism's tendency to photosynthesize, to eat or be eaten, to compete, or to signal sexually, socially, and to other species. In many real cases of adaptation, small numbers of loci have major effects on traits such as resistance to chemicals, parasites, diseases, or predators (e.g. Fig. 2.1; see also Macnair, 1991), although there are good theoretical reasons for expecting that most adaptation should be gradual and polygenic (Fisher, 1958; Barton & Turelli, 1989). In Batesian mimicry, polymorphisms within palatable populations can only be maintained if there are no intermediates between mimetic forms, requiring single, dominant genes (or collections of tightly linked genes known as 'supergenes') that switch between different mimetic patterns; non-mimetic intermediates are selected against (Plate 2.1, facing p. 214; Clarke *et al.*, 1968; Clarke & Sheppard, 1971; Turner, 1984). In this case, there is strong selection for major gene inheritance. But in other cases of major gene inheritance, for example, monomorphic Müllerian mimicry, there is no selection for single gene inheritance because populations are not normally polymorphic. Instead, the simple inheritance of *Heliconius* racial diversity at a handful of loci is most plausibly explained by the existence of selectively advantageous loci that affect wing colour in major ways (Fig. 2.1 & Plate 2.2; Turner, 1984; Mallet, 1989; Nijhout, 1991).

Complex or continuous traits, such as height, speed, or overall fitness,

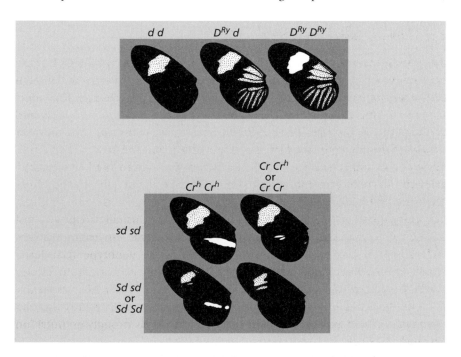

Fig. 2.1 Action of three loci affecting the Müllerian mimetic colour pattern of *Heliconius erato emma* and *H. e. favorinus* in a hybrid zone in the Huallaga valley, N.E. Peru. D^{Ry} changes the forewing band from red to yellow, and adds orange forewing basal patches and hindwing rays. *Cr* and *Sd* interact to remove the yellow hindwing bar found in $Cr^h Cr^h sd sd$ genotypes, and *Sd* also changes the shape of the forewing band from broad to narrow (after Mallet *et al.*, 1990). Races are shown in Plate 2.2; hybrid index in Fig. 2.3.

cannot be studied as easily as discrete differences inherited at single genes. The method of quantitative genetics allows a partitioning of the total phenotypic variance ($V_P = V_G + V_E$) of a quantitative trait consisting of additive genetic variance (V_G) plus environmental and non-additive genetic variance (V_E), to give an estimate of heritability, $h^2 = V_G/V_P$, which varies between 0 and 1. Heritabilities can be estimated from phenotypic correlations between relatives (Falconer, 1989). Heritability is useful because, in spite of ignorance of the underlying inheritance, it predicts the approximate success of selection on any continuous trait. For example, supposing the heritability of pig weight is 0.75, if pigs 12 kg larger than the population mean are selected as parents, their progeny would be expected to attain weights about 12×0.75 (= 9 kg) more than the mean. These approximations work provided that the environment (V_E) remains relatively constant, and that many independent genes, each of individually small effect, act on the selected trait.

Within populations

Polymorphisms in morphological characters have been much studied by ecological geneticists, but it is perhaps not generally realized how extremely common they are. The most obvious example is the existence of separate sexes, or dioecy, in most higher animals and many plants. Other examples are polymorphisms in flower colour, mating types such as pin and thrum flowers in the hermaphroditic primrose *Primula veris*, banding patterns in the snail *Cepaea nemoralis*, jaw morphology affecting feeding habits in the cichlid genus *Cichlasoma*, the blue and 'snow' phase of the snow goose *Anser caerulescens*, black and brown phases of the brown bear *Ursus arctos*, and polymorphic Batesian mimicry in females of butterflies such as *Papilio memnon* (Plate 2.1). These forms are usually referred to as 'varieties' (still a valid taxonomic category for botanists, but in zoology now used mainly by amateurs) or 'phases' in birds and mammals.

All organisms are variable morphologically; we can gauge the prevalence of discrete polymorphisms by counting listed colour-pattern polymorphisms in field guides. Excluding rare aberrations, seasonal, maturity, and sexual phases, about 4% of over 600 resident North American birds have clearly marked polymorphisms (based on Robbins *et al.*, 1983). The amount of polymorphism is highly taxon dependent: 10 out of 11 British grasshoppers (Acrididae) have presumably genetic colour varieties, and each species has an average of 5.1 varieties; in contrast, only half the 10 British bush-crickets (Tettigoniidae) are polymorphic, and none has more than two varieties, giving 1.5 varieties on average per species (Ragge, 1965). Eleven per cent of 63 British butterflies have more than one common morph or major variation (Thomas & Lewington, 1991), and many British moths are highly variable, both as larvae and as adults (Skinner & Wilson, 1984). It is normally accepted, but by no means proved, that most of this morphological diversity is maintained by frequency-dependent balancing, or 'apostatic' selection (Poulton, 1890; Clarke, 1962). For example, predators can form searching images of common, palatable insect prey (Tinbergen, 1960), leading to an advantage for rare phenotypes.

There is also plenty of heritable variation in quantitative traits such as height, weight or size. Some of the best examples of quantitative inheritance come from agricultural experiments; for instance, selection over 76 years caused an increase in oil content in corn of nearly fourfold, from 5% to 19% (Dudley, 1977). On average, yield in agricultural crops and domestic animals improves at a rate of 1% or more per year. In contrast, the winning times of racehorses, which have an extremely narrow genetic base, have improved at less than 0.1% per year (Hill, 1988). Fewer studies have been done in natural populations, but published reports show similarly high heritabilities. For instance, heritability of beak characters under strong selection in the Darwin's finch *Geospiza fortis* were beak length $h^2 = 0.65$, beak depth $h^2 = 0.79$, beak width $h^2 = 0.90$ (Grant, 1986). Heritabilities of male traits under directional sexual selection include: call length in field crickets *Gryllus integer* $h^2 = 0.7$ (Hedrick, 1986, 1988) and tail length in the male barn swallow *Hirundo rustica* $h^2 = 0.59$ (Møller, 1994). These values are as high as that for chicken egg weight, $h^2 = 0.55$ (Falconer, 1989). Why such high levels of genetic variation of strongly selected traits are maintained in the wild is very poorly understood, because V_G is expected to decline as genes for suboptimal traits are purged. Various possibilities for the maintenance of quantitative variation are under discussion: frequency-dependent or other balancing selection, high mutation rates in quantitative traits, antagonistic genetic correlations, and spatial variation in selection (Barton & Turelli, 1989). This paradox of the maintenance of quantitative genetic variation strongly affects debates on conservation (see Section 2.4.1).

Between populations and races

Visible morphological differences are commoner between than within populations. Good examples are the melanic forms of otherwise pale moths found around many industrial British cities (Kettlewell, 1973; Bishop *et al.*, 1976). The tree *Eucalyptus urnigera* has glaucous, waxy leaves which resist freezing above 1000 m and bright green leaves below 800 m; between the two there is a narrow cline of glaucousness (Thomas & Barber, 1974). In other cases, variation consists of broad clines in traits such as morphology of plants (Clausen *et al.*, 1947) or skin colour in humans.

Divergent morphological races found over wide geographic areas are often taxonomically described as subspecies. Interracial variation leads to high levels of intrapopulation variation only in narrow hybrid zones where subspecies or races meet, consisting of sharp clines at several characters. Some examples of subspecific variation are the shape of penis-like intromittent organs in woodlice *Oniscus asellus* (Bilton, 1994), colour patterns in *Limenitis* (Platt, 1983) and *Heliconius* butterflies (Plate 2.2; Brown *et al.*, 1974; Mallet, 1993), toads of the genus *Bombina* (Szymura & Barton, 1991), and birds such as the toucans (Ramphastidae; Haffer, 1974), and flickers (*Colaptes auratus*; Moore & Price, 1993). Racial variability depends on the taxon examined, but it is certainly frequent in many groups (Hewitt, 1988). Each of the world's 8600 known bird species has an average of over three subspecies considered valid, with many species having 10 or more (Mayr, 1970). Of the 65 heliconiine butterflies in tropical America, 74% have more than one subspecies considered valid; on average there are 8.6 subspecies per species (Brown, 1979). In contrast, there is virtually no geographic variation in morphology of neotropical hairstreak butterflies (Theclinae; R. Robbins, pers. comm., 1994). This is also true for better known thecline faunas: the 17 European species have only about 1.2 named subspecies, while 75 North American theclines have only 1.5 subspecies per species (Scott, 1986; Higgins & Riley, 1993). These morphological subspecies counts probably underestimate racial variation, some of which must be due to cryptic factors like chromosomal rearrangements (White, 1978; and see below). Morphological differences between populations are often due to quantitative variation between populations, such as in height and life history in plants (Clausen *et al.*, 1947) or skin colour in humans. Racial variation as well as this may also be due to some single genes of major effect as in Fig. 2.1 and Plate 2.2.

Between species

Morphological differences between individuals or populations form a continuum with morphological differences between species. Good examples are found in herons and egrets (Ardeidae): some species are polymorphic for white and dark phases; in others, white or dark colouring is fixed. Some good species can be crossed, allowing estimation of the numbers of genes affecting morphological differences. For example, Val (1977), Templeton (1977) and Lande (1981) found that there were at least 6–8 genes (probably more) controlling the difference in sexually selected head shape between the sympatric *Drosophila heteroneura* and *D. silvestris* of Hawaii.

2.3.2 Other phenotypic variation

Above I have used variation in external morphology as an example of phenotypic diversity, but there are many equally important behavioural, physiological or biochemical phenotypic traits which also vary between individuals, populations, and species. Examples include phenology, secondary biochemistry (see Appendix 2.1) and adaptations to soil conditions in plants, and behaviour such as host plant choice in insects. These traits are often continuous and polygenic, such as the hatching phenology of the true fruit fly *Rhagoletis pomonella* adapted to particular hosts (Smith, 1988), but single genes once again seem common, e.g. in host choice and seasonality in the lacewing genus *Chrysopa* (Tauber & Tauber, 1977), and in female pheromone blend and male response in moths (Löfstedt, 1993).

2.3.3 Karyotypic variation

The chromosomal arrangement of genes (or karyotype; Appendix 2.1) is often under strong selection. This is not usually because of direct effects on the phenotype, although there is some evidence that the positions of genes may affect their function, but because rearrangements affect the mechanics of meiosis and recombination (Dobzhansky, 1937; White, 1978; King, 1993). Chromosomal variation consists of rearrangements such as inversions, translocations (including centric fissions and fusions of nearly complete chromosomes) and polyploidy. Rearrangements often cause chromosome breakage, or duplications and deletions in the meiotic products of heterozygotes (heterokaryotypes) leading to unbalanced gametes and reduced fertility. This produces purifying selection against heterokaryotypes which leads to eventual fixation of one rearrangement within populations.

Within populations

In spite of the likelihood of purifying selection, chromosomal polymorphisms are fairly common within species. Dobzhansky (1937) found many *Drosophila* species to be polymorphic for paracentric (single arm) inversions. Crossing-over between breakpoints in inversion heterokaryotypes would be expected to generate unbalanced gametes and reduce fitness, leading to selection for homozygosity. However, in males there is no crossing-over at all, and in females unbalanced products of meiosis are shunted into polar bodies rather than eggs; inversions in *Drosophila* act as cross-over suppressors which reduce fertility very little, if at all. Inversion polymorphisms in *D. pseudoobscura* are maintained by heterozygote advantage (heterosis), often with very strong selection against homokaryotypes (~90%; Dobzhansky, 1937). Equilibrium frequencies of polymorphic forms can depend on altitude, season and latitude, and on temperature in the laboratory. Dobzhansky (e.g. 1970) believed that inversions trapped beneficial combinations of coadapted genes involved in heterosis and temperature adaptation, and that inversions are favoured because they preserve these adaptive combinations. However, as Maynard Smith (1989: p. 68) points out, it is still unknown whether inversions trap some single loci that are themselves heterotic, or whether heterosis at the chromosomal level is caused by different deleterious recessive mutations accumulating on each inversion type. If the latter, inversions could be pathological in that they prevent the recombination-led purging of deleterious alleles. Strong chromosomal heterosis also results in virtually lethal homokaryotypes in evening primroses (*Oenothera*; see Dobzhansky, 1937), frogs (Schmidt, 1993), newts (Macgregor & Horner, 1980), and *Anopheles* mosquitoes (Seawright *et al.*, 1991).

Between populations and species

Where there is selection against heterokaryotypes, local populations should lack chromosomal variation. Chromosomal diversity is thus expected mostly between rather than within populations. In nature, chromosomal polymorphism does indeed seem rarer than geographic variation between chromosomal races in diverse organisms such as flowering plants (Grant, 1981), grasshoppers (White, 1978) and mammals (Searle, 1993). When chromosomal races abut, they often form stable narrow clines or hybrid zones maintained by selection against heterokaryotypes (Bazykin, 1969; Hewitt, 1988; Barton & Gale, 1993).

The partial meiotic sterility of heterokaryotypes is reminiscent of sterility between species, and chromosomal evolution is therefore often linked to speciation. White (1978) proposed that chromosomal evolution, initiated by genetic drift or biases in meiosis, could trigger speciation directly (see also Bush *et al.*, 1977; King, 1993). Species often differ in chromosomal morphology: for example, humans differ from chimpanzees by 16 pericentric inversions (i.e. involving the centromere) and one chromosomal fusion (Yunis & Prakash, 1982). However, White's 'stasipatric' chromosomal speciation theory and its derivatives are controversial because it seems that speciation is rarely due to a single chromosomal change. The existence of many chromosomal hybrid zones in which polymorphic forms are effectively panmictic shows that instant speciation is not a normal result of chromosomal rearrangement. Most evolutionists now believe that chromosomal evolution contributes to sterility between species, but that genic sterility is more important. There is, however, one exception: polyploidy. When newly formed tetraploids hybridize with ancestral diploids, triploid hybrids, which are usually sterile, are produced, so that chromosome doubling can lead to instant speciation. In organisms such as flowering plants, where selfing is possible, speciation often does result from single, or even several parallel, mutations to polyploidy (Grant, 1981).

2.3.4 Soluble enzyme loci

In the 1960s, techniques for revealing the genetic diversity of proteins became available (Harris, 1966; Appendix 2.1). It was possible for the first time (apart from immunological studies on proteins, for instance on human blood groups) to assess polymorphisms in protein sequence, a direct translation of DNA sequence.

Within populations

An enormous amount of protein diversity was quickly found in natural populations, whether the studies were done in humans, vertebrates, invertebrates, or flowering plants. Two explanations seemed possible (Lewontin, 1974): (i) some pointed out that if the alleles were effectively neutral, polymorphism would result because stochastic losses of alleles due to genetic drift in large populations could be balanced by gains from new mutation; and (ii) others believed that this enormous variation was maintained by means of balancing selection. The debate is still not fully resolved. Certain patterns fit neutrality: for instance bigger enzymes have more variation, as might be expected if larger enzymes have more neutral sites (Ward *et al.*, 1992). But strong variations in the rate of base pair and amino acid substitution along different lineages indicate that non-random processes, presumably selection, are responsible for some of this evolution (Gillespie, 1991). Molecular patterns of polymorphism at maybe the best-known enzyme locus, alcohol dehydrogenase *(Adh)* in *Drosophila melanogaster*, show that it is almost certainly selected (Kreitman & Hudson, 1991). Regardless of the exact causes of diversity at allozyme loci, they are extremely useful genetic markers.

Between populations and species

In contrast to morphology and karyotype, soluble enzymes usually have a high proportion of variability within populations compared to that between populations (Lewontin, 1974). However, in hybrid zones between morphological or karyotypic races, there are usually corresponding allozyme differences, showing that strong genetic barriers often exist: 20% of enzymes tested showed fixed or nearly fixed differences across these zones (Barton & Hewitt, 1983). These genetic findings confirm that races and subspecies, at least when justified using several characters, may be evolutionarily significant entities close to the level of species (Avise & Ball, 1990; Moritz, 1994), contrary to what had been assumed (Wilson & Brown, 1953; Mayr, 1970). Although allozyme differences are usually somewhat greater between species than between subspecies, there is plenty of overlap as we go from local populations to subspecies to full species (Ayala *et al.*, 1974; Emelianov *et al.*, 1995).

2.3.5 DNA markers

The ability to investigate DNA sequences directly became available to population biologists only during the late 1970s. Recent advances such as the polymerase chain reaction (PCR) have put molecular techniques within the abilities and budgets of population biologists. We can now use an enormous variety of tools to assess sequence variation in nuclear or organelle DNA (Appendix 2.1). With the automation of data collection made possible by robotic PCR and DNA sequencing, data is now being generated more rapidly than it can be analysed.

Within populations

As we proceed from the phenotype down to the level of the DNA sequence, so the resolution, or amount of detectable diversity within populations, increases enormously. For instance, there are no visible morphological effects of the fast and slow *Adh* alleles in *Drosophila melanogaster*, and only a single amino acid difference affects the charge on the protein; however, a small sample of 11 alleles showed 98 polymorphic sites in the 4800 bp region around *Adh*, of which 36 are silent mutations (i.e. do not alter amino acid sequence) within the 1900 bp *Adh* locus itself; the only amino acid difference is generated by a single base pair change (Kreitman & Hudson, 1991). Some of this copious variation must be effectively neutral; especially in non-coding DNA outside the gene or in introns, or at synonymous (silent) differences at the third base pairs of codons in exons.

From a practical point of view, the high resolution of molecular techniques means that they are often useful where allozymes fail. For example, Queller *et al.* (1993) needed polymorphic genetic markers to give relatedness estimates in behavioural studies of social wasps. Unfortunately, social insects have notoriously low genetic diversity at enzyme loci. Microsatellites (2–5 base pair repeats; Appendix 2.1), which proved highly polymorphic, were used instead. In studies of paternity in birds and mammals, hypervariable minisatellite or 'fingerprint' loci have been used for the same reason (Appendix 2.1; Burke & Bruford, 1987; Bruford & Wayne, 1994). The only drawback is that most variation at the DNA level is so remote from the phenotypic diversity in which we are actually interested that molecular techniques often do little more than provide genetic markers.

Between populations

Patterns of DNA differences between populations and species will be generated by the same processes (drift, selection, mutation and gene flow) as those seen at allozymes, so we would expect similar results. But this argument may be incorrect. *Drosophila melanogaster* worldwide has very similar allozyme diversity, suggesting that gene flow is global. However, nuclear DNA shows much higher levels of diversity in Africa, the probable site of origin of *D. melanogaster*, than in other parts of the world (Begun & Aquadro, 1993). The mtDNA of *D. melanogaster* also has more local variation in Africa than elsewhere (Singh & Hale, 1994). One possible explanation is that protein polymorphisms are much more strictly selected than hitherto imagined, whereas DNA markers show a strong genetic diversity gradient set up by the original emigration from Africa. It is important to know whether these results are general, and, if so, to understand why such discrepancies exist between markers. More studies investigating joint patterns of DNA and allozyme variation in this and other species would be valuable.

Between species

DNA differences between species are similar to, although usually more marked than, allozyme differences. DNA gives abundant, high resolution markers which are especially useful in distinguishing sibling species or in estimating phylogenies (see e.g. Avise, 1994; Moritz, 1994).

2.4 Examples of analyses of genetic diversity in natural populations

2.4.1 Within populations

Measurement

As with species diversity, there are three components to genetic diversity: the numbers (richness) of alleles, their relative abundance (evenness), and their distinctiveness. The distinctiveness, or genetic divergence of alleles due to evolutionary genealogy is a fascinating, and growing area of evolutionary study with many potential uses in biodiversity and conservation (Hudson, 1990; Milligan *et al.*, 1994). Here I concentrate on the first two measures. The simplest measure of diversity at a gene is the total number of alleles, but new mutants arise in every generation, so the lower limit to the frequency of an allele in a diploid is given by unique alleles, with frequency $1/2N$, where N is the population size. With new mutations in large populations, there are likely to be many such very rare alleles. It is therefore virtually impossible to estimate the number of alleles with reasonable sample sizes. A more complete measure of allelic diversity is the heterozygosity, H, which incorporates both number and frequencies of alleles, and which stabilizes with reasonable sample sizes. The heterozygosity at a locus can be measured either from direct counts of heterozygotes (H_{dc}) or as the heterozygosity expected under the Hardy–Weinberg law of random union of gametes (H_e – see Box 2.1a). H_e is in fact identical to Simpson's (1949) measure of species diversity in community ecology. In most local populations, we can by the neighbourhood definition (see Section 2.2) assume random mating, at least with respect to random marker loci. With random mating H_{dc} and H_e will be approximately equivalent, but the latter, which has a smaller variance since the sample size of alleles is twice that of genotypes, is normally used as an estimate of genetic diversity.

However, deviations from Hardy–Weinberg are frequently found, and there are various options for dealing with them. If the data are molecular or allozymic, the first thing to worry about is scoring errors. These are common when the genetics of allozymes have not been carefully investigated (see Mallet *et al.* (1993) for documentation of examples in *Heliothis* moths). Otherwise the pattern might be caused by selection, inbreeding, or statistical artifacts. If there is systematic inbreeding, for example partial selfing in plants, the deviation from Hardy–Weinberg can be interpreted as an inbreeding coefficient (Box 2.1a) and should affect all loci approximately equally. If the deviation is caused by selection, it will vary between loci. In practice, it is hard to measure small selection coefficients, and consistent patterns of inbreeding are rare, at least in animals. In outbreeding sexual diploids, many deviations from Hardy–Weinberg are probably artifacts due to lumping divergent populations, small sample sizes, or scoring difficulties.

We expect a random association of alleles between genes in the same way that we expect genotype frequencies at single loci to obey the Hardy–Weinberg law. The expected frequency of the gametic type *AB*, π_{AB}, is given by the combined probability of A and of B, which is $\pi_A\pi_B$, i.e. the product of the gene frequencies. A deviation from random association is called gametic (or 'linkage') disequilibrium, D_{AB}: this can be standardized to give a correlation coefficient between loci, R_{AB} (Box 2.1a). Gametic disequilibria are found in natural populations when the rate of accumulation of associations via selection or drift is large compared with their destruction by recombination. Gametic disequilibria are more likely between tightly linked loci; but if selection is intense, they will form even between unlinked genes. Supergenes in Batesian mimicry (Plate 2.1) and the association of particular allozymes with crossover-inhibiting inversions in *Drosophila* are examples of gametic disequilibria in natural populations (Hedrick *et al.*, 1978; Maynard Smith, 1989). Disequilibria also occur when there is gene flow between divergent populations. Disequilibria produced are proportional to gene flow, with the useful result that gene flow between two partially isolated populations, or along a cline, can be estimated from disequilibria (Barton & Gale, 1993).

Theory

In small diploid populations, allelic diversity will decline due to drift by about $1/2N$ per generation on average, where N is the population size (Wright, 1931). The expected heterozygosity after t generations is therefore $H_t \approx H_0(1 - 1/[2N])^t$, where H_0 is the initial heterozygosity. For large populations, this equation shows that drift will have very little effect, but in fact the approximation assumes an idealized population of hermaphrodites mating randomly, including with themselves, and with Poisson-distributed family

26

sizes. For real populations, we need to substitute an effective size, N_e, the idealized population size in the model that gives the same rate of loss of alleles as in the actual population size N under study. For instance, if a few bull elephant seals monopolize all available females, N_e will be much nearer the number of dominant males than the total number of individuals in the population. Typically, $N_e < N$ within any generation (Nunney, 1993). Over time, N_e approximates to the harmonic mean of the per generation effective sizes, and so will be strongly influenced by the lowest population sizes (Wright, 1931), or 'bottlenecks'.

Reduction in heterozygosity and its effects

Wright's theory has become very important in conservation because low heterozygosity could cause inbreeding depression in small populations. But in recent years, the emphasis on maintenance of genetic diversity has been questioned (Lande, 1988; Caughley, 1994). Baur and Schmid (Chapter 7) discuss large-scale patterns of genetic variation with respect to overall population size. In this section, I will discuss evidence for local reductions in heterozygosity in natural and managed populations and ask whether reduction in genetic diversity compromises long-term fitness.

Some now abundant species are much less diverse genetically than close relatives, for example the Northern vs. Southern elephant seal (Bonnell & Selander, 1974) and the moths *Yponomeuta rorellus* vs. other *Yponomeuta* spp. (Menken, 1987) and *Helicoverpa zea* vs. *H. armigera* (Mallet *et al.*, 1993). These examples of low genetic diversity were presumably caused by relatively recent founder events or 'bottlenecks' in population size. Peripheral populations, especially those on islands, often have lower heterozygosities and unusual frequencies of alleles when compared with mainland populations: for instance, island races of mammals have on average half the genetic diversity found on the mainland (Kirkpatrick, 1981; and see Chapter 7). In most cases, the historical reasons for reductions in genetic diversity, although presumably caused by bottlenecks, are not known. However, the turkey oak gallwasp *Andricus quercuscalicis* has spread in historical times across Europe from the Mediterranean region, following introduction of its host-plant as an ornamental tree. This spread was accompanied by a strong reduction in variability by the time it reached the edge of its range, presumably caused by repeated bottlenecks (Stone & Sunnucks, 1993).

Heterozygosity at genes affecting quantitative traits is directly proportional to genetic variance and therefore to heritability (Falconer, 1989). A lack of genetic variance may cause extinction of a species if it cannot adapt to new environmental challenges; heritability is very important for long-term fitness (Lande & Barrowclough, 1987). In conservation, we might therefore

27

like to maximize heritable variation of traits likely to be advantageous. How is this best achieved, and what is the minimum viable population size for effective long-term conservation?

A naive model of this situation would assume a linear relationship between long-term population fitness and heterozygosity among populations within a species (Fig. 2.2, **A**). However, actual heritabilities, for example of yield in inbred lines of domestic animals and plants, remain very high in small populations suggesting that reduced heterozygosity has few ill effects. This implies that genes important for fitness are redundant, and that populations must lose variability at a great many individual loci before losing much genetic variance at phenotypic traits, or alternatively, that balancing selection is strong, or that mutation quickly replenishes variation. Under this scenario, loss of heritability and fitness will occur only at very small population sizes (Fig. 2.2, **B**); demography and ecology of endangered species are therefore deemed considerably more important than their genetic welfare (Lande & Barrowclough, 1987; Lande, 1988).

Theories of quantitative genetic variation do not yet enable us to predict whether the effect of bottlenecks on long-term fitness will look more like Fig. 2.2, **A** or 2.2, **B**; in particular, it is unclear whether the heritable quantitative

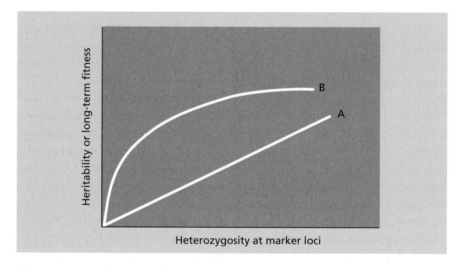

Fig. 2.2 Two possible relationships between heterozygosity within a species and heritability or long-term fitness after a population bottleneck. **A.** Heritability and long-term fitness are linearly related to heterozygosity at marker loci. **B.** Heritability and long-term fitness are lost only when heterozygosity at marker loci becomes very low, for example after an extended bottleneck. The actual relationship between heritability and long-term fitness depends strongly on how quantitative variation is maintained, which is not well understood (Lande & Barrowclough, 1987; Barton & Turelli, 1989). For the purposes of this diagram, heritability is assumed directly proportional to long-term fitness in that it provides genetic buffering against environmental change. It is also assumed that high-frequency polymorphic deleterious alleles have been purged, so that the remaining deleterious alleles are either fixed or at mutation/selection balance.

variation maintained in small populations is of the 'wrong kind', i.e. largely deleterious (Barton & Turelli, 1989; Houle, 1989; Kondrashov & Turelli, 1993). However, we can get some clues from natural populations with reduced genetic diversity, for example on islands. For example, a population of house mice on the Isle of May, off the coast of Scotland (presumably originally colonized by a small founder population from the mainland) was genetically divergent and impoverished. Berry *et al.* (1991) and Scriven (1992) introduced an inoculum of genetically different mice from the Isle of Eday in the Forth of Firth, and tracked changes in frequency of allozyme, DNA and chromosomal markers. The experiment was performed to test whether the unusual allozyme frequencies on the Isle of May were due to local adaptation. If selection controlled the local enzyme frequencies, it was expected that each marker would behave differently, and that the introduced alleles would mostly not increase.

The results were unexpected. All nuclear markers studied had similar dynamics; foreign alleles spread very rapidly until they reached about 50% of the population all over the island. Introduced markers were apparently 'hitchhiking' on strongly favoured foreign genomes; their spread had little of the heterogeneity expected due to selection at individual marker loci. More importantly for conservationists, any local adaptation the May population had achieved was far outweighed by the large number of less favourable alleles it carried. The reduced genetic diversity on the Isle of May was associated with invasibility to foreign alleles. These genetic data were backed by other observations. Native Isle of May mice were docile and moved slowly when confined in a handling tin; mice trapped after the introduction repeatedly jumped as well as ran, and often bit the handlers' gloves when the live traps were opened. There was also evidence for a substantial increase in abundance of mice on the Isle of May (G. Triggs, P. Scriven, P. King, pers. comm., 1994). Size also increased: mandibles were about 6–8% longer post-introduction compared with either May or Eday native mice, suggesting heterosis (Scriven & Bauchau, 1992).

Vrijenhoek (1994) performed a very similar introduction into a low heterozygosity population of the desert topminnow *Poeciliopsis monacha*. The results were similar: allozyme variation rapidly recovered after the introduction of fish from a more variable population. In this study, measures of fitness (population size, freedom from parasitism, interspecific competition) were all shown to increase after the introduction. Natural populations of the plants *Scabiosa* and *Salvia* with low genetic diversity also have strongly reduced fitness (Bijlsma *et al.*, 1994).

These experiments show that fitness can decline in natural populations with low genetic diversity. It would be worth performing many more field experiments to test whether natural populations with low marker heterozygosity are less fit than populations with normal genetic diversity. Two recent experiments have shown that inbreeding depression is expressed much more strongly under stressful natural conditions than under relaxed regimes: in white-footed mice (*Peromyscus*), inbreeding strongly reduced survival in the field, but only weakly in the laboratory; in song sparrows (*Geospiza*), progeny of outcrossed parents survived a population crash much better than naturally inbred progeny, whereas in times of plenty inbred individuals survived well. At the moment we cannot reject even the extreme idea that long-term fitness and persistence are linearly related to genetic variability (Fig. 2.2, **B**). Under this possibility, there is no easily quantifiable minimum viable population for the maintenance of fitness-related genetic diversity; the bigger and more variable the population the more genetically resistant it would be to extinction.

The current results, if confirmed by further experiments, also suggest that genetic management of populations of endangered species should be given a higher priority than is currently fashionable. Species with low heterozygosity are not necessarily less viable (e.g. the cheetah or other carnivores – see Caughley, 1994), but low diversity populations within high diversity species may be. We should preserve larger chunks of habitat in which to maintain as much of the existing genetic diversity of endangered species as possible. In cases where several isolated populations of a species show strongly divergent reduced genetic diversity, genetic management might include cross-population introductions to maintain high levels of fitness in each one, although this must be balanced against the risk of introducing disease and homogenizing locally adaptive variation.

2.4.2 Between populations

Measurement

Genetic diversity of subpopulations in a metapopulation implies that frequencies of each allele, π, differ between subpopulations. The difference in frequencies can simply be measured by the variance of frequencies for each allele, or Wahlund variance σ_π^2. Standardizing this by maximum possible variance in allele frequency, $\bar{\pi}(1 - \bar{\pi})$ gives Wright's familiar statistic $F_{st} = \sigma_\pi^2/\bar{\pi}(1 - \bar{\pi})$. F_{st} is very convenient because it also measures the deviation from Hardy–Weinberg caused by population subdivision, a kind of inbreeding coefficient already in use for studies within populations. To generalize this statistic to all genes sampled, some sort of averaging must be done across alleles and loci. The between population analogue of F_{is} (Box 2.1b) is used for this, which also shows that an equivalent way of understanding this statistic is as the proportion of total allelic diversity, H_e, found between as opposed to within populations. (These formulations ignore difficulties in making estimates of F_{st} with limited sample sizes, a problem treated by Slatkin & Barton (1989) and Weir (1990)). F_{st} measures the variance of gene frequency fluctuations across space; it would also be useful to measure their spatial extent. This can be done by measuring correlations of allele frequencies between populations at different distances (see below).

The problem of estimating F_{st} and other parameters in genetic population structure is not easily solved. Very few computer packages exist for calculating the relevant estimates (e.g. Swofford & Selander, 1981; Weir, 1990; Goudet, in prep.; Raymond & Rousset, in prep.). With small sample sizes, different estimators give wildly different values. One of the programs, BIOSYS (no longer supported by its writer, David Swofford, but still widely used), has two routines which estimate F_{st}, but the answers do not agree: one subroutine (FSTAT) takes an unweighted average of the allele frequency variation and yields a value that amplifies stochastic errors of estimating the frequencies of rare alleles, while another subroutine (WRIGHT78) takes a weighted average which gives a lower estimate of F_{st}. Good programs with multiple options and clear explanations would be greatly valued by practical population geneticists!

Theory

F_{st} measures the standardized variance in gene frequency, and is useful because it can be related directly to genetic drift in metapopulations. For instance, under a simple 'island model' where an infinite number of subpopulations of size N_e receive migration at rate m per generation from the mainland, the variation of neutral allele frequencies will rather quickly stabilize at a stochastic equilibrium given by $F_{st} \approx 1/(1 + 4N_e m)$ (Wright, 1951). Luckily, the result holds approximately in more realistic population structures, such as two-dimensional stepping-stone models, and even (still more approximately) in continuous populations when the neighbourhood size N_b is substituted for Nm (Slatkin & Barton, 1989). Allowance can also be made for extinction of local subpopulations (Whitlock, 1992). Slatkin (1985) has found empirically that the conditional frequency of a small subset of the rarest alleles, those 'private' (peculiar) to particular samples, can be used to estimate Nm, and he has developed other methods based on gene genealogies obtained in sequencing studies (see Chapter 7; Slatkin, 1989, 1991; Slatkin & Maddison, 1989; Milligan *et al.*, 1994).

If the population is more or less continuous, allele frequencies will fluctuate with distance due to genetic drift and selection. The correlation of allele frequencies should decline with distance in a more or less predictable way. When populations are close together, so that there is plenty of gene exchange, there should be few differences in gene frequency: allele frequencies should be highly correlated. If populations are very far apart (i.e. much greater than the gene flow distance, σ_x), there should be strong differences in allele frequency: allele frequencies should be uncorrelated. The decline in correlations of allele frequencies with distance should, in effect, allow us to estimate the spatial scale (σ_x) of drift. This is the basis of the 'spatial

31

autocorrelation' method for studying allele frequencies (Sokal *et al.*, 1989). Unfortunately, apart from this simple argument, there is as yet no published theory to use declines of correlation with distance to estimate σ_x, the gene flow distance. It may not even be possible to achieve a workable theory if the correlation depends on hard-to-estimate parameters like mutation rates and low levels of selection. As a result, spatial autocorrelation has been characterized as a technique for exploratory data analysis rather than a fully developed tool (Slatkin & Arter, 1991a,b; see also reply by Sokal & Oden, 1991).

Spatial patterns in nature

Studies of population structure potentially give estimates of the tendency towards genetic drift in terms both of local population size N_b, and of migration rates m and migration distance σ_x. These would be of enormous use for two reasons. First, we could use estimates of σ_x to estimate the minimum patch size to which populations can adapt. Insecticide treatments of disease vectors or other pests, for instance, could be designed to minimize local evolution of insecticide resistance. In conservation, this information would be useful in assessing population structure and tendency to lose genetic diversity. Second, the information can give an estimate of the importance of genetic drift in evolution, in particular the likelihood of the shifting balance (Wright, 1980), in which drift initiates a shift between two adaptive peaks. In practice, as indicated above, we are still a long way from being able to use genetic information in this way. A review is given in Chapter 7; I give only two examples of attempts to use spatial information in understanding population structure.

Wright *et al.* (1942) found that the frequency of allelism of lethal recessives on chromosome III of *Drosophila pseudoobscura* was 0.0213 within sampling stations (<0.1 km apart), 0.0088 between stations within localities (0.2–3.5 km apart), and 0.0041 between localities (16–21 km apart). Clearly, there is a reduction in allelism of these lethal alleles with distance, and this is most likely due to a higher identity of alleles by descent within sampling stations and localities than between. However, in the island model formulation used by Wright *et al.*, spatial information was not adequately sampled. Barton and Clark (1990: 158–9) applied a correlation model to data on allozymes and fluctuations of chromosome frequency in a hybrid zone of the grasshopper *Podisma pedestris* in which overall $F_{st} \approx 0.04$. The correlations of allele frequencies declined to an insignificant level beyond about 30 m, suggesting that dispersal was less than 30 m. Previous mark-recapture studies, which indicated that $\sigma_x \approx 20$ m for *Podisma* (Barton & Hewitt, 1982), are in good approximate agreement.

2.4.3 Diagnosis of species

Most existing genetic diversity is found between species. This section shows how genetic techniques are being used to refine Darwin's morphological cluster idea of species, while leaving intact his finding of continuity between individual variation, populations, races and species. Adopting the emerging 'genotypic cluster' definition of species can clarify problems in evolution, taxonomy and conservation.

A brief history of species definitions

There is now more confusion in the literature on species concepts even than in Darwin's day. For instance, Ridley (1993) discusses no less than seven concepts (phenetic, biological, recognition, ecological, cladistic, pluralistic and evolutionary), and concludes that a confusing combination of four of these (biological, recognition, ecological and cladistic) is ideal. This confusion is extraordinary, because Darwin (1859) thought that he had solved the problem of defining species when he stated:

> Hereafter, we shall be compelled to acknowledge that the only distinction between species and well-marked varieties is, that the latter are known, or believed, to be connected at the present day by intermediate gradations, whereas species were formerly thus connected.

Darwin, who invented the idea of evolutionary species, had to adopt a more fluid definition of species than previous creationist attempts. After the acceptance of evolution, species were simply varieties that had diverged further than other varieties, and which, when in contact, lacked intermediates.

Darwin's definition was generally accepted for about 80 years. However, by the 1930s and 1940s, two problems had arisen. First, discrete polymorphisms, as in mimetic females of the butterfly *Papilio memnon* (Plate 2.1), could be classified under Darwin's definition as separate species, whereas they in fact formed part of a single interbreeding population (Poulton, 1904). Second, Darwin's 'morphological species concept' could not distinguish sibling species like *Drosophila persimilis* and *D. pseudoobscura*, which, though morphologically nearly identical, were intersterile (Dobzhansky, 1937). In response to these difficulties, Dobzhansky (1937) and Mayr (1940, 1942) adopted the 'biological species concept' based on interbreeding: groups of individuals were separate species if they did not interbreed when in contact, or, if not in contact, were presumed unable to interbreed. Mayr (1942, 1970) in particular was highly influential in uniting two previous ideas on species. The first was that geographical replacement taxa could often be classed as races within 'polytypic species' (first realized by American ornithologists in the 1880s; see Stresemann, 1975). By the turn of the century, most taxonomists accepted this polytypic species definition, in which geographic populations were demoted from species to subspecies within wide-ranging species (Rothschild & Jordan, 1903).

The second idea, already mentioned, was that separate species do not interbreed in sympatry, whereas varieties or races do (Poulton, 1904; Dobzhansky, 1937). These ideas arose in the infancy of molecular genetics; genes were at that time not known to consist of smaller units, and as late as 1951 evolutionists doubted that proteins were the immediate gene products (see the discussion by Dobzhansky (1951: p. 13) of the one-gene-one-enzyme hypothesis). It is perhaps not surprising that the biological species concept is an idea which did not depend explicitly on state-of-the-art genetics, unlike the rest of the evolutionary synthesis (Mallet, 1995). Modern genetic knowledge now makes a genetic definition much simpler.

Species as genotypic clusters

Darwin (1859, 1871) saw species as morphological clusters that did not intergrade with others where they overlapped. In genetic terms, species are non-intergrading genotypic clusters within local areas. In areas of overlap, separate species must differ in frequency at some loci, or they will remain undetected, but it is the way in which the genetic diversity is partitioned into genotypic clusters rather than the amount of difference which allows us to distinguish species. For loci that differ, a sample of genotypes from sympatric genotypic clusters will show two patterns: first, there will be fewer heterozygotes than expected under Hardy–Weinberg at each locus; second, there will be correlations between loci (or deviations from gametic equilibrium, see Box 2.1a). This is in the spirit of Darwin's definition because morphological intermediates can be defined both trait by trait (equivalent to heterozygotes at single loci) and on the basis of combinations between traits (equivalent to recombinant gametic types). Note that geographic races within a species must also differ genetically, and so may form separate genotypic clusters in allopatry, but will blend to form a single genotypic cluster where they overlap.

In practice, genotypic distributions are already used to detect sibling species. Dobzhansky (1951), who promoted the idea that *D. persimilis* was a different species from *D. pseudoobscura*, felt it necessary to show that mating incompatibilities were correlated with differences in the frequency of chromosomal rearrangements as well as slight differences in body size and genitalia morphology. The synergy between mating studies and genotypic approaches is well demonstrated by a recent set of discoveries of mosquito sibling species. Two sympatric forms of the 'common malaria mosquito' of North America (*Anopheles quadrimaculatus*), when hybridized, produced sterile or reduced numbers of offspring. This suggested that there were two separate species, now designated A and B (Lanzaro *et al.*, 1988; Kaiser *et al.*,

1988a). This finding was confirmed when strong chromosomal and allozyme differences were also found between the forms (Kaiser *et al.*, 1988c; Lanzaro *et al.*, 1990). A further sympatric species, C, was discovered virtually simultaneously via both crossing experiments and studies of genetic markers (Kaiser *et al.*, 1988b), and a fourth sympatric species, D, was detected solely via allozyme studies (Narang *et al.*, 1989). The understanding of *Anopheles* species complexes is not merely idle philosophy; in Africa, sibling species of the *A. gambiae* complex differ strongly in their tendency to carry malaria. In North America, malaria has now been eradicated, but it is unknown which *A. quadrimaculatus* sibling species carried the disease.

Sexual isolation and hybrid inviability are bad characters for defining species because they are common within species (Darwin, 1859). For example, insect populations differing in heritable bacterial symbionts (Rousset & Raymond, 1991; Turelli & Hoffmann, 1991) or transposable elements (Engels, 1983) which cause hybrid dysgenesis are considered conspecific because they lack other genetic differences in sympatry. Mate choice is common within as well as between species (Andersson, 1994), and is also a poor candidate trait for defining species. In general, mating incompatibilities and mate choice can be used as evidence for species status, but must be accompanied by genetic evidence for separate clusters.

As an example of the purely genotypic approach, I choose a system where colour pattern differences in a butterfly are known to be genetic. *Heliconius erato* has many races which abut, and which usually intergrade freely (Plate 2.2; Brown *et al.*, 1974; Turner, 1984). Near Tarapoto, Peru, there is a narrow hybrid zone between an Andean race from the upper Huallaga and Mayo river valleys, and an Amazonian race from the lowlands near Yurimaguas. The differences between the colour patterns are inherited at three unlinked major loci (Fig. 2.1), of which one is codominant and the other two are dominant (Mallet *et al.*, 1990). A hybrid index can be constructed by counting the number of Amazonian phenotypic contributions out of a total of four (one for each allele at the codominant gene and one for each dominant gene). The codominant locus conforms to Hardy–Weinberg (Mallet *et al.*, 1990), and the distribution of the hybrid index shows that there is only a single genotypic cluster (Fig. 2.3a). There are rather strong gametic correlations between the unlinked genes ($R_{AB} \approx 35\%$; Mallet *et al.*, 1990), but since most individuals are recombinants, the distribution of the hybrid index is broadened rather than made bimodal. These *Heliconius* races, which do not differ at allozyme loci, are considered members of the same species because single and multilocus intermediates are common in the centre of the hybrid zone. The probable cause of these intermediates, is, of course, near-random mating among the forms and near-full viability of hybrids; however, these extra inferences about gene flow and selection need not enter into the definition of races vs. species.

In contrast, where *Heliconius erato cyrbia* meets its close relative *H. himera*

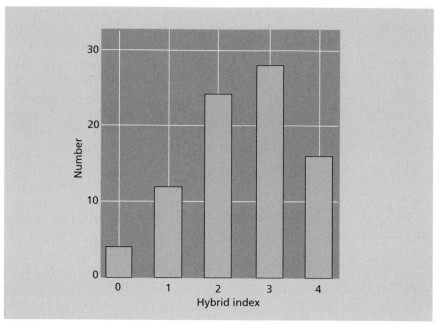

(a)

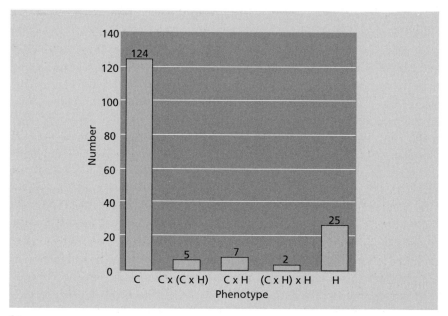

(b)

Fig. 2.3 The frequency of hybrids in two hybrid zones involving *Heliconius erato*. (a) Hybrid index scores of individuals collected near Pongo de Cainarache, San Martin, Peru, in a single population in the centre of a hybrid zone between *H. erato emma* and *H. e. favorinus* (Plate 2.2). (b) Frequency of F_1 (C×H) and backcross [C×(C×H)] and [H×(C×H)] hybrids in a single population of a hybrid zone between *H. erato* (C) and *H. himera* (H; Plate 2.2). The collection was from a single population near Guayquichuma, S. Ecuador.

36

(see Plate 2.2) at Guayquichuma in southern Ecuador, hybrids are rare and most individuals are apparently pure *cyrbia* or pure *himera* (Fig. 2.3b). At least six loci determine the colour pattern differences, so the rarity of intermediates is highly non-random. Allozyme loci also show several almost fixed differences between sympatric *cyrbia* and *himera*. Hybrids are also mostly F_1 or simple backcrosses, suggesting pre- or post-mating incompatibilities (Descimon & Mast de Maeght, 1984; Jiggins *et al.*, in press), but again, these incompatibilities are secondary to the genotypic definition. The existence of two peaks in the genotypic frequency distribution argues for separate species status of the *erato* and *himera*. In the same way, we maintain species of duck, even though natural hybrids are regularly found (Grant & Grant, 1992).

A number of recent evolutionists explicitly adopt the genotypic cluster approach. Darwin's finches on Isla Daphne Major in the Galapagos hybridize frequently, but are considered separate species by Grant and Grant (Grant & Grant, 1992; Grant, 1993), on the grounds that intermediates in heritable beak morphology are rare in spite of gene flow. Avise and Ball (1990), Sbordoni (1993), Patton (1993) and Patton and Smith (1994) use explicit genotype-based definitions of species applied to allozyme and molecular genotypes in their discussions of hybridization in species complexes. Recently Cohan (1994) showed that gene flow between bacteria did not preclude the maintenance of separate genetic clusters, and suggested that bacterial species are therefore best defined as genotypic clusters rather than via gene flow. The genotypic method is thus a general way of defining speices.

Species as testable clusters, not concepts

A genotypic definition is more easily used than species 'concepts' because it emphasizes an easily testable *pattern* of genotype frequencies rather than a hypothesized *process* of the origin or maintenance of that pattern. Genotypic clusters thus avoid the common, but risky practice of incorporating evolutionary hypotheses into definitions of the objects of these same hypotheses. For instance, the biological species concept emphasizes reproductive isolation or lack of gene flow as a way of defining species: but it is actually the results of reproductive isolation, rather than the isolation itself, that we can observe and use to classify species. The phylogenetic species concept (Cracraft, 1989), although in a sense pattern-based, again emphasizes an estimated phylogenetic process rather than the tangible results of that process (Mallet, 1995).

To compare hypotheses for the origin or maintenance of species, it is obviously best to have a definition that assumes nothing about species maintenance or evolution. A 'concept', which includes an idea of the process of species maintenance (e.g. reproductive isolation, ecological differentiation), is not what is needed in taxonomy, evolutionary studies, or conservation. For example, the biological species concept, by requiring a unifying force of gene flow within species, might tempt one more towards an allopatric explanation of speciation, involving an external disruption of gene flow, than a simple definition based on the pattern of genotypes. In conservation, we would like to preserve actual morphological and genetic diversity, rather than groupings that depend on evolutionary concepts or ideas about this diversity.

Continuity between species and races

One perceived disadvantage of the biological species concept is that isolated races, for example on islands, cannot easily be distinguished from species. It is well known that mating can occur in captivity between sympatric forms that do not mate in the wild, so there is no way to distinguish between geographic races and species unless they overlap (Wallace, 1865; Mayr, 1942, 1970). The genotypic approach has the same 'problem', and taxonomists typically adopt the null hypothesis of 'same species' when dealing with island populations that are similar. However, because races as well as species are by definition separate genetic clusters (see Section 2.2), the genotypic definition emphasizes the continuity of races and species, and that species and races really are the same kind of thing. In contrast, the biological concept tries to construct an unnecessary and untrue Aristotelian 'essence' of species that is applicable to such cases.

The biological species concept fails even on its own territory, when two forms overlap and hybridize in the wild. Sometimes we call them separate species because intermediates are rare, as in ducks; at other times we call them races because intermediates are common, as in the flicker *Colaptes auratus* (Moore & Price, 1993). A dividing line, where the two forms are considered independent, exists but cannot be quantified in terms of mate choice, inviability or infertility of hybrids (Sokal & Crovello, 1970). Nor is hybridization rare; under the above definition, at least 9% of the world's birds and 12% of European butterflies, as well as many plants, are known to hybridize naturally with closely related species (Guillaumin & Descimon, 1976; Grant & Grant, 1992; H. Descimon, pers. comm., 1995). It is much easier to distinguish between races and species using a genotypic definition. If there are two overlapping but distinct clusters of genotypes with few intermediates, they are separate species; when there is but a single genotypic cluster in a local area, there is only one species. The null hypothesis should clearly be a single species; a test comparing the more complex hypothesis of two independent genotypic clusters can be used to reject the simpler model (Mallet, 1995). Hybrid inviability or assortative mating can still provide evidence for separate genotypic clusters, but genetic or morphological data must be used to confirm the diagnosis, as took place with *Drosophila* and *Anopheles*. As we have seen, in practice the genotypic approach is already used in the study of sibling species and and in hybrid zones. We may use hybrid inviability or assortative mating as *evidence* that two forms are separate genotypic clusters, but not as the primary *definition* of those species. The continuing existence of separate species is dependent on whether inviability selection is powerful enough compared with gene flow to maintain separate genotypic clusters when two populations are in contact; but this dependency should not be built into the definition of species. The genotypic approach to species definition is both testable, and allows a clear appreciation of Darwin's insight: the continuity between varieties, races and full species.

Species, genetics and conservation

Biodiversity and conservation studies typically use the species level as their basis. But without a species definition that gives accurate diagnosis, the numbers of species are virtually arbitrary (O'Hara, 1994). This arbitrariness is in part due to the difficulty of what to define in allopatry, which, as explained above, cannot easily be solved except by agreement among taxonomists. The arbitrariness is due to the fact of evolutionary continuity of species and races, though a strictly genotypic approach applied to continental faunas and floras should help. The species category should perhaps be de-emphasized in conservation (Moritz, 1994). Many geographic subspecies are more divergent, both morphologically and genetically, than some sibling species pairs; if we must choose, which should we conserve? My own preference would be for the more divergent subspecies. Whether we look at morphology, chromosomes, allozymes or DNA, races and species overlap in their levels of divergence. To conserve nature, we must save populations containing representative genetic diversity. These representative populations may often be below the level of species.

Acknowledgments

I am very grateful to Bruno Baur, Martin Brookes, Sarah Durant, Owen McMillan, Jan De Ruiter, Chris Thomas, Mike Whitlock, and an anonymous reviewer for helpful comments on this manuscript. I am grateful to Bernard D'Abrera for his photographic skills on Plates 2.1 and 2.2, and to Dick Vane-Wright, Phil Ackery and the Natural History Museum, London, for help. This work was supported by grants from NERC and BBSRC.

References

Andersson M. (1994) *Sexual Selection*. Princeton University Press, New Jersey. [2.4.2]
Avise J.C. (1994) *Molecular Markers, Natural History and Evolution*. Chapman and Hall, London. [2.3.4] [Appendix 2.1]
Avise J.C. & Ball R.M. (1990) Principles of genealogical concordance in species concepts and

biological taxonomy. In: *Oxford Surveys in Evolutionary Biology*, Vol. 7 (eds D.J. Futuyma & J. Antonovics), pp. 45–67. Oxford University Press. [2.3.4] [2.4.2]

Avise J.C., Arnold J., Ball R.M., Bermingham E., Lamb T., Neigel J.E., Reeb C.A. & Saunders N.C. (1987) Intraspecific phylogeography: the mitochondrial DNA bridge between population genetics and systematics. *Annu. Rev. Ecol. Syst.* **18**, 489–522. [Appendix 2.1]

Ayala F.J., Tracey M.L., Hedgecock D. & Richmond R. (1974) Genetic differentiation during the speciation process in *Drosophila*. *Evolution* **28**, 576–592. [2.3.4]

Barton N. & Clark A.G. (1990) Population structure and processes in evolution. In: *Population Genetics and Evolution* (eds K. Wohrmann & S. Jain), pp. 115–174. Springer-Verlag, Berlin. [2.4.2]

Barton N.H. & Gale K.S. (1993) Genetic analysis of hybrid zones. In: *Hybrid Zones and the Evolutionary Process* (ed. R.G. Harrison), pp. 13–45. Oxford University Press, New York. [2.2] [2.3.3] [2.4.1]

Barton N.H. & Hewitt G.M. (1982) A measure of dispersal in the grasshopper *Podisma pedestris* (Orthoptera: Acrididae). *Heredity* **48**, 237–249. [2.4.2]

Barton N.H. & Hewitt G.M. (1983) Hybrid zones as barriers to gene flow. In: *Protein Polymorphism: Adaptive and Taxonomic Significance* (eds G.S. Oxford & D. Rollinson), pp. 341–359. Academic Press, London. [2.3.4]

Barton N.H. & Turelli M. (1989) Evolutionary quantitative genetics: how little do we know? *Annu. Rev. Genet.* **23**, 337–370. [2.3.1] [2.4.1]

Bazykin A.D. (1969) Hypothetical mechanism of speciation. *Evolution* **23**, 685–687. [2.3.3]

Begun D.J. & Aquadro C.F. (1993) African and North American populations of *Drosophila melanogaster* are very different at the DNA level. *Nature* **365**, 548–550. [2.3.4]

Berry R.J., Triggs G.S., King P., Nash H.R. & Noble L.R. (1991) Hybridization and gene flow in house mice introduced into an existing population on an island. *J. Zool.* **225**, 615–632. [2.4.1]

Bijlsma R., Ouborg N.J. & van Treuren R. (1994) On genetic erosion and population extinction in plants: a case study in *Scabiosa columbaria* and *Salvia pratensis*. In: *Conservation Genetics* (eds V. Loeschcke, J. Tomiuk & S.K. Jain), pp. 255–272. Birkhäuser, Basel. [2.4.1]

Bilton D.T. (1994) Intraspecific variation in the terrestrial isopod *Oniscus asellus* Linnaeus, 1758 (Crustacea: Isopoda: Oniscidea), with a description of a new subspecies from Western Europe, and a preliminary account of an extensive mosaic hybrid zone. *Zool. J. Linn. Soc.* **110**, 325–354. [2.3.1]

Bishop J.A., Cook L.M. & Muggleton J. (1976) Variation in some moths from the industrial northwest of England. *Zool. J. Linn. Soc.* **58**, 273–296. [2.3.1]

Bonnell M.L. & Selander R.K. (1974) Elephant seals: genetic variation and near-extinction. *Science* **184**, 908–909. [2.4.1]

Brakefield P.M. & Larsen T.B. (1984) The evolutionary significance of dry and wet season forms in some tropical butterflies. *Biol. J. Linn. Soc.* **22**, 1–12. [2.3.1]

Brown K.S. (1967) Chemotaxonomy and chemomimicry: the case of 3-hydroxy-kynurenine. *Syst. Zool.* **16**, 213–216. [Appendix 2.1]

Brown K.S. (1979) *Ecologia Geográfica e Evolução nas Florestas Neotropicais*. Universidade Estadual de Campinas, Campinas, Brazil. Livre de Docencia. [2.3.1]

Brown K.S., Sheppard P.M. & Turner J.R.G. (1974) Quaternary refugia in tropical America: evidence from race formation in *Heliconius* butterflies. *Proc. Roy. Soc., Lond. B* **187**, 369–378. [2.3.1] [2.4.2]

Bruford M.W. & Wayne R.K. (1994) The use of molecular genetic techniques to address conservation questions. In: *Molecular Environmental Biology* (ed. S.J. Garte), pp. 11–28. Lewis Publishers (CRC Press), Boca Raton. [2.3.4] [Appendix 2.1]

Burke T. (1989) DNA fingerprints and other methods for the study of mating success. *Trends Ecol. Evol.* **4**, 139–144 [Appendix 2.1]

Burke T. & Bruford M.W. (1987) DNA fingerprinting in birds. *Nature* **327**, 149–152. [2.3.4]

Bush G.L., Case S.M., Wilson A.C. & Patton J.L. (1977) Rapid speciation and chromosomal evolution in mammals. *Proc. Natl. Acad. Sci., USA* **74**, 3942–3946. [2.3.3]

Caughley G. (1994) Directions in conservation biology. *J. Anim. Ecol.* **63**, 215–244. [2.4.1]

Chase M.W., Soltis D.E., Olmstead R.G., Morgan D., Les D.H., *et al.* (1993) Phylogenetics of seed plants: an analysis of nucleotide sequences from the plastid gene *rbcl*. *Ann. Missouri Bot. Gard.* **80**, 528–580. [Appendix 2.1]

Clarke B. (1962) Balanced polymorphism and the diversity of sympatric species. In: *Taxonomy and Geography* (ed. D. Nichols), pp. 47–70. (Systematics Association Publication 4) Systematics Association, Oxford. [2.3.1]

Clarke C.A. & Sheppard P.M. (1971) Further studies on the genetics of the mimetic butterfly *Papilio memnon*. *Phil. Trans. Roy. Soc., Lond. B* **263**, 35–70. [2.3.1]

Clarke C.A., Sheppard P.M. & Thornton I.W.B. (1968) The genetics of the mimetic butterfly *Papilio memnon*. *Phil. Trans. Roy. Soc., Lond. B* **254**, 37–89. [2.3.1]

Clausen J., Keck D.D. & Hiesey W.M. (1947) Heredity of geographically and ecologically isolated races. *Am. Nat.* **81**, 114–133. [2.3.1]

Cohan F.M. (1994) The effects of rare by promiscuous genetic exchange on evolutionary divergence in prokaryotes. *Am. Nat.* **143**, 965–986. [2.4.2]

Cracraft J. (1989) Speciation and its ontology: the empirical consequences of alternative species concepts for understanding patterns and processes of differentiation. In: *Speciation and its Consequences* (eds D. Otte & J.A. Endler), pp. 28–59. Sinauer Associates, Sunderland, Massachusetts. [2.4.2]

Darwin C. (1859) *On the Origin of Species by Means of Natural Selection, or the Preservation of Favoured Races in the Struggle for Life* (1st edn). John Murray, London. [2.4.2]

Darwin C. (1871) *The Descent of Man, and Selection in Relation to Sex* (2nd edn). John Murray, London. [2.4.2]

Dawkins R. (1976) *The Selfish Gene*. Oxford University Press, London. [2.1.2]

Dawkins R. (1982) *The Extended Phenotype*. WH Freeman, Oxford. [2.1.2]

Descimon H. & Mast de Maeght J. (1984) Semispecies relationships between *Heliconius erato cyrbia* Godt. and *H. himera* Hew. in southwestern Ecuador. *J. Res. Lepid.* **22**, 229–239. [2.4.2]

Dobzhansky T. (1937) *Genetics and the Origin of Species*. Columbia University Press, New York. [2.3.3] [2.4.2]

Dobzhansky T. (1951) *Genetics and the Origin of Species* (3rd, revised edn). Columbia University Press, New York. [2.4.2]

Dobzhansky T. (1970) *Genetics of the Evolutionary Process*. Columbia University Press, New York. [2.3.3]

Dudley J.W. (1977) Seventy-six generations of selection for oil and protein percentage in maize. In: *Proceedings of the International Conference on Quantitative Genetics* (eds E. Pollack, O. Kempthorne & T.B. Bailey), pp. 459–473. Iowa State University Press, Ames, Iowa. [2.3.1]

Emelianov I., Mallet J. & Baltensweiler W. (1995) Genetic differentiation in the larch budmoth *Zeiraphera diniana* (Lepidoptera: Tortricidae): polymorphism, host races or sibling species? *Heredity* **75**, in press. [2.3.4]

Endler J.A. (1977) *Geographic Variation, Speciation, and Clines*. Princeton University Press, New Jersey. [2.2]

Engels W.R. (1983) The P family of transposable elements in *Drosophila*. *Annu. Rev. Genet.* **17**, 319–344. [2.4.2]

Falconer D.S. (1989) *Introduction to Quantitative Genetics* (3rd edn). Longman, Harlow, Essex. [2.3.1] [2.4.1]

Fisher R.A. (1958) *The Genetical Theory of Natural Selection* (2nd edn). Dover, New York. [2.3.1]

Ford E.B. (1941) Studies on the chemistry of pigments in the Lepidoptera, with reference to their bearing on systematics. (1)The anthoxanthins. *Proc. Roy. Entomol. Soc. Lond. A* **16**, 65–90. [Appendix 2.1]

Forrest G.I. (1994) Biochemical markers in tree improvement programmes. *Forestry Abstracts* **55**, 123–153. [Apendix 2.1]

Gabriel W. & Lynch M. (1992) The selective advantage of reaction norms for environmental tolerance. *J. Evol. Biol.* **5**, 41–59. [2.3.1]

Ghiselin M.T. (1975) A radical solution to the species problem. *Syst. Zool.* **23**, 536–544. [2.1.2]

Gillespie J.H. (1991) *The Causes of Molecular Evolution.* Oxford University Press. [2.3.3] [2.3.4]

Grant P.R. (1986) *Ecology and Evolution of Darwin's Finches.* Princeton University Press, New Jersey. [2.3.1]

Grant P.R. (1993) Hybridization of Darwin's finches on Isla Daphne Major, Galápagos. *Phil. Trans. Roy. Soc., Lond. B* **340**, 127–139. [2.4.3]

Grant P.R. & Grant B.R. (1992) Hybridization of bird species. *Science* **256**, 193–197. [2.4.2]

Grant V. (1981) *Plant Speciation* (2nd edn). Columbia University Press, New York. [2.3.3]

Guillaumin M. & Descimon H. (1976) La notion d'espèce chez les lépidoptères. In: *Les Problèmes de l'Espèce dans le Règne Animal*, Vol. 1 (eds C. Bocquet, J. Génermont & M. Lamotte), pp. 129–201. Société zoologique de France, Paris. [2.4.2]

Haffer J. (1974) Avian speciation in tropical South America. *Publ. Nuttall Ornith. Club* **14**, 1–390. [2.3.1]

Harris H. (1966) Enzyme polymorphisms in man. *Proc. Roy. Soc. Lond. B* **164**, 298–310. [2.3.4]

Harrison R.G. (ed.) (1993) *Hybrid Zones and the Evolutionary Process.* Oxford University Press, New York. [2.2]

Hedrick A.V. (1986) Female preferences for male calling bout duration in a field cricket. *Behav. Ecol. Sociobiol.* **19**, 73–77. [2.3.1]

Hedrick A.V. (1988) Female choice and the heritability of attractive male traits: an empirical study. *Am. Nat.* **132**, 267–276. [2.3.1]

Hedrick P., Jain S. & Holden L. (1978) Multilocus systems in evolution. *Evol. Biol.* **11**, 101–182. [2.4.1]

Hewitt G.M. (1988) Hybrid zones — natural laboratories for evolutionary studies. *Trends Ecol. Evol.* **3**, 158–167. [2.2] [2.3.1] [2.3.3]

Higgins L.G. & Riley N.D. (1993) *Butterflies of Britain and Europe.* HarperCollins, London. [2.3.1]

Hill W.G. (1988) Why aren't horses faster? *Nature* **332**, 678. [2.3.1]

Hillis D.M. & Moritz C. (eds) (1990) *Molecular Systematics.* Sinauer, Sunderland, Massachusetts. [Appendix 2.1]

Hoelzel A.R. (ed.) (1992) *Molecular Genetic Analysis of Populations.* IRL Press at Oxford University Press [Appendix 2.1]

Houle D. (1989) The maintenance of genetic variation in finite populations. *Evolution* **43**, 1767–1780. [2.4.1]

Howard R.W. & Blomquist G.J. (1982) Chemical ecology and biochemistry of insect hydrocarbons. *Annu. Rev. Entomol.* **27**, 149–172. [Appendix 2.1]

Hudson R. (1990) Gene genealogies and the coalescent process. In: *Oxford Surveys in Evolutionary Biology*, Vol. 7 (eds D. Futuyma & J. Antonovics), pp. 1–44. Oxford University Press. [2.4.1]

Jiggins C., McMillan O., Neukirchen W. & Mallet J. What can hybrid zones tell us about speciation? The case of *Heliconius erato* and *H. himera* (Lepidoptera: Nymphalidae). *Biol. J. Linn. Soc.*, in press. [2.4.3]

Kaiser P.E., Mitchell S.E., Lanzaro G.C. & Seawright J.A. (1988a) Hybridization of laboratory strains of sibling species A and B of *Anopheles quadrimaculatus. J. Am. Mosq. Control Assoc.* **4**, 34–38. [2.4.2]

Kaiser P.E., Narang S.K., Seawright J.A. & Kline D.L. (1988b) A new member of the *Anopheles quadrimaculatus* complex, species C. *J. Am. Mosq. Control Assoc.* **4**, 494–499. [2.4.2]

Kaiser P.E., Seawright J.A. & Birky B.K. (1988c) Chromosome polymorphism in natural populations of *Anopheles quadrimaculatus*. Say species A and B. *Genome* **30**, 138–146. [2.4.2]

Kettlewell H.B.D. (1973) *The Evolution of Melanism: the study of a recurring necessity*. Blackwell Scientific, Oxford. [2.3.1]

King M. (1993) *Species Evolution: the role of chromosome change*. Cambridge University Press. [2.3.3]

Kirkpatrick C.W. (1981) Genetic structure of insular populations. In: *Mammalian Population Genetics* (eds M.H. Smith & J. Joule), pp. 28–59. University of Georgia Press, Athens, Georgia. [2.4.1]

Kondrashov A.S. & Turelli M. (1992) Deleterious mutations, apparent stabilizing selection and the maintenance of quantitative variation. *Genetics* **132**, 603–618. [2.4.1]

Kreitman M. & Hudson R.R. (1991) Inferring the evolutionary histories of the Adh and Adh-dup loci in *Drosophila melanogaster* from patterns of polymorphism and divergence. *Genetics* **127**, 565–582. [2.3.4]

Lande R. (1981) The minimum number of genes contributing to quantitative variation between and within populations. *Genetics* **99**, 541–553. [2.3.1]

Lande R. (1988) Genetics and demography in biological conservation. *Science* **241**, 1455–1460. [2.4.1]

Lande R. & Barrowclough G.F. (1987) Effective population size, genetic variation, and their use in population management. In: *Viable Populations for Conservation* (ed. M.E. Soulé), pp. 87–123. Cambridge University Press. [2.4.1]

Lanzaro G.C., Narang S.K., Mitchell S.E., Kaiser P.E. & Seawright J.A. (1988) Hybrid male sterility in crosses between field and laboratory strains of *Anopheles quadrimaculatus* Say (Diptera: Culicidae). *J. Med. Entomol.* **25**, 248–255. [2.4.2]

Lanzaro G.C., Narang S.K. & Seawright J.A. (1990) Speciation in an anopheline (Diptera: Culicidae) mosquito: enzyme polymorphism and the genetic structure of populations. *Ann. Entomol. Soc. Am.* **83**, 578–585. [2.4.2]

Lewontin R.C. (1974) *The Genetic Basis of Evolutionary Change*. Columbia University Press, New York. [2.3.4]

Löfstedt C. (1993) Moth pheromone genetics and evolution. *Phil. Trans. Roy. Soc., Lond. B* **340**, 167–177. [2.3.2]

Lorkovic Z. (1990) The butterfly chromosomes and their application in systematics and phylogeny. In: *Butterflies of Europe*, Vol. 2 (ed. O. Kudrna), pp. 332–396. Aula, Wiesbaden. [Appendix 2.1]

Macgregor H.C. & Horner H.A. (1980) Heteromorphism for chromosome 1, a requirement for development in crested newts. *Chromosoma* **76**, 111–122. [2.3.3]

Macgregor H. & Varley J. (1988) *Working with Animal Chromosomes* (2nd edn). Wiley, Chichester. [Appendix 2.1]

Macnair M.R. (1991) Why the evolution of resistance to anthropogenic toxins normally involves major gene changes: the limits to natural selection. *Genetica* **84**, 213–219. [2.3.1]

Mallet J. (1989) The genetics of warning colour in Peruvian hybrid zones of *Heliconius erato* and *H. melpomene*. *Proc. Roy. Soc. Lond. B* **236**, 163–185. [2.3.1]

Mallet J. (1993) Speciation, raciation, and color pattern evolution in *Heliconius* butterflies: evidence from hybrid zones. In: *Hybrid Zones and the Evolutionary Process* (ed. R.G. Harrison), pp. 226–260. Oxford University Press, New York. [2.4.1]

Mallet J. (1995) A species definition for the Modern Synthesis. *Trends Ecol. Evol.* **10**, 294–299 [2.4.2]

Mallet J., Barton N., Lamas G., Santisteban J., Muedas M. & Eeley H. (1990) Estimates of selection and gene flow from measures of cline width and linkage disequilibrium in *Heliconius* hybrid zones. *Genetics* **124**, 921–936. [2.4.2]

Mallet J., Korman A., Heckel D.G. & King P. (1993) Biochemical genetics of *Heliothis* and

Helicoverpa (Lepidoptera: Noctuidae) and evidence for a founder event in *Helicoverpa zea*. *Ann. Entomol. Soc. Am.* **86**, 189–197. [2.3.1]

Marks J. (1994) Black, white, other. *Nat. Hist.* **103** (December), 32–35. [2.2]

Maxson L.R. & Maxson R.D. (1990) Proteins II: immunological techniques. In: *Molecular Systematics* (eds D.M. Hillis & C. Moritz), pp. 127–155. Sinauer, Sunderland, Massachusetts. [Appendix 2.1]

May B. (1992) Starch gel electrophoresis of allozymes. In: *Molecular Genetic Analysis of Populations* (ed. A.R. Hoelzel), pp. 1–27, 271–280. IRL Press at Oxford University Press. [Appendix 2.1]

Maynard Smith J. (1989) *Evolutionary Genetics*. Oxford University Press. [2.3.3] [2.4.1]

Mayr E. (1940) Speciation phenomena in birds. *Am. Nat.* **74**, 249–278. [2.4.2]

Mayr E. (1942) *Systematics and Origin of Species*. Columbia University Press, New York. [2.4.2]

Mayr E. (1970) *Populations, Species, and Evolution*. Harvard University Press, Cambridge, Massachusetts. [2.3.1] [2.3.4] [2.4.2]

Menken S.B. (1987) Is the extremely low heterozygosity level in *Yponomeuta rorellus* caused by bottlenecks? *Evolution* **41**, 630–637. [2.4.1]

Milligan B.G., Leebens-Mack J. & Strand A.E. (1994) Conservation genetics: beyond the maintenance of marker diversity. *Molec. Ecol.* **3**, 423–435. [2.4.1] [2.4.2]

Møller A.P. (1994) *Sexual Selection and the Barn Swallow*. Oxford University Press. [2.3.1]

Moore W.S. & Price J.T. (1993) Nature of selection in the northern flicker hybrid zone and its implications for speciation theory. In: *Hybrid Zones and the Evolutionary Process* (ed. R.G. Harrison), pp. 196–225. Oxford University Press, New York. [2.3.1] [2.4.2]

Moritz C. (1994) Defining 'Evolutionarily Significant Units' for conservation. *Trends Ecol. Evol.* **9**, 373–375. [2.3.4] [2.4.2]

Murphy R.W., Sites J.W., Buth D.G. & Haufler C.H. (1990) Proteins I: isozyme electrophoresis. In: *Molecular Systematics* (eds D.M. Hillis & C. Moritz), pp. 45–126. Sinauer, Sunderland, Massachusetts. [Appendix 2.1]

Narang S.K., Kaiser P.E. & Seawright J.A. (1989) Identification of species D, a new member of the *Anopheles quadrimaculatus* complex: a biochemical key. *J. Am. Mosq. Control Assoc.* **5**, 317–324. [2.4.2]

Nijhout H.F. (1991) *The Development and Evolution of Butterfly Wing Patterns*. Smithsonian Institution Press, Washington DC. [2.3.1]

Nunney L. (1993) The influence of mating system and overlapping generations on effective population size. *Evolution* **47**, 1329–1314. [2.4.1]

O'Hara R.J. (1994) Evolutionary history and the species problem. *Am. Zoologist* **34**, 12–22. [2.4.2]

Palumbi S.R., Martin A, Romano S., McMillan W.O., Stice L. & Grabowski G. (1991) *The Simple Fool's Guide to PCR* (2nd edn.). Department of Zoology, University of Hawaii, Honolulu. [Appendix 2.1]

Pasteur N., Pasteur G., Bonhomme F., Catalan J. & Britton-Davidian J. (1988) *Practical Isozyme Genetics*. Ellis Horwood, Chichester. [Appendix 2.1]

Patton J.L. (1993) Hybridization and hybrid zones in pocket gophers (Rodentia, Geomyidae). In: *Hybrid Zones and the Evolutionary Process* (ed. R.G. Harrison), pp. 290–308. Oxford University Press, New York. [2.4.2]

Patton J.L. & Smith M.F. (1994) Paraphyly, polyphyly and the nature of species boundaries in pocket gophers (genus *Thomomys*). *Syst. Biol.* **43**, 11–26. [2.4.2]

Platt A.P. (1983) Evolution of North American admiral butterflies (*Limenitis*: Nymphalidae). *Bull. Entomol. Soc. Am.* **29**, 10–22. [2.3.1]

Poulton E.B. (1890) *The Colours of Animals*. Trübner & Co Ltd, London. [2.3.1]

Poulton E.B. (1904) What is a species? *Proc. Entomol. Soc. Lond.* **1903**, lxxvii–cxvi. [2.4.2]

Queller D.C., Strassmann J.E. & Hughes C.R. (1993) Microsatellites and kinship. *Trends Ecol. Evol.* **8**, 285–288. [2.3.4] [Appendix 2.1]

Ragge D.R. (1965) *Grasshoppers, Crickets and Cockroaches of the British Isles*. F. Warne, London. [2.3.1]

Reznick D.A. (1990) Plasticity in age and size at maturity in male guppies (*Poecilia reticulata*): an experimental evaluation of alternative models. *J.Evol. Biol.* **33**, 185–203. [2.3.1]

Richardson B.J., Baverstock P.R. & Adams M. (1986) *Allozyme Electrophoresis: a handbook for animal systematics and population studies*. Academic Press, London. [Appendix 2.1]

Ridley M. (1993) *Evolution*. Blackwell Scientific, Oxford. [2.4.3]

Robbins C.S., Bruun B., Zim H.S. & Singer A. (1983) *A Guide to Field Identification. Birds of North America* (Revised, expanded edn). Golden Press, New York. [2.3.1]

Rooney D.E. & Czepulkowski B.H. (eds) (1992) *Human Cytogenetics: a practical approach* (2nd edn), *Vol. I. constitutional analysis*. IRL Press, Oxford. [Appendix 2.1]

Rothschild W. & Jordan K. (1903) A revision of the lepidopterous family Sphingidae. *Novitat. Zool.* **9** (supplement), **i–cxxv**: 1–813. [2.4.2]

Rousset F. & Raymond M. (1991) Cytoplasmic incompatibility in insects: why sterilize females? *Trends Ecol. Evol.* **6**, 54–57. [2.4.2]

Sbordoni V. (1993) Molecular systematics and the multidimensional concept of species. *Biochem. Syst. Ecol.* **21**, 39–42. [2.4.2]

Schmid B. (1992) Phenotypic variation in plants. *Evol. Trends Plants* **6**, 45–60. [2.3.1]

Schmidt B.R. (1993) Are hybridogenetic frogs cyclical parthenogens? *Trends Ecol. Evol.* **8**, 271–273. [2.3.3]

Scott J.A. (1986) *Butterflies of North America*. Stanford University Press, Stanford. [2.3.1]

Scriven P.N. (1992) Robertsonian translocations introduced into an island population of house mice. *J. Zool.* **227**, 493–502. [2.4.1]

Scriven P.N. & Bauchau V. (1992) The effect of hybridization on mandible morphology in an island population of the house mouse. *J. Zool.* **226**, 573–583. [2.4.1]

Searle J.B. (1993) Chromosomal hybrid zones in eutherian mammals. In: *Hybrid Zones and the Evolutionary Process* (ed. R.G. Harrison), pp. 309–353. Oxford Universtiy Press, New York. [2.3.3]

Seawright J.A., Kaiser P.E. & Narang S.K. (1991) A unique chromosomal dimorphism in species A and B of the *Anopheles quadrimaculatus* complex. *J. Hered.* **82**, 221–227. [2.3.3]

Sessions S.K. (1990) Chromosomes: molecular cytogenetics. In: *Molecular Systematics* (eds D.M. Hillis & C. Moritz), pp. 156–203. Sinauer, Sunderland, Massachusetts. [Appendix 2.1]

Sharma A.K. & Sharma A. (1965) *Chromosome Techniques: theory and practice*. Butterworths, London. [Appendix 2.1]

Sibley C.G. & Ahlquist J.E. (1986) Reconstructing bird phylogenies by comparing DNA. *Sci. Amer.* **254**, 82–93. [Appendix 2.1]

Simpson E.H. (1949) Measurement of diversity. *Nature* **163**, 688. [2.4.1]

Singh R.S. & Hale L.R. (1994) Regional variation in fruitflies. *Nature* **369**, 450. [2.3.4]

Skinner B. & Wilson D. (1984) *Colour Identification Guide to Moths of the British Isles (Macrolepidoptera)*. Viking Penguin, Harmondsworth, Middlesex. [2.3.1]

Slatkin M. (1985) Rare alleles as indicators of gene flow. *Evolution* **39**, 53–65. [2.4.2]

Slatkin M. (1989) Detecting small amounts of gene flow from phylogenies of alleles. *Genetics* **121**, 609–612. [2.4.2]

Slatkin M. (1991) Inbreeding coefficients and coalescence times. *Genet. Res.* **58**, 167–175. [2.4.2]

Slatkin M. & Arter H.E. (1991a) Reply to Sokal and Oden. *Am. Nat.* **138**, 522–523. [2.4.2]

Slatkin M. & Arter H.E. (1991b) Spatial autocorrelation methods in population genetics. *Am. Nat.* **138**, 499–517. [2.4.2]

Slatkin M. & Barton N.H. (1989) A comparison of three indirect methods for estimating average levels of gene flow. *Evolution* **43**, 1349–1368. [2.2] [2.4.2]

Slatkin M. & Maddison W.P. (1989) A cladistic measure of gene flow inferred from the phylogenies of alleles. *Genetics* **123**, 603–614. [2.4.2]

Smith D.C. (1988) Heritable divergence of *Rhagoletis pomonella* host races by seasonal asynchrony. *Nature* **336**, 66–67. [2.3.2]

Sokal R.R. & Crovello T.J. (1970) The biological species concept: a critical evaluation. *Am. Nat.* **104**, 107–123. [2.4.2]

Sodal R.R. & Oden N.L. (1991) Spatial autocorrelation analysis as an inferential tool in population genetics. *Am. Nat.* **138**, 518–521. [2.4.2]

Sokal R.R., Jacquez G.M. & Wooten M.C. (1989) Spatial autocorrelation analysis of migration and selection. *Genetics* **121**, 845–855. [2.4.2]

Soltis P.S., Soltis D.E. & Doyle J.J. (eds) (1990) *Molecular Systematics of Plants*. Chapman & Hall, New York. [Appendix 2.1]

Stearns S.C. (1989) The evolutionary significance of phenotypic plasticity. *BioScience* **39**, 436–445. [2.3.1]

Stone G.N. & Sunnucks P. (1993) Genetic consequences of an invasion through a patchy environment: the cynipid gallwasp *Andricus quercuscalicis* (Hymenoptera: Cynipidae). *Molec. Ecol.* **2**, 251–268. [2.4.1]

Stresemann E. (1975) *Ornithology: from Aristotle to the Present*. Harvard University Press, Cambridge, Massachusetts Reprinted edition of Stresemann's 1951 'Entwicklung der Ornithologie'. Trans. H.J. & C. Epstein; ed. G.W. Cottrell; foreword and epilogue by E. Mayr. [2.4.2]

Swofford D.L. & Selander R.B. (1981) BIOSYS-1: a FORTRAN program for the comprehensive analysis of electrophoretic data in population genetics and systematics. *J. Hered.* **72**, 281–283. [2.4.2]

Szymura J.M. & Barton N.H. (1991) The genetic structure of the hybrid zone between the fire-bellied toads *Bombina bombina* and *B. variegata*: comparisons between transects and between loci. *Evolution* **45**, 237–261. [2.3.1]

Tauber C.A. & Tauber M.J. (1977) Sympatric speciation based on allelic changes at three loci: evidence from natural populations in two habitats. *Science* **197**, 1298–1299. [2.3.2]

Templeton A.R. (1977) Analysis of head shape differences between two interfertile species of Hawaiian *Drosophila*. *Evolution* **31**, 630–641. [2.3.1]

Thomas D.A. & Barber H.N. (1974) Studies on leaf characteristics of a cline of *Eucalyptus urnigera* from Mount Wellington, Tasmania. I. Water repellancy and the freezing of leaves. *Aust. J. Bot.* **22**, 501–512. [2.3.1]

Thomas J.A. & Lewington R. (1991) *The Butterflies of Britain and Ireland*. Dorling Kindersley, London. [2.3.1]

Tinbergen L. (1960) The natural control of insects in pinewoods. I. Factors influencing the intensity of predation in songbirds. *Archs. Neerl. Zool.* **13**, 265–343. [2.3.1]

Trask B.J. (1991) Fluorescence *in-situ* hybridization applications in cytogenetics and gene mapping. *Trends Genet.* **7**, 149–154. [Appendix 2.1]

Turelli M. & Hoffmann A.A. (1991) Rapid spread of an inherited incompatibility factor in California *Drosophila*. *Nature* **353**, 440–442. [2.4.2]

Turner J.R.G. (1984) Mimicry: the palatability spectrum and its consequences. In: *The Biology of Butterflies* (eds R.I. Vane-Wright & P.R. Ackery), pp. 141–161. (Symposia of the Royal Entomological Society of London No. 11.) Academic Press, London. [2.3.1] [2.4.2]

Val F.C. (1977) Genetic analysis of the morphological differences between two interfertile species of Hawaiian *Drosophila*. *Evolution* **31**, 611–629. [2.3.1]

Vrijenhoek R.C. (1994) Genetic diversity and fitness in small populations. In: *Conservation Genetics* (eds V. Loeschcke, J. Tomiuk & S.K. Jain), pp. 37–53. Birkhäuser, Basel. [2.4.1]

Wallace A.R. (1865) On the phenomena of variation and geographical distribution as illustrated by the Papilionidae of the Malayan region. *Trans. Linn. Soc. Lond.* **25**, 1–71. [2.4.2]

Ward R.D., Skibinski D.O.F. & Woodwark M. (1992) Protein heterozygosity, protein structure, and taxonomic differentiation. *Evol. Biol.* **26**, 73–159. [2.3.4]

Weir B.S. (1990) *Genetic Data Analysis*. Sinauer, Sunderland, Massachusetts. [2.4.2]

White M.J.D.(1978) *Modes of Speciation*. W.H. Freeman, San Francisco. [2.3.1] [2.3.3]

Whitlock M.C.(1992) Temporal fluctuations in demographic parameters and the genetic variance among populations. *Evolution* **46**, 608–615. [2.4.2]

Whittaker R.H. & Feeny P.P. (1971)Allelochemics: chemical interactions between species. *Science* **171**, 757–770. [Appendix 2.1]

Williams J.G.K., Kubelik A.R., Livak K.J., Rafalski J.A. & Tingey S.V. (1990) DNA polymorphisms amplified by arbitrary primers are useful as genetic markers. *Nucl. Acids Res.* **18**, 6531–6535. [Appendix 2.1]

Wilson A.C. (1985) The molecular basis of evolution. *Sci. Am.* **253**, 164–173. [Appendix 2.1]

Wilson E.O. & Brown W.L. (1953) The subspecies concept and its taxonomic application. *Syst. Zool.* **2**, 97–111. [2.3.4]

Wright S. (1931) Evolution in Mendelian populations. *Genetics* **10**, 97–159. [2.4.1]

Wright S. (1943) Isolation by distance. *Genetics* **28**, 114–138. [2.2]

Wright S. (1951) The genetical structure of populations. *Ann. Eugenics* **15**, 323–354. [2.2] [2.4.2]

Wright S. (1969) *Evolution and the Genetics of Populations, Volume 2. The theory of gene frequencies*. University of Chicago Press, Chicago. [2.2]

Wright S. (1980) Genic and organismic selection. *Evolution* **34**, 825–843. [2.4.2]

Wright S., Dobzhansky T. & Hovanitz W. (1942) The allelism of lethals in the third chromosome of *Drosophila pseudoobscura*. *Genetics* **27**, 363–394. [2.4.2]

Yunis J.J. & Prakash O. (1982) The origin of man: a chromosomal pictorial legacy. *Science* **215**, 1525–1530. [2.3.3]

Appendix 2.1

A quick guide to genetic and molecular markers used in biodiversity studies.

(By Martin Brookes, Igor Emelianov, Owen McMillan, Owen Rose and James Mallet.)

Marker	Type of data	Cost × effort	Resolution	Taxonomic level within which useful	Brief description
CHROMOSOMES					
Karyotype	Chromosome counts	L	L	≥Genus − population	Squash cell, observe stained chromosomes under microscope. Cytological stains reveal banding patterns (Sharma & Sharma, 1965; MacGregor & Varley, 1988; Lorkovic, 1990; Sessions, 1990; Rooney & Czepulkowski, 1992). FISH visualizes chromosomal location of sequences using fluorescer-marked DNA diagnostic probe*. 'Chromosome paints' use FISH probes to mark whole chromosomes (Trask, 1991)
C- and G-banding	Banding patterns −genotypes	L	L/M	(Note: polyteny only in certain groups/tissues, e.g. larval salivary glands of Diptera)	
Polytene chromosomes	Banding patterns −genotypes	L	M		
In situ hybridization (ISH, FISH)	Physical location of DNA sequences on chromosome	H	M		
'SECONDARY' BIOCHEMISTRY					
Phenolics (inc. tannins), terpenoids, flavonoids, glycosides, pigments, cuticular hydrocarbons, etc.	Presence/absence or quantitative measures	L	L	≥Genus (rarely, within species)	Secondary (non-essential to metabolism, often defensive or sexual) chemicals are extracted and identified. This technique is especially useful in plants (Whittaker & Feeny, 1971; Forrest, 1994), but also useful in some animals (e.g. pigments in butterflies; Ford, 1941; Brown, 1967), and cuticular hydrocarbons in insects (Howard & Blomquist, 1982)
PROTEINS					
Immunological methods					
General	'Immunological distance'	M	L	≥Species	Antibodies to purified proteins (antigens) react more strongly to proteins of similar amino acid sequence than to proteins with divergent sequence.
Blood groups	Dominant genotypes	L	M	Population (humans only)	

continued

Marker	Type of data	Cost × effort	Resolution	Taxonomic level within which useful	Brief description
					Strength of antibody–antigen interaction gives a value for immunological distance (Maxson & Maxson, 1990)
Electrophoresis Allozymes and other soluble proteins	Genotypes	L	M	Genus –population	Crude non-denatured protein extracts are separated by electrophoresis†. Stains specific to enzyme activity or other protein characteristic visualize the position of protein on gel (Richardson *et al.*, 1986; Pasteur *et al.*, 1988; Murphy *et al.*, 1990; May, 1992)
ORGANELLE DNA *Mitochondrial, mtDNA* Total, restriction analysis	Haplotypes	H	H	≥Genus – population	Restriction analysis‡ of purified organelle or total DNA, or direct sequencing§ of PCR‖ amplified regions can be used to assay variation within organelle DNA. Because organelle DNA is haploid and mostly non-recombining, sequence or restriction site differences can be used to construct single gene phylogenetic trees, or genealogies. MtDNA (≈16kb in mammals) evolves very fast, thereby producing much variation even within species. Evolution within the much larger (≥200kb) cpDNA is slower, so that analyses are usually restricted to the species level and above. (Avise *et al.*, 1987; Wilson *et al.*, 1985; Soltis *et al.*, 1990; Chase *et al.*, 1993)
Partial sequences	Sequence	VH	H		
Chloroplast, cpDNA (as for mtDNA)	Haplotypes, sequences	H, VH	H	≥Genus –species	

continued on p. 50

Marker	Type of data	Cost × effort	Resolution	Taxonomic level within which useful	Brief description
REPEATED NUCLEAR DNA					
Middle repetitive DNA					
Ribosomal DNA, rDNA	Dominant multilocus genotypes	H	L	≥ Genus – population	Restriction digest‡ of total DNA is electrophoresed†, Southern blotted†, and probed* with rDNA. Banding pattern reveals variation in length and restriction sites of the rDNA repeat units. Alternatively, the repeats can be sequenced. There are often problems in interpretation because of multiple copies, which cannot be separated, and concerted evolution which tends to homogenize ribosomal genes within individuals (Avise, 1994)
Highly repetitive DNA, variable number tandem repeats (VNTRs)					
Multilocus minisatellite, DNA 'fingerprint'	Dominant multilocus genotypes	H	F	Population	Minisatellites are repeated sequences about 10–50 bp/repeat, and are present in tandem arrays scattered through the genome. Total cellular DNA is digested with a restriction enzyme‡, electrophoresed and blotted†. The filter is probed with a part of the repeat to reveal banding patterns. The number of minisatellite repeats evolves rapidly by 'slippage' during replication, resulting in length variation. DNA 'fingerprints' therefore differ greatly between individuals in a population. Relatedness can be estimated from percent band-sharing

continued

50

Marker	Type of data	Cost × effort	Resolution	Taxonomic level within which useful	Brief description
Single locus minisatellite	Genotype	H	H	Population	PCR‖ and electrophoresis† or electrophoresis followed by stringent hybridization* can reveal length variation at single loci. Resulting genotypic data is more easily interpreted than multilocus fingerprints (Burke, 1989)
Microsatellites	Genotype	H	H	Species – population	Microsatellites are short sequence repeats (usually 2–5 bp/repeat). A genomic library¶ is probed using the simple repeat. The cloned fragments positive for microsatellites are then sequenced§ to provide flanking primer sequences for PCR. PCR‖ is then used to amplify the specific repeat locus in different individuals; subsequently electrophoresis† reveals length variation, depending on the number of repeats. Obtaining primers by cloning, probing and sequencing is very time-consuming and costly; however, once primers are known, PCR analysis is relatively cheap. Evolution by slippage is very fast, yielding numerous alleles per locus (Queller *et al.*, 1993)

continued on p. 52

Marker	Type of data	Cost × effort	Resolution	Taxonomic level within which useful	Brief description
SINGLE COPY NUCLEAR DNA					
Multiple sequences					
DNA–DNA hybridization	Differences in 'melting temperature' (temperature at which dissociation occurs); can be used to obtain genetic distance measures between taxa	H	L	⩾Species	Total DNA is purified to remove repetitive fraction, made single-stranded, labelled, then incubated with excess of sample DNA from same or different species. This allows reannealing* with heterologous strands. Mixture is then slowly heated, giving rates of 'melting'* inversely related to sequence similarity (Sibley & Ahlquist, 1986)
Randomly amplified polymorphic DNA, RAPD	Multilocus dominant genotypes	L	F	Population	Short random primers (8–10 bp) are used in PCR‖ to amplify random DNA fragments which are then separated by electrophoresis† (Williams et al., 1990)
Single copy sequences					
RFLP	Genotypes	H	M/H	⩾Genus – population	A particular sequence is obtained by probing* or PCR‖, then subjected to restriction analysis‡ or sequencing§. Rapidly evolving regions such as introns can be assayed to examine diversity below the species level (Avise, 1994)
Gene sequence	Sequence	VH	H	⩾Genus – population	

Notes Cost and effort: L, low; M, medium; H, high; VH, very high; this column attempts to give an extremely rough idea of the cost of a 'typical study', which may vary widely. Resolution: L, low, unclear inheritance, often at multiple loci; M, medium, clear single locus (or chromosome, organelle) inheritance, usually with under 5 alleles per locus; H, high, single locus inheritance, often with many more than 5 alleles per locus; F, 'fingerprint', multiple discrete bands on a gel, each presumably encoded by a single locus with dominant expression.

Important techniques used in molecular biology

DNA denaturation/renaturation or hybridization. Heating to about 90°C denatures DNA into single strands, renaturation occurs below about 60°C. This reaction is the basis for an enormous variety of DNA techniques. The conditions of renaturation can be varied to give variable stringency (specificity), depending on the application. In DNA–DNA hybridization, the temperature of denaturing (melting) is used to assess sequence similarity.

Hybridization (or probing) after blotting uses a single-stranded probe sequence to mark the position of homologous single-stranded on electrophoretic gels or other media. *In situ* hybridization (ISH, from which FISH or fluorescent *in situ* hybridization, GISH or genomic ISH) is a method for locating probed sequences on cytological specimens.

†*Electrophoresis*. Method that separates molecules by charge as well as by molecular weight. Aqueous samples are placed on a matrix across which an electric field is applied. Molecules then migrate through the matrix depending on their charge, molecular weight and conformation. Proteins are usually separated on starch or polyacrylamide (polyacrylamide gel electrophoresis, or PAGE) gels, or on plates coated with a thin layer of cellulose acetate. DNA separation is usually done on agarose (300–20 000 bp) or polyacrylamide (10–1000 bp). After electrophoresis, blotting is often performed to transfer DNA (Southern blotting), RNA (Northern blotting) or protein (Western blotting) to other media so that it can be visualized, often by autoradiography after probing (see ‖ below).

‡*Restriction site analysis*. Bacterial enzymes, called 'restriction endonucleases', cut DNA at specific sites, 'restriction sites', which are often small inverted repeat 'palindromes'; for example GAATTC is cut by *Eco*RI, an enzyme found in *Escherichia coli*. Restriction endonucleases in nature defend bacteria against invading DNA of phage viruses. DNA cut with restriction enzymes may be polymorphic for presence or absence of restriction sites, or due to variation in length between sites. Both produce variation in the length of restriction fragments seen on electrophoretic gels, giving rise to the term 'restriction fragment length polymorphism' (RFLP).

§*DNA sequencing*. In the Sanger method, *in vitro* DNA synthesis is interrupted repeatedly during DNA extension at one of the four nucleotides (A, G, C, or T). For each nucleotide, a series of DNA chains is produced that correspond to each position of interruption. Radioactively labelled products of four such reactions, one for each nucleotide, are run side-by-side on long polyacrylamide gels, allowing the sequence to be deduced from the four-lane pattern of bands. The procedure has now been mechanized; nucleotide-specific fluorescent labels allow single reactions to be run, followed by single-lane electrophoresis giving four-colour (one for each base) ladders. Modern robotic sequencing and gel-reading techniques, together with PCR, has allowed an order-of-magnitude increase in the rate of production of highly accurate sequences, so that it is rapidly becoming feasible to do large scale sequence surveys with hundreds of samples.

‖*Polymerase chain reaction* (*PCR*). A revolutionary *in vitro* technique for specific amplification of up to about 5 kb of DNA. A pair of primers, usually about 20 bp long, are constructed which are homologous to the 5′ end of each flanking sequence. These primers, together with template DNA, DNA polymerase, and free deoxyribonucleotides A,G,C,T, and other essential ingredients are added together in a mixture. The reaction is run in an apparatus, called a thermal cycler, which alternately allows denaturation at high temperatures, and primer annealing followed by DNA synthesis (primer extension) at lower temperatures. Most reactions use a polymerase (*Taq* polymerase) from *Thermophilus aquaticus*, a bacterium which lives in hot springs, because this enzyme is resistant to the high temperatures needed to denature DNA. Each new single stranded copy can itself act as a template, so the copy number should, in theory, double with every thermal cycle. Because copies increase exponentially, the term 'chain reaction' is used. Extremely small quantities of template DNA can be analysed, leading to applications in forensic and even fossil DNA from highly degraded samples; however, this sensitivity can also lead to problems caused by contamination.

¶*Cloning and vectors*. Vectors are the means of cloning (amplifying) DNA sequences of varying length *in vivo*. Typical vectors are plasmids (small circular non-chromosomal DNA fragments found in bacteria), phages, cosmids (plasmids which contain phage packaging sequences which enable purification of the sequence of interest), or YACs (yeast artificial chromosomes). Cloning preceded, and has been partially superseded by, PCR. However, much longer sequences can be cloned, especially in YACs (20–300 kb), than amplified in PCR (up to k for most purposes). Cloning may be non-specific, for example when a genomic library is obtained by cloning random pieces of DNA from all over a species' genome, or highly specific, as in the cloning of a specific cDNA, i.e. DNA obtained by reverse transcription from the mRNA of a specific structural gene.

General references: Hillis & Moritz (1990), Palumbi *et al.* (1991), Hoelzel (1992), Forrest (1994), Bruford & Wayne (1994) and Avise (1994).

3: Comparing character diversity among biotas

PAUL H. WILLIAMS and CHRIS J. HUMPHRIES

3.1 Introduction

A measure of biodiversity must solve two problems. First, what, ultimately, is to be measured? Second, how, realistically, can appropriate data be obtained?

What are we trying to measure? In general, measures are used to assess the diversity of one or more subsets of objects in comparison either with the entire set (what proportion of total biodiversity is represented in taxon A, vegetation class B, or the biota of area C?), or with other subsets (are they more diverse than taxon D, vegetation class E, or the biota of area F?). Biodiversity can be viewed as the irreducible complexity of all life. Unfortunately not all aspects can be captured in any single measure (Chapter 1; McNeely et al., 1990; Reid et al., 1992; Groombridge, 1992; Wilson, 1992). Choosing one particular aspect to measure is, in effect, choosing which aspect of biodiversity is seen to hold the *value* that needs to be quantified. Consequently, deciding upon which particular aspect of biodiversity to value is a deeper problem than, for example, adjudicating among any inherent preferences of molecular biologists to count substitutions in molecular sequences, of taxonomists to count species, or of ecologists to count vegetation formations. The particular set of values adopted in this chapter does not preclude the use of others, but rather provides one systematic approach, from which it should be possible to compare the consequences of using other kinds of values.

How are we going to measure it? Measuring biodiversity directly as some aspect of the variety of all life will usually be impractical or too expensive (May, 1990; Ehrlich, 1992). For example, not only does a substantial proportion of species remain undescribed, but it is near impossible to count all of the species within a temperate suburban garden, let alone a tropical rain forest. A more realistic approach requires the best quick approximations. This does not mean, however, that measures based on detailed information about organisms are irrelevant. On the contrary, they are essential, because they provide a baseline for assessing which rapid and low-cost approximations are most effective.

The approach of this chapter is to consider where value lies in biodiversity in order to formulate measures, interpreting value in terms of character differences (Section 3.2); which kinds of characters hold biodiversity value (Section 3.3); which assumptions of character distribution among organisms are required for comparisons of character diversity (Section 3.4); how character diversity measures may be calculated using taxonomic surrogates (Section 3.5); and how other surrogate approaches may be used at larger, more practical scales for measuring more of overall biodiversity (Section 3.6).

3.2 Characters as a currency of difference and diversity

If biodiversity is seen as the irreducible complexity of all life, then there may be no single objective definition or measure of biodiversity, only measures appropriate for particular, restricted purposes (Norton, 1994). Consequently, multiple interpretations of biodiversity value are possible (such as attempts to consider diversity in pattern and process separately, e.g. Noss, 1990), and any single formulation cannot include all other possible value systems. However, if formulating a biodiversity measure demands making a choice about values, then an explicit formulation using one value can provide a basis from which to compare the consequences and trade-offs with other kinds of value.

So how might we choose one broadly appropriate kind of value as a starting point for a biodiversity measure? At the most basic level, dictionaries (e.g. OED, 1990) define diversity in terms of 'difference' or 'unlikeness.' Applying the idea of difference to biological diversity, one approach has viewed a dandelion and a fern as representing more diversity than any two individuals or species of dandelions, because the more distantly related organisms in some sense represent more variety or differences (Vane-Wright *et al.*, 1991; Williams *et al.*, 1991; Harper & Hawksworth, 1994).

Differences among organisms are commonly recognized in their constituent biochemicals, morphological structures and behavioural traits. Such differences can be construed as the obverse of the pattern of special similarities or *homologies* that is used to classify organisms. The description of homologies, or *characters*, for characterizing groups of organisms, is the particular subject of taxonomic studies. Homologies may be recognized by tests of conjunction (single occurrence within the same organism), topological correspondence (similarity of position within the organism) and congruence (shared distribution among organisms) (Patterson, 1982, 1987). Homologies are important because of their use in reconstructing estimates of phylogeny (by interpreting characters as sharing a single evolutionary origin through common ancestry). Estimates of phylogeny can then be used to predict the distribution among organisms of the many more characters that have not been sampled. Thus characters or homologies can be seen as one candidate for an underlying *currency* of difference: greater biodiversity value might be possessed by those sets of organisms or biotas that have larger collections of different characters.

The link between biodiversity value and characters of organisms has been

55

explored further in the context of evaluation for biodiversity conservation (Faith, 1992a; Williams *et al.* 1994a). Aside from aesthetics and the intrinsic rights of species, the most generally convincing ethical arguments for biodiversity conservation depend on utility values, not in the narrow sense of the use value of the few species presently exploited, but for representing the potential value of the broadest range of biodiversity. This idea is encapsulated in *option value*, an economic justification for conservation as a form of insurance (Weisbrod, 1964; Hanemann, 1984, 1989), for ensuring possible future uses or services from some organisms. Reid (1994) describes the option value of biodiversity as maximizing human capacity to adapt to changing ecological conditions, which in turn requires maximizing the rest of life's capacity to adapt to change.

Accepting the notion that the value of biodiversity resides in the capacity of organisms to adapt to change redirects the focus of biodiversity value away from species (or higher taxonomic ranks) down to species' characters, where selection is likely to act. Which characters or what their future values will be cannot be known at present. Ignorance of how an organism may prove to be of value in the future means there can be no justification for attempting to weight differentially the units of diversity value. The inevitable consequence of equal weighting of characters is the differential weighting of the organisms that own them, because some organisms can add more currency to the conservation 'bank' than others.

3.3 Kinds of characters

Are all characters equal, be they genetic, morphological, or functional? In the context of identifying sets of entities with the greatest capacity to adapt to change (i.e. the greatest option value), characters are likely to contribute value only if they are expressed in the phenotype of an organism, or in the phenotypes of its progeny. Consequently, for 'genetic' characters, recessive alleles may have value, but unexpressed substitutions in DNA base sequences may not (Williams *et al.*, 1994a).

This distinction between classes of characters (e.g. genetic, phenotypic, functional) would be unimportant were it not that they can show different distributions of diversity among organisms. Models for changes in selectively neutral base differences may yield very different patterns of evolutionary change among organisms from changes in characters under strong selective constraint. Therefore it cannot be assumed that different kinds of character diversity are necessarily distributed in similar ways. Increased selection for richness in any one class of characters may to some extent deselect for another (at least in principle), even if scores for the two are highly correlated among individuals. Not only does selecting a currency require taking responsibility for the choice of what is of most value, but it also implies accepting a lesser value for the remaining characters or attributes.

56

3.3.1 Genetic characters

Generally, because phenotypic characters are expected to be (to a large extent) expressions of parts of the genetic code, genetic characters have been seen as a more fundamental currency of diversity among organisms. Thus, although Wilson (1992) identified the biological species concept as the 'pivotal unit' for the description of biodiversity, he considered the ultimate source of this diversity to be at the level of genes.

Utilitarian justifications for valuing intra-specific genetic characters may have originated from selective breeding programmes. Value has been extended to inter-specific genetic characters, for their potential for creating transgenic organisms and for chemical (e.g. pharmaceutical) prospecting among little-known species. In a recent review of biodiversity by Groombridge (1992), most references to the value of inter-specific genetic diversity are connected with plant breeding or with pharmaceutical prospecting. Therefore it is arguably the phenotypic products of genes, including molecules, that are directly valued.

To the extent that phenotypic characters are expressions of genetic characters at the level of entire transcribed genes or recessive alleles, phenotypic diversity and genetic diversity can be seen as being equivalent views of the level at which diversity value may be seen to reside. The only purpose in distinguishing the term character diversity from genetic diversity here is that the former deliberately excludes variety at lower levels of organization, such as within DNA base sequences, where these differences have no effect on the phenotype and selection (Chapter 2).

3.3.2 Phenotypic characters

It is phenotypic diversity, and perhaps especially morphological diversity, that people respond to first and foremost, because it is most directly amenable to the senses (and so perhaps linked to any emotional response). For many today, just as for Linnaeus (1758: pp. 12–13), the diversity of life is divided into mammals, birds, reptiles and amphibians, fish, insects, 'worms' (or for some, 'animals', a category which excludes those previously listed), and plants. Thus when Groombridge (1992: p. iv) reviewed species diversity, it was divided into micro-organisms, lower plants, higher plants, nematodes, deep-sea invertebrates, soil macrofauna, fishes, higher vertebrates and island species. While habitat criteria clearly influenced this classification, morphological similarity is predominant, and the genetic relationship of natural, monophyletic groups (Farris, 1977) is conspicuously lacking.

Phenotypic characters, or their underlying genetic characters in the sense of whole alleles and transcribed genes, have been suggested to be closest to the currency of most direct use to people (Williams *et al.*, 1994a). All of the utilitarian justifications for whole-gene characters can be applied to the phenotypic products of those genes, including molecules. More directly than at least the unexpressed variation in DNA base sequences, phenotypic characters may be the actual stuff in which value resides (for evolution or utility, e.g. selecting for disease resistance).

3.3.3 Functional characters

Functional diversity among organisms is expected to have value for maintaining the integrity of ecosystems, and hence the services that they provide (Walker, 1992; Schulze & Mooney, 1993).

There are difficulties in deciding how a currency based on functional characters might be measured (Chapter 5). Function includes both the levels of functional anatomy and trophic relationships of organisms. At least part of the option value of morphological characters may lie in their relationship to functional characters. When defined broadly, individual functional groups tend to comprise entire higher taxa, or at least it has often been found convenient to categorize species in this way (Moran & Southwood, 1982; Stork, 1987). Thus there is probably some relationship between functional-group richness of an area and its higher-taxon richness, as counted across a particular rank of the nomenclatural hierarchy (Gaston, 1994; Linder & Midgley, 1994). The relationship is not simple, because even disparate organisms (e.g. from different kingdoms) may sometimes belong to the same guild (Eggleton & Gaston, 1990; Hochberg & Lawton, 1990). In that sense, organisms may be functionally equivalent, and yet they could be of very different (and not consistent) 'desirability' among conservationists (e.g. nematode and hymenopteran parasitoids, hawkmoths and hummingbirds, seed-eating ants and rodents).

3.4 Predicting character ownership

If complete knowledge of relevant character differences were available, then in principle this could be used directly in diversity measures, without recourse to evolutionary models (Williams *et al.*, 1991; Weitzman, 1992, 1993; Solow *et al.*, 1993). Unfortunately, characters cannot usually be counted directly because the number of characters is potentially vast and complete knowledge of characters in the sense used here is seldom available.

In practice, information on the distribution of all characters among organisms usually has to be extrapolated from very small samples of characters. The relative distribution of characters among organisms can be predicted from their genealogy—the phylogenetic tree—using the general evolutionary model of 'descent with modification' (character changes are inherited by all descendants on a tree), together with the appropriate special process model for how character distributions may be related to a genealogical pattern. Cladistic methods can be used to distinguish the divergent (rather than convergent) character changes in samples in order to build trees (classifications) as estimates of genealogical pattern. These trees should then be more generally predictive of all of the other kinds of character difference (Farris, 1983).

Reconstruction of the phylogenetic tree is not a trivial problem. There is difficulty in recognizing homology, the evidence for natural genealogical groups, which is at least as bad for molecular data (because of the problems of alignment, paralogy and xenology) as for morphological data (Patterson *et al.*, 1993). Many different methods for building trees are available (>100: Huelsenbeck & Hillis, 1993), some of which are believed to be better than others, although this depends on the degree to which different assumptions are admitted. Resulting trees for the same organisms rarely agree entirely, whether because different data samples or different tree-building methods were used. The one historically correct phylogenetic tree can never be recognized, only hypothesized (Eggleton & Vane-Wright, 1994). These problems are beyond the scope of this chapter (but see Kluge & Wolfe, 1993).

3.4.1 The problem

If either absolutely all kinds of characters are of value, or if it is completely uncertain which characters are likely to be of value, then the only criterion with which to compare the diversity of sets of organisms is by using representative samples across all kinds of characters. As previously mentioned, the consequences of currency choice for diversity measures are likely to diverge strongly because the different classes of characters show different patterns of distribution among organisms. These different patterns of ownership of characters by organisms are described to some extent by models of evolutionary change in characters. All of these models require contentious, but inescapable assumptions.

3.4.2 Models of character evolution

Predicting character distributions among organisms requires not just a strongly corroborated estimate of the phylogenetic tree, but also selection of an explicit, special evolutionary process model for the way in which characters change across the tree. The possibilities include an 'empirical' model, an anagenetic 'clock' model, and a cladogenetic model. The choice of special model is not necessarily linked to the method used to estimate the tree, but will be governed instead by how the class of characters chosen as a currency of biodiversity value is believed to change with time.

The 'empirical' model of character change, in the present sense, uses counts of character differences from a sample as an estimate of the relative overall numbers of character differences among organisms. Counts of character differences are represented by scaling the relative branch lengths of a tree

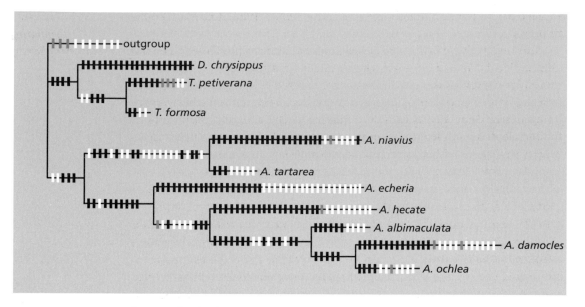

Fig. 3.1 Metric tree for African species of milkweed butterflies (family Nymphalidae) of the genera *Danaus* (1 species), *Tirumala* (2 species) and *Amauris* (7/12 species). The branching order (topology) of the tree is derived by cladistic analysis of 217 morphological and chemical characters (volatile components extracted from male scent organs) from Ackery & Vane-Wright (1984), Vane-Wright *et al.* (1992) and Schulz *et al.* (1993). The branch lengths between each branching point on the tree are then scaled by the number of character changes within this same character sample, which are shown as vertical bars: black bars show non-homoplasious 'forward' changes; light grey bars show homoplasious 'forward' changes and non-homoplasious 'reversals'; and dark grey bars show homoplasious 'reversals'.

(Fig. 3.1). Such a tree, which often has different path lengths from the root to the different terminals, is termed a metric tree.

Even though the vast majority of characters remain unsampled (e.g. Patterson *et al.*, 1993), branch-length estimates from differences within a sample of characters used to reconstruct the tree must be accepted as representative of the overall relative distribution of character changes (Faith, 1992a). This requires the bold assumption that a given sample of characters is unbiased, not only in the sense that sampling effort should be even among branches (Williams & Humphries, 1994), but also in that patterns of both sampled and unsampled characters should behave as though they obeyed the same evolutionary model. This could present severe difficulties, for example in a situation where selectively neutral DNA base sequence data were used in an attempt to predict the distribution of characters of functional morphology that have been selectively constrained. Thus the maidenhair tree *Ginkgo biloba* and the coelacanth *Latimeria chalumnae* may both have diverged from their closest surviving relatives a long time ago, and may both have undergone great molecular change (*Ginkgo*: Hamby & Zimmer, 1992; *Latimeria*: Hillis *et al.*, 1991), but morphological or ecological change does not appear to have been commensurately great (*Ginkgo*: Crane, 1988; *Latimeria*: Forey, in prep.).

Another model of evolutionary change that has been particularly popular with molecular biologists is the 'clock' model. This is usually interpreted as molecular changes (such as DNA base substitutions) accumulating with increasing time elapsed along the branches of trees between branching events (anagenesis). Generally, because the number of character changes is related to time elapsed, path lengths from the root to the different terminals tend to be very similar (or with a strict clock, identical, Fig. 3.2). A tree with terminals all at the same distance from the root is termed 'ultrametric'.

The regularity of this molecular 'clock' remains contentious (Scherer, 1990; Gillespie, 1991), but perhaps because a substantial proportion of the genetic code has been considered to be selectively neutral in evolution (Kimura, 1983), it is more often accepted that numbers of changes in these attributes (usually base-pair substitutions rather than whole-gene characters) are related to time elapsed within a lineage (as opposed to changes merely accumulating with time at widely varying rates) and so to the age of lineages. Clock models are also assumed by some tree-building methods (UPGMA and some distance methods).

If necessary, even the nomenclatural hierarchy can be used to infer branch-length information (Faith, 1992b, 1994c). For example, one interpretation might be to place all modern organisms on trees at an equal distance

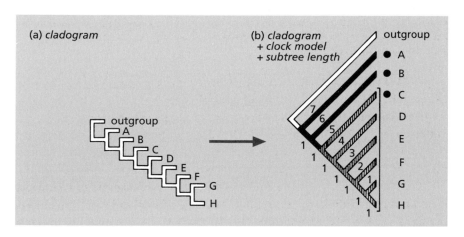

Fig. 3.2 Genealogical trees used to infer the distribution of clock-like character changes and biodiversity value as character richness. For a genealogical classification (a) of eight modern species, A–H, and assuming a correlation between time elapsed and character divergence (so that numbers of character changes associated with each terminal branch are predictable from the position of the neighbouring nodes), then character richness is maximized for three species by (b) representing the earliest-diverging (highest) taxa (species A + B + (any one from C to H)). Species essential to any maximally-scoring set are shown by spots at the end of black branches, whereas species that remain equivalent as choices are shown bracketed across the ends of alternative grey branches, with branches to species contributing lower value to these sets shown in white.

from the root (e.g. converting a cladogram into an ultrametric tree diagram) not only on a time axis, but also in terms of numbers of character changes (Fig. 3.2). This is a special case of the clock model, in which a sample of characters may be seen as having been used in the past to estimate a branch length on an ultrametric, nomenclatural tree. It assumes not only that the nomenclatural hierarchy is a reasonable representation of the branching pattern of the phylogenetic tree, but also that nomenclatural rank reflects age and character divergence. It is likely to be useful only within the scope of classifications by single authors, because of the bias of heterogeneous splitting or lumping of higher taxa among authors. It also depends on deciding which nomenclatural ranks are informative and what their relative branch-length contributions should be.

In contrast, if no assumptions about clock-like correlations between numbers of character changes and time elapsed are made, then it may not be straightforward to predict from the position of a branching point on a tree how many character changes are likely to be associated either with that branching event (cladogenetic changes) or with the neighbouring branch lengths (anagenetic changes). However, it remains possible (for morphological diversity in particular) that the predictive value of the tree's structure may be high, particularly if character changes still tend to accumulate with increasing numbers of branching points along a lineage (Fig. 3.3).

For example, one key difference in the interpretation of classifications based on morphological characters compared with molecular characters has been that a relationship between numbers of character changes and time

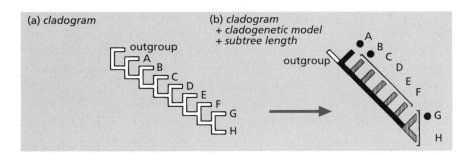

Fig. 3.3 Genealogical trees used to infer the distribution of non-clock-like character changes and biodiversity value as character richness. For a genealogical classification (a) of eight modern species, A–H, and assuming no correlation between time and character divergence (so that numbers of character changes associated with each internode are either not predictable from the position of the neighbouring nodes, or alternatively character changes are associated primarily with nodes), then character richness is maximized for three species by (b) the longest subtree linking the species (species A + (any one from B to F) + (G or H)). Species essential to any maximally-scoring set are shown by spots at the end of black branches, whereas species that remain equivalent as choices are shown bracketed across the ends of alternative grey branches, with branches to species contributing lower value to these sets shown in white.

elapsed is generally less widely accepted in the former. One explanation could be that most morphological changes may be associated with cladogenesis (changing at the branching of lineages, with speciation) (Gould & Eldredge, 1977, 1993; Vrba, 1980; Ax, 1987). This is not to say that there need be a constant number of morphological changes at cladogenesis, but it would be useful to know whether there is a correlation between the numbers of changes and the numbers of cladogenetic events among lineages within a particular group. Of course, any assessment of whether this pattern were general would be greatly strengthened if it were possible to take account of any evidence (e.g. Nee *et al.*, 1994) for 'hidden' cladogenetic events involving extinct lineages.

If whole-gene or phenotypic characters are preferred as a currency of diversity value, then the empirical model is preferred over the extreme clock or cladogenetic models, providing that the predicted phylogeny is based on a large sample of whole-gene or phenotypic characters. In the absence of this information, the cladogenetic model is preferred to the clock model as a predictor of whole-gene or phenotypic character distribution among organisms.

3.5 Measures for comparing character diversity

3.5.1 An overview

Many broadly genealogical or phylogenetic measures of diversity have been proposed. The earliest taxonomic approach was a measure of hierarchical diversity by Pielou (1967; for a critique see Faith, 1994a). This was followed by attempts to explore diversity as patterns of cladistic relationship (Vane-Wright *et al.* in May, 1990; Vane-Wright *et al.*, 1991; Williams *et al.*, 1991; Nixon & Wheeler, 1992; Williams & Humphries, 1994). This approach explicitly attempts to free diversity measures from measuring difference in numbers of characters from samples, which are inevitably biased to some degree; searching instead for concepts of diversity expressed in terms of patterns of relationship. But in so doing, the approach became divorced from the justification of maximizing the character-based currency of option value (Faith, 1992a; Williams *et al.*, 1994a). In response, approaches aiming explicitly to maximize character representation were proposed by Altschul and Lipman (1990), Crozier (1992), Faith (1992a,b), Weitzman (1992, 1993), Solow *et al.* (1993), Faith and Walker (1994), Krajewski (1994) and Crozier and Kusmierski (in press). For critiques of these measures see Faith (1992a, 1993, 1994b). A distinctive approach by Erwin (1991) and by Brooks *et al.* (1992) was to value differentially those biotas that include many particularly closely related (recently diverged) species, because these were considered to reside in likely 'species-dynamo' areas for future evolution. This 'rats and sparrows' approach has been criticized as deliberate selection

63

for *low* character diversity and for its strong assumptions about the future course of evolution (Williams, 1993; Williams *et al.*, 1994a).

In discussing a range of topological (tree-shape) diversity measures, Williams *et al.* (1991) identified two incompatible criteria for measuring the diversity in subsets of terminals on genealogical trees (aside from simple richness): higher-taxon diversity and dispersion (regularity) diversity. Clarification in the interpretation of diversity value for conservation in terms of option value (Faith, 1992a) and how this can be related to evolutionary models (Williams *et al.*, 1994a) has shown that these criteria select for character richness and character combinations respectively. Thus, after selecting an appropriate currency of valued characters and a special evolutionary model that will predict its distribution over genealogical trees, the major remaining choice is whether greatest value is seen as residing in richness of individual characters, or with their integration as richness in combinations of characters.

3.5.2 Character richness

If value is placed on different characters independent of one another, such as might be the case when prospecting for disease resistance in plants, pharmaceuticals etc., then the aim of a biodiversity measure would be to give a higher score to the set of organisms that maximizes the number of different character states represented.

Overall character richness for subsets of organisms can be maximized on metric trees with branch lengths proportional to character changes (Fig. 3.1) by selecting the set of organisms with the maximum *subtree length* (Figs 3.2, 3.3) ('phylogenetic diversity' of Faith, 1992a,b; Williams, 1993, 1994; Faith & Walker, 1994). Maximum character richness is obtained by maximizing subtree length as the sum of the lengths of the branches in the path along the tree connecting all the terminals in the subset (Faith, 1992a). Subtrees should be treated as unrooted if diversity is to be maximized within a subset of taxa (Williams & Humphries, 1994). Ultrametric trees, which result from perfectly clock-like character change, are a special case, for which diversity is maximized by selecting the set of organisms that is richest in the earliest-diverging taxa. A measure of higher-taxon richness is available to give greater weight to sets of organisms richer in higher taxa when branch-length information is lacking (Williams *et al.*, 1991). Even for this case, subtree length remains a consistent general approach (Faith, 1992a,b; Faith & Walker, 1994), but requiring that the branch lengths implied by ultrametric trees should be added explicitly.

3.5.3 Character combinations

Much of the foregoing discussion has been based on the premise that the greatest diversity value will result from maximizing richness in individual or unique character states. However, it may be broader differences in *combinations* of unique states among characters that are seen to have greater option value. These combinations may be required as integrated, functional suites of characters in order to perform certain ecosystem services (e.g. a biocontrol agent may need to be specific to a particular target organism, in a particular habitat, at a particular time of year, and with particular properties of dispersal). An alternative approach for conservation evaluation would therefore be to maximize richness in different combinations of characters.

Choosing organisms that are regularly spaced across trees samples most evenly the topology of the tree, in the sense that it represents most evenly taxa at all levels in the hierarchy, and therefore represents evenly the different character combinations that diagnose these taxa (Fig. 3.4) (Williams, 1993; Williams & Humphries, 1994). A measure of regularity based on Poisson's work on rare events (Greig-Smith, 1983) was used as the basis for a cladistic dispersion measure, which favours sets of organisms with the largest and most even numbers of nodes between them (Williams *et al.,* 1991). This measure had no clearly justified balance between regularity, numbers of organisms, (terminals), and subtree length (Williams *et al.,* 1991; Faith, 1992b). Recently, Faith and Walker (1994) have drawn on a family of *p*-median procedures from the Operations Research literature (Tansel *et al.,*

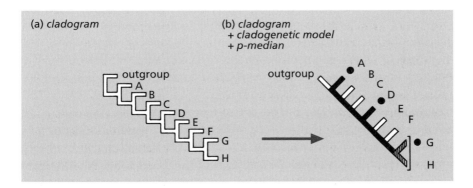

Fig. 3.4 Genealogical trees used to infer the distribution of non-clock-like character changes and biodiversity valued as richness of character combinations. For a genealogical classification (a) of eight modern species, A–H, and assuming no correlation between time and character divergence (so that numbers of character changes associated with each internode are either not predictable from the position of the neighbouring nodes, or alternatively character changes are associated primarily with nodes), then character combinations are maximized for three species by (b) a subtree that is an even or regular representation of the overall tree shape (species A + D + (G or H)). Species essential to any maximally-scoring set are shown by spots at the end of black branches, whereas species that remain equivalent as choices are shown bracketed across the ends of alternative grey branches, with branches to species contributing lower value to these sets shown in white.

1983; Love *et al.*, 1988) that also attempt to locate objects in a regular pattern on a network (such as a metric tree). Using a formal disc model for the distribution of character combinations on a metric tree, Faith and Walker (1994) provide an improved measure of the representation of character combinations in *p*-median diversity.

The general model represents different combinations of characters as 'discs' of differing 'radius' along the branches of metric trees (Faith & Walker, 1994). If particular combinations of unique character states come and go over time, any point along a tree may be the centre of distribution for a character-state combination. The distance that can be travelled along the tree while still finding a particular combination defines the combination radius. All points are equally likely to be the centres for a combination, and combinations can have any radius with equal probability. The *continuous p-median* then maximizes all historical character combinations by maximizing intersection of all discs anywhere on a tree with the selected terminals, by minimizing the sum or average distances from all points on a tree to their nearest selected terminal among the subset of *p* terminals. The *discrete p-median* maximizes extant character combinations by maximizing intersection of discs centred on unselected terminals with the selected terminals, by minimizing the sum or average distances from all unselected terminals on a tree to their nearest selected terminal.

Faith (1994c) prefers the continuous *p*-median over the discrete *p*-median because several similar species forming unresolved 'bushes' (polytomies) on trees contribute less to the score with the former. The effect of these unresolved parts of trees on the discrete *p*-median manifests itself as an inconsistency in highest scoring subsets between species in the polytomy and single species at the end of longer branches, as numbers of species increase evenly across the entire tree (a very similar problem to that described by Faith, 1992b, for the earlier dispersion measure of Williams *et al.*, 1991). The same inconsistency would also be shown by the continuous *p*-median if the number of species in the unresolved bush on the tree were sufficiently high. Williams (1994, 1995b) suggests that the discrete *p*-median may be more appropriate to the conservation goal of seeking representation of only those character combinations actually found in extant (terminal) species on trees.

3.6 Measuring overall biodiversity

The taxonomic diversity measures described so far in this chapter are formulated for maximizing diversity value within a small taxonomic group of organisms, for which estimates of genealogy and geographical distribution exist. In this situation it gives the most precise measure of diversity. This is a realistic approach for particularly well-known groups of organisms (e.g. Mickleburgh *et al.*, 1992; Williams & Humphries, 1994), although it is not practical for overall biodiversity (all life). Taxonomic diversity measures are particularly likely to differ from measures of species richness in the distribution of values among areas (Figs 3.5, 3.6) when the numbers of species are low, and when faunas or floras differ strongly in the degree to which the species are taxonomically clumped (i.e. when the most species-rich biotas are relatively character-poor and vice versa) (Williams & Humphries, 1994).

3.6.1 The problem

Biodiversity concerns all life, so measures of biodiversity to predict character diversity should take account of all organisms. Unfortunately, just as it is not practical to count all of the characters for a species, so it is not practical to count all species for an area (although the reasons differ). Information on distribution and genealogical relationships is unlikely to be available for all species in the foreseeable future, even within very local studies.

3.6.2 Surrogacy as a solution

The conclusion drawn from the present interpretation of diversity value in terms of characters is not that biodiversity must always be scored using taxonomic measures, but rather that other less direct approaches using more practical surrogates are also justified. Without wishing to view it as the only possible unifying scheme, the consequence of identifying biodiversity value with characters could be seen as placing the traditional three levels of biodiversity (genes, species and ecosystem, e.g. Noss, 1990) within a scale of surrogacy for measuring overall biodiversity, arranged from the more direct (e.g. taxonomic measures) to the more remote (e.g. landscape measures) (Table 3.1; Williams, 1995b).

Surrogate measures for character diversity can be chosen for the practical advantage that information on their distribution should be more widely available or more easily acquired. In using genealogical pattern to predict character distributions, the taxonomic approach to diversity recognizes that characters of organisms cannot be counted directly, so that indirect, surrogate approaches are inevitable. It is then a natural extension to recognize a scale of surrogacy for character diversity, within which indirect measures appropriate to a particular situation can be chosen. The use of surrogates in general relies on a predictive relationship between the surrogate variable and the target variable, which in this case is biodiversity interpreted as character diversity. This predictive relationship should always be demonstrated and not just assumed (Landres *et al.*, 1988; Noss, 1990).

Species richness is likely to become a reasonable surrogate for character richness when dealing with large numbers of species (Fig. 3.7: in particular Fig. 3.7a shows a stronger relationship than that shown in Fig. 3.6 for fewer species of the same group). The estimates of character richness in Fig. 3.7 are

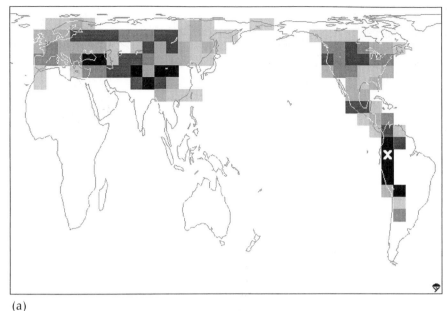

(a)

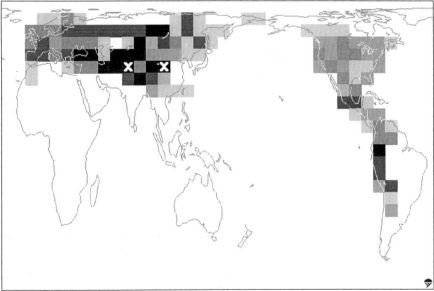

(b)

Fig. 3.5

crude in that they are derived from the available taxonomic classifications rather than directly from any large samples of characters. However, for this relationship to fail would require that the more species-rich faunas and floras become progressively more highly taxonomically clumped (i.e. hypo-diverse or character-poor, in the sense that additional species contribute few complementary characters) with increasing numbers of species (Williams, 1993;

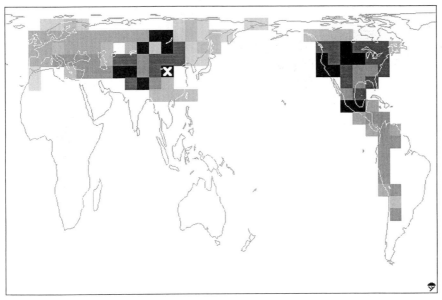

(c)

Fig. 3.5 Comparison of the consequences of different diversity measures for the distribution of biodiversity value, for an example with some species-rich, but unusually highly taxonomically clumped (character-poor), faunas in South America. Maps show regional diversity scores for the *sibiricus*-group of 43 species of bumble bees (genus *Bombus*, family Apidae; distribution data and classification from Williams, 1991, 1995a) on an equal-area grid map (grid-cell area c. 611 000 km², for intervals of 10° longitude) (Williams, 1994): (a) (*opposite*) assuming value is approximated by species richness; (b) (*opposite*) assuming value lies in richness of individual characters, without assuming clock-like patterns of character change, then diversity is measured as subtree-length diversity with equal branch lengths (cf. Fig. 3.3); or (c) (*above*) assuming value lies in richness of character combinations, without assuming clock-like character change, then diversity is measured as discrete *p*-median diversity with equal branch lengths (cf. Fig. 3.4). Scores are represented by logarithmic grey-scale intensities, in classes of approximately equal size by the frequency of values between minimum (light grey) and maximum (black with white 'X'), with white for no records of these bees.

Williams & Humphries, 1994). Any relationship between species richness and character diversity would be fortunate because it greatly reduces the demands on data (because estimates of genealogical relationship are not required), and therefore reduces survey costs in comparison with the use of tree-based measures.

In some circumstances, higher-taxon richness can provide a surrogate for the ultimate value of character richness. This approach reduces the massive extrapolation invoked when using the indicator-group approach (reviewed by Kremen *et al.*, 1993; Reid *et al.*, 1993). It also avoids the assumptions of environmental surrogates for character diversity (Faith & Walker, 1994) of a uniform random distribution of species in niche space (or simple graded transformations thereof), and of an equilibrium distribution of organisms among patches of suitable habitat. The principle behind using higher taxa is

69

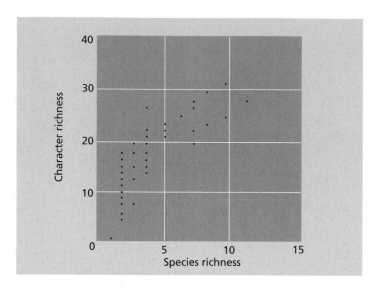

Fig. 3.6 Species richness and predictions of character richness for the 43 species of bumble bees of the *sibiricus*-group among the equal-area grid cells of the grid from Fig. 3.5 (Spearman rank correlation coefficient 0.922) (distribution data and classification from Williams, 1991, 1995a). Character richness is predicted from a cladistic classification, the cladogenetic model of evolutionary character change (with all branches set to the same 'length') and subtree-length diversity.

that mapping 1000 genera or families represents more of total biodiversity than mapping 1000 species, but without a commensurate increase in costs. Higher-taxon richness has already been used as a surrogate for species richness (Fig. 3.8; Chapter 4, *this volume*; Gaston & Williams, 1993; Williams & Gaston, 1994; Willams *et al.*, 1994b). Higher taxa can also be used for estimates of character diversity with tree-based diversity measures, if higher taxa are monophyletic (Williams, 1993). This latter interpretation is likely to be most robust for the character richness approach with the clock model (because the diversity measure is most dependent on the highest-level, most ancient relationships), and least robust for the character combination approach with the cladogenetic model (because the diversity measure is more dependent on complete sampling of the lowest-level taxa, which are obscured within higher taxa). A particular strength of the higher-taxon surrogate for character diversity in place of species data is that, because unlike many other surrogates it retains information on the identity of taxa within each area, there is some knowledge of the spatial turnover of taxa among areas (Williams *et al.*, 1994b).

More remote surrogates must always be used with caution because of decreased precision in their representation of character diversity (Table 3.1). For example, with vegetation, some classes (e.g. on highly disturbed land)

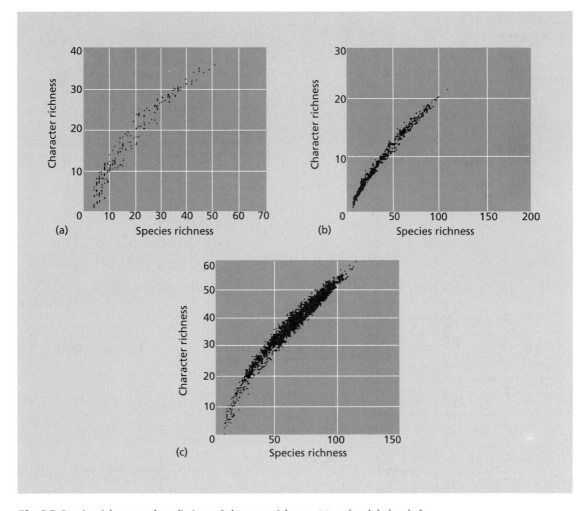

Fig. 3.7 Species richness and predictions of character richness: (a) at the global scale for 241 species of bumble bees (genus *Bombus*, family Apidae) among cells of an equal-area grid (Fig. 3.5) (Spearman rank correlation coefficient 0.986; distribution data and classification from Williams, 1991, 1995a); (b) at the regional scale for the 729 species of the families Dichapetalaceae, Lecythidaceae, Caryocaraceae, Chrysobalanaceae and the genus *Panopsis* among the 1° × 1° grid cells of the *Flora Neotropica* grid (Spearman rank correlation coefficient 0.994; distribution data and classification from Williams, Prance *et al.*, in press, and references therein); and (c) at the national scale for 218 species of breeding birds (class Aves) among 10 × 10 km grid cells of the British National Grid (Spearman rank correlation coefficient 0.952; distribution data from Gibbons *et al.*, 1993; Williams/Gibbons *et al.*, in press; classification based on Sibley & Ahlquist, 1990). The distribution of character richness is predicted from existing classifications, using the cladogenetic model of evolutionary character change (but ignoring all species outside the area considered and with all branches set to the same 'length') and subtree-length measure of diversity.

may be defined by having a simple subset of the flora of other classes, so that choosing to represent these depauperate classes in preference to the more inclusive classes could actually reduce species and character diversity as an

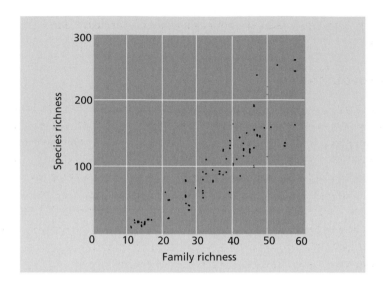

Fig. 3.8 Species richness of plants plotted against family richness, based on plants of ≥2.5 cm stem diameter on 74 plots (53 from the neotropics) of 0.1 ha for which counts were indicated to be complete (Spearman rank correlation coefficient 0.929) (from Williams *et al.*, 1994b).

Table 3.1 If biodiversity value is associated with characters as a fundamental currency, then units at higher levels of biological organization may be arranged on a scale of surrogacy, with advantages of precision for more proximate measures, and advantages of low survey cost for more remote measures.

Advantage: precision as a measure of character diversity	A scale of surrogacy for character diversity	Advantage: ease (/cost) of measurement
Low	(Ecosystems)	High
↓	Landscapes	↑
↓	Land classes	↑
↓	Species assemblages	↑
↓	Higher taxa	↑
↓	Species	↑
High	(Characters)	Low

opportunity cost. Similarly with land classes, some classes (e.g. different classes of deserts) may show low complementarity for characters in comparison with choices made more directly by character-based selection.

Running counter to the scale of precision in surrogacy, approaches at higher levels have other crucial strengths (Table 3.1; Williams, 1995b). Surrogates may become far more practical for measuring overall character diversity (and more information may already be available) at the progressively higher scales of species, higher taxa, and classifications or ordinations of species assemblages, land patches, landscapes and ecosystems. These higher surrogates may also integrate more of the functional processes that are important for maintaining both ecosystem services and ecosystem viability.

3.7 Conclusion

A problem for biodiversity measurement is that only selected aspects can be valued in any one measure. Character diversity is close to one popular understanding of diversity as the difference among organisms, although clarification is needed of the significance of different kinds of characters. Generally it is not possible to quantify even selected aspects (such as character diversity) directly, so that the use of surrogacy in some sense is inevitable and much work needs to be done to explore possible surrogates as the best quick approximations for biodiversity measurement. Fortunately, species richness appears to provide a good first approximation to character richness, at least when applied to large natural assemblages of species.

Acknowledgments

P.H.W. thanks Kevin Gaston, Dan Faith, Sandra Mitchell, Katrina Brown and Dick Vane-Wright for discussion.

References

Ackery P.R. & Vane-Wright R.I. (1984) *Milkweed Butterflies, their Cladistics and Biology, Being an Account of the Natural History of the Danainae, a Subfamily of the Lepidoptera*. British Museum (Natural History) and Cornell University Press, London. [3.4.2]

Altschul S.F. & Lipman D.J. (1990) Equal animals. *Nature* **348**, 493–494. [3.5.1]

Ax P. (1987) *The Phylogenetic System*. Wiley, Chichester. [3.4.2]

Brooks D.R., Mayden R.L. & McLennan D.A. (1992) Phylogeny and biodiversity: conserving our evolutionary legacy. *Trends Ecol. Evol.* **7**, 55–59. [3.5.1]

Crane P.R. (1988) Major clades and relationships in the 'higher' gymnosperms. In: *Origin and Evolution of Gymnosperms* (ed. C.B. Beck), pp. 218–272. Columbia University Press, New York. [3.4.2]

Crozier R.H. (1992) Genetic diversity and the agony of choice. *Biol. Conserv.* **61**, 11–15. [3.5.1]

Crozier R.H. & Kusmierski R.M. (1995) Genetic distances and the setting of conservation priorities. *Biol. Conserv.*, in press. [3.5.1]

Eggleton P. & Gaston K.J. (1990) 'Parasitoid' species and assemblages: convenient definitions or misleading compromises? *Oikos* **59**, 417–421. [3.3.3]

Eggleton P.J. & Vane-Wright R.I. (1994) Some principles of phylogenetics and their implications for comparative biology. In: *Phylogenetics and Ecology* (eds P.J. Eggleton & R.I. Vane-Wright), pp. 345–366. Academic Press, London. [3.4]

Ehrlich P.R. (1992) Population biology of checkerspot butterflies and the preservation of global biodiversity. *Oikos* **63**, 6–12. [3.1]

Erwin T.L. (1991) An evolutionary basis for conservation strategies. *Science* **253**, 750–752. [3.5.1]

Faith D.P. (1992a) Conservation evaluation and phylogenetic diversity. *Biol. Conserv.* **61**, 1–10. [3.2] [3.4.2] [3.5.1] [3.5.2]

Faith D.P. (1992b) Systematics and conservation: on predicting the feature diversity of subsets of taxa. *Cladistics* **8**, 361–373. [3.4.2] [3.5.1] [3.5.2] [3.5.3]

Faith D.P. (1993) Biodiversity and systematics: the use and misuse of divergence information in assessing taxonomic diversity. *Pacific Conserv. Biol.* **1**, 53–57. [3.5.1]

Faith D.P. (1994a) Phylogenetic diversity: a general framework for the prediction of feature diversity. In: *Systematics and Conservation Evaluation* (eds P.L. Forey, C.J. Humphries & R.I. Vane-Wright), pp. 251–268. Oxford University Press. [3.5.1]

Faith D.P. (1994b) Genetic diversity and taxonomic priorities for conservation. *Biol. Conserv.* **68**, 69–74. [3.5.1]

Faith D.P. (1994c) Phylogenetic pattern and the quantification of organismal biodiversity. *Phil. Trans. Roy. Soc., Lond. B* **345**, 45–58. [3.4.2] [3.5.3]

Faith D.P. & Walker P.A. (1994) *DIVERSITY: Reference and User's guide. Version 2.0.* CSIRO, Canberra. [3.5.1] [3.5.2] [3.5.3] [3.6.2]

Farris J.S. (1977) On the phenetic approach to vertebrate classification. In: *Major Patterns in Vertebrate Evolution* (eds M.K. Hecht, P.C. Goody & B.M. Hecht), pp. 823–850. Plenum, New York. [3.3.2]

Farris J.S. (1983) The logical basis of phylogenetic analysis. In: *Advances in Cladistics II* (eds N.I. Platnick & V.I. Funk), pp. 7–36. Columbia University Press, New York. [3.4]

Gaston K.J. (1994) Biodiversity—measurement. *Prog. Phys. Geog.* **18**, 565–574. [3.3.3]

Gaston K.J. & Williams P.H. (1993) Mapping the world's species—the higher taxon approach. *Biodiv. Lett.* **1**, 2–8. [3.6.2]

Gibbons D.W., Reid J.B. & Chapman R.A. (1993) *The New Atlas of Breeding Birds in Britain and Ireland: 1988–91.* Poyser, London. [3.6.2]

Gillespie J.H. (1991) *The Causes of Molecular Evolution.* Oxford University Press. [3.4.2]

Gould S.J. & Eldredge N. (1977) Punctuated equilibria: the tempo and mode of evolution reconsidered. *Paleobiology* **3**, 115–151. [3.4.2]

Gould S.J. & Eldredge N. (1993) Punctuated equilibrium comes of age. *Nature* **366**, 223–227. [3.4.2]

Greig-Smith P. (1983) *Quantitative Plant Ecology.* Blackwell Scientific, Oxford. [3.5.2]

Groombridge B. (ed.) (1992) *Global Biodiversity: status of the Earth's living resources.* Chapman & Hall, London. [3.1] [3.1.1] [3.3.2]

Hamby R.K. & Zimmer E.A. (1992) Ribosomal RNA as a phylogenetic tool in plant systematics. In: *Molecular Systematics in Plants* (eds P.S. Soltis, D.E. Soltis & J.J. Doyle), pp. 50–91. Chapman & Hall, London. [3.4.2]

Hanemann W.M. (1984) *On Reconciling Different Concepts of Option Value.* Working Paper No. 295, Department of Agricultural and Resource Economics, University of California, Berkeley. [3.2]

Hanemann W.M. (1989) Information and the concept of option value. *J. Environ. Econ.* **16**, 23–37. [3.2]

Harper J.L. & Hawksworth D.L. (1994) Biodiversity: measurement and estimation. *Phil. Trans. Roy. Soc., Lond. B* **345**, 5–12. [3.2]

Hillis D.M., Dixon, M.T. & Ammerman, L.K. (1991) The relationships of the coelacanth *Latimeria chalumnae*: evidence from sequences of vertebrate 28S ribosomal RNA genes. *Environ. Biol. Fishes* **32**, 119–131. [3.4.2]

Hochberg M.E. & Lawton J.H. (1990) Competition between kingdoms. *Trends Ecol. Evol.* **5**, 367–371. [3.3.3]

Huelsenbeck J.P. & Hillis D.M. (1993) Success of phylogenetic methods in the four-taxon case. *Syst. Biol.* **42**, 247–264. [3.4]

Kimura M. (1983) *The Neutral Theory of Molecular Evolution.* Cambridge University Press. [3.4.2]

Kluge A.G. & Wolfe A.J. (1993) Cladistics: what's in a word? *Cladistics* **9**, 183–199. [3.4]

Krajewski C. (1994) Phylogenetic measures of biodiversity: a comparison and critique. *Biol. Conserv.* **69**, 33–39. [3.5.1]

Kremen C., Colwell R.K., Erwin T.L., Murphy D.D., Noss R.F. & Sanjayan M.A. (1993) Terrestrial arthropod assemblages: their use in conservation planning. *Conserv. Biol.* **7**, 796–806. [3.6.2]

Landres P.B., Verner J. & Thomas J.W. (1988) Ecological uses of vertebrate indicator species: a critique. *Conserv. Biol.* **2**, 316–328. [3.6.2]

Linder H.P. & Midgley J.J. (1994) Taxonomy, compositional biodiversity and functional biodiversity of fynbos. *South Afr. J. Sci.* **90**, 329–333. [3.3.3]

Linnaeus C. (1758) *Systema Naturae per Regna Tria Naturae, Secundum Classes, Ordines, Genera, Species, cum Characteribus, Differentiis, Synonymis, Locis* (Vol. 1, edn. 10). Holmiae. [3.3.2]

Love R.F., Morris J.G. & Wesolowsky G.O. (1988) *Facilities Location: models and methods.* North-Holland, London. [3.5.3]

McNeely J.A., Miller K.R., Reid W.V., Mittermeier R.A. & Werner T.B. (1990) *Conserving the World's Biological Diversity.* IUCN, WRI, CI, WWF and World Bank, Washington DC. [3.1]

May R.M. (1990) Taxonomy as destiny. *Nature* **347**, 129–130. [3.1] [3.5.1]

Mickleburgh S.P., Hutson A.M. & Racey P.A. (1992) *Old World Fruit Bats: an action plan for their conservation.* IUCN, Gland, Switzerland. [3.6]

Moran V.C. & Southwood T.R.E. (1982) The guild composition of arthropod communities in trees. *J. Anim. Ecol.* **51**, 289–306. [3.3.3]

Nee S., Holmes E.C., May R.M. & Harvey P.H. (1994) Extinction rates can be estimated from molecular phylogenies. *Phil. Trans. Roy. Soc., Lond.* B **344**, 77–82. [3.4.2]

Nixon K.C. & Wheeler Q.D. (1992) Measures of phylogenetic diversity. In: *Extinction and Phylogeny* (eds M.J. Novacek & Q.D. Wheeler), pp. 216–234. Columbia University Press, New York. [3.5.1]

Norton B.G. (1994) On what we should save: the role of culture in determining conservation targets. In: *Systematics and Conservation Evaluation* (eds P.L. Forey, C.J. Humphries & R.I. Vane-Wright), pp. 23–39. Oxford University Press. [3.2]

Noss R.F. (1990) Indicators for measuring biodiversity: a hierarchical approach. *Conserv. Biol.* **4**, 355–364. [3.2] [3.6.2]

Oxford English Dictionary (1990) *The Shorter Oxford English Dictionary.* (3rd edn) Clarendon Press, Oxford. [3.2]

Patterson C. (1982) Morphological characters and homology. In: *Problems of Phylogenetic Reconstruction* (eds K.A. Joysey & A.E. Friday), pp. 21–74. (Systematics Association Special Volume No. 21). Systematics Association, London. [3.2]

Patterson C. (ed.) (1987) *Molecules and Morphology in Evolution: conflict or compromise?* Cambridge University Press. [3.2]

Patterson C., Williams D.M. & Humphries C.J. (1993) Congruence between molecular and morphological phylogenies. *Annu. Rev. Ecol. Syst.* **24**, 153–188. [3.4] [3.4.2]

Pielou E.C. (1967) The use of information theory in the study of the diversity of biological populations. *Proceedings of the 5th Berkeley Symposium on Mathematical and Statistical Problems* **4**, 163–177. [3.5.1]

Reid W., Barber C. & Miller K. (1992) *Global Biodiversity Strategy: guidelines for action to save, study and use Earth's biotic wealth sustainably and equitably.* WRI, IUCN and UNEP, Washington DC. [3.1]

Reid W.V. (1994) Setting objectives for conservation evalution. In: *Systematics and Conservation Evaluation* (eds P.L. Forey, C.J. Humphries & R.I. Vane-Wright), pp. 1–13. Oxford University Press. [3.2]

Reid W.V., McNeely J.A., Tunstall D.B., Bryant D.A. & Winograd M. (1993) *Biodiversity Indicators for Policy-makers.* WRI and IUCN, Washington DC. [3.6.2]

Scherer S. (1990) The protein molecular clock. Time for a reevaluation. *Evol. Biol.* **24**, 83–106. [3.4.2]

Schulz S., Boppré M. & Vane-Wright R.I. (1993) Specific mixtures of secretions from male scent organs of African milkweed butterflies (Danainae). *Phil. Trans. Roy. Soc., Lond.* B **342**, 161–181. [3.3.3] [3.4.2]

Schulze E.-D. & Mooney H.A. (eds) (1993) *Biodiversity and Ecosystem Function.* Springer-Verlag, Berlin. [3.3.3]

Sibley C.G. & Ahlquist J.E. (1990) *Phylogeny and Classification of Birds: a study in molecular evolution.* Yale University Press, New Haven. [3.6.2]

Solow A., Polasky S. & Broadus J. (1993) On the measurement of biological diversity. *J. Environ. Econ. Manag.* **24**, 60–68. [3.4] [3.5.1]

Stork N.E. (1987) Guild structure of arthropods from Bornean rain forest trees. *Ecol. Entomol.* **12**, 69–80. [3.3.3]

Tansel B.C., Francis R.L. & Lowe T.J. (1983) Location on networks: a survey. Part I: the *p*-center and *p*-median problems. *Management Sci.* **29**, 482–497. [3.5.3]

Vane-Wright R.I., Humphries C.J. & Williams P.H. (1991) What to protect?–Systematics and the agony of choice. *Biol. Conserv.* **55**, 235–254. [3.2] [3.5.1]

Vane-Wright R.I., Schulz S. & Boppré M. (1992) The cladistics of *Amauris* butterflies: congruence, consensus and total evidence. *Cladistics* **8**, 125–138. [3.4.2]

Vrba E.S. (1980) Evolution, species and fossils: how does life evolve? *South Afr. J. Sci.* **76**, 61–84. [3.4.2]

Walker B.H. (1992) Biodiversity and ecological redundancy. *Conserv. Biol.* **6**, 18–23. [3.3.3]

Weisbrod B.A. (1964) Collective-consumption services of individual-consumption services of individual-consumption goods. *Quart J. Econ.* **78**, 471–477. [3.2]

Weitzman M.L. (1992) On diversity. *Quart. J. Econ.* **107**, 363–405. [3.4] [3.5.1]

Weitzman M.L. (1993) What to preserve? An application of diversity theory to crane conservation. *Quart. J. Econ.* **108**, 157–183. [3.4] [3.5.1]

Williams P.H. (1991) The bumble bees of the Kashmir Himalaya (Hymenoptera: Apidae, Bombini). *Bull. Brit. Mus. (Nat. Hist.), Entom.* **60**, 1–204. [3.2] [3.4] [3.5.1] [3.5.3]

Williams P.H. (1993) Measuring more of biodiversity for choosing conservation areas, using taxonomic relatedness. In: *International Symposium on Biodiversity and Conservation* (ed. T.-Y. Moon), pp. 194–227. Korean Entomological Institute, Seoul. [3.5.1] [3.5.2] [3.6.2]

Williams P.H. (1994) *WORLDMAP priority areas for biodiversity. Using version 3.* Privately distributed, London. [3.5.2] [3.6]

Williams P.H. (1995a) Phylogenetic relationships among the bumble bees (*Bombus* Latr)· a re-appraisal of morphological evidence. *Syst. Entom.,* **19**, 327–344. [3.6] [3.6.2]

Williams P.H. (1995b) Biodiversity value and taxonomic relatedness. In: *The Genesis and Maintenance of Biological Diversity* (eds M.E. Hochberg, J. Clobert & R. Barbault). Oxford University Press, in press. [3.5.3] [3.6.2]

Williams P.H. & Gaston K.J. (1994) Measuring more of biodiversity: can higher-taxon richness predict wholesale species richness? *Biol. Conserv.* **67**, 211–217. [3.6.2]

Williams P.H. & Humphries C.J. (1994) Biodiversity, taxonomic relatedness and endemism in conservation. In: *Systematics and Conservation Evaluation* (eds P.L. Forey, C.J. Humphries & R.I. Vane-Wright), pp. 269–287. Oxford University Press. [3.4.2] [3.5.1] [3.5.2] [3.6] [3.6.2]

Williams P.H., Humphries C.J. & Vane-Wright R.I. (1991) Measuring biodiversity: taxonomic relatedness for conservation priorities. *Aust. Syst. Bot.* **4**, 665–679. [3.4] [3.5.1] [3.5.2]

Williams P.H., Gaston K.J. & Humphries C.J. (1994a) Do conservationists and molecular biologists value differences between organisms in the same way? *Biodiv. Lett.* **2**, 67–78. [3.3] [3.5.1]

Williams, P., Gibbons, D., Margules, C., Rebelo, A., Humphries, C. & Pressey, R. A comparison of richness hotspots, rarity hotspots, and complementary areas for conserving diversity using British birds. *Cons. Biol.*, in press. [3.6.2]

Williams P.H., Humphries C.J. & Gaston K.J. (1994b) Centres of seed-plant diversity: the family way. *Proc. Roy. Soc. Lond. B* **256**, 67–70. [3.6.2]

Williams, P.H., Prance, G.T., Humphries, C.J. & Edwards, K.S. Promise and problems in applying quantitative complementary areas for representing the diversity of some Neotropical plants (families Dichapetalaceae, Lecythidaceae, Caryocaraceae, Chrysobalanaceae and Proteaceae). *Biol. J. Lin. Soc.*, in press. [3.6.2]

Wilson E.O. (1992) *The Diversity of Life.* Penguin Press, London. [3.1] [3.5.2]

4: Species richness: measure and measurement

KEVIN J. GASTON

4.1 Introduction

Notwithstanding the breadth of the concept of biodiversity, number of species remains its most frequently and widely applied measure. There are several possible reasons, of differing levels of contention and empirical support. First, species richness is thought by many to capture much of the essence of biodiversity. Second, the meaning of species richness is apparently widely understood, and there is no need to derive complex indices to express it. Third, species richness is considered in practice often to be a measurable parameter. Fourth, much data on species richness already exist.

This chapter serves as a broad review both of species richness as a measure of biodiversity, and of the measurement of species richness.

4.2 Species richness as a measure of biodiversity

The species is regarded in many quarters as the fundamental unit of biodiversity, species richness as the fundamental meaning of biodiversity, and the high level of species extinction as the main manifestation of the biodiversity crisis. Indeed, the connection between species richness and biodiversity is sufficiently well developed that many authors move freely between the two terms without identifying any of the assumptions that they are making in so doing.

This said, it is obvious that biodiversity, as commonly defined (Chapter 1), and species richness are far from synonymous. Species richness is at best a measure of one aspect of biodiversity. At its simplest, biodiversity is the variety of life. In that species are by definition in some sense different, species richness must capture some facet of this variety. Its weakness in so doing has typically been illustrated with reference to the issue of whether an assemblage of two closely related species, say two species of pierid butterflies, is more or less biodiverse than an assemblage of two distantly related species, say a species of butterfly and a species of hoverfly. While, intuitively, the latter assemblage would seem to be the more biodiverse, in terms of species richness the assemblages are equally biodiverse. Moreover, this remains true when one alters the scenario in various ways, such as distinguishing the assemblages on the basis of the numbers of individuals (and hence potentially the genetic diversity) of each of the species rather than their degree of relatedness, or on the basis of their functional diversity.

The extent to which this is a weakness depends perhaps less on the outcomes of such simple scenarios than on those which concern many studies of biodiversity, which commonly involve assemblages numbering at least tens, if not hundreds or thousands of species. It is not necessarily obvious how simple arguments based on two species will extrapolate to these situations. In fact, there is evidence that differences in the species richness of assemblages may parallel differences in some other measures of their biodiversity, and diversity more generally. For example:

1 Species richness can be correlated positively with some measures of ecological diversity (Hurlbert, 1971; Paasivirta, 1976; Cousins, 1977, 1994; Smith *et al.*, 1979; Magurran, 1988), and has been regarded as one useful measure of that diversity (Magurran, 1988). Indeed, many standard measures of ecological diversity are measures of the numbers of certain types of species (e.g. common species) in an assemblage (Brewer & Williamson, 1994). Indices of ecological diversity calculated at different taxonomic levels can be strongly correlated amongst samples (Kaesler *et al.*, 1978).

2 Species richness and numbers of higher taxonomic units tend to be related positively (a point of potential value in the estimation of species richness itself, see below; Gaston & Williams, 1993; Williams & Gaston, 1994). This implies that species richness may give some broad indication of diversity at a higher level, and perhaps of general morphological diversity.

3 When species richness becomes moderately high, those areas which are recognized as having high biodiversity on the basis of measures of phylogenetic disparity (Chapter 3) tend to converge on those having high species richness (Vane-Wright & Rahardja, 1993; Williams, 1993; Kershaw *et al.*, 1994; Williams & Humphries, 1994). Williams and Humphries (1994) find that for 22 of 25 data-sets the highest scoring biota by taxonomic dispersion is among the highest scoring by unweighted species richness.

4 Various descriptors of the structure of food webs (e.g. connectance, food chain length, number of trophic links) have been reported to be correlated with species richness. Although debate continues about the reality and form of some of these patterns (e.g. Schoener, 1989; Warren, 1990; Martinez, 1991, 1993; Havens, 1992, 1993a, 1993b; Hall & Raffaelli, 1993; Yodzis, 1993; Bengtsson, 1994), they suggest that species richness captures some elements of functional diversity.

5 Relatively high species richness is often associated with relatively high topographic diversity (e.g. Holland, 1978; Schall & Pianka, 1978; Miller, 1986; Miller *et al.*, 1987; Martin & Gurrea, 1990). This suggests that species richness may capture some elements of the diversity of landscapes, although this is a rather unconventional way to view the relationship and interpretation is complicated by the difficulty of controlling for differences in area (see below).

In sum, species richness not only provides a measure of the variety of life as represented by a count of species, it tends also to capture a number of other facets of that variety. In arguing that species richness provides a reasonably useful measure of biodiversity we need to be clear that this does not equate to the desirability, or otherwise, of identifying areas of high species richness as high conservation priorities. Much of the contention surrounding species richness as a measure of biodiversity derives from a failure to recognize this distinction.

4.3 Issues in the measurement of species richness

Accepting that species richness serves as but one measure of biodiversity, we need then to consider some practical problems of its measurement.

4.3.1 Species concepts

A severe but frequently ignored complication to the measurement of species richness is the lack of an agreed and universally applied definition of what a species is, or indeed of agreement as to the utility of the term at all. A variety of broad species concepts exist (variously termed biological, cladistic, cohesion, ecological, evolutionary, phylogenetic, and taxonomic; for references see Otte & Endler, 1989; Rojas, 1992; Vane-Wright, 1992). These may emphasize either pattern or process, are of varying conceptual and practical (pragmatic) appeal, and overlap to varying degrees. For any given assemblage, the numbers of species recognized and the level of congruence in species limits may potentially differ dramatically, dependent on the species concept which is applied. For example, Cracraft (1992) contrasts the 40–42 species of birds-of-paradise (Paradisaeidae) recognized using the biological species concept, with the approximately 90 species diagnosable using a phylogenetic species concept.

From the viewpoint of biodiversity studies, the problem of differing species concepts gives rise to two primary concerns. First, there is the need to ensure that for any given higher taxon the same concept is being applied when comparisons are made of species numbers in different areas or at different times. The fact that species concepts have tended to develop, and that techniques of discrimination change makes this hard to achieve. For example, the taxonomy of Lepidoptera has been revolutionized through the use of dissections to reveal genitalic characters, but this technique has been widely applied only since the second quarter of this century, and has been used to a greater extent in the investigation of some faunas than of others. Whilst in practice the species concept has remained essentially a morphological one, the details of this concept undoubtedly have changed. Likewise, the

79

invention of more powerful means of observation has increased the usage of morphology in recognition of algal species. The number of species described of the synurophyte *Mallomonas* rose dramatically when the details of scale morphology were discovered, and when electron microscopy revealed further detail (Andersen, 1992).

The second concern is the need to apply similar species concepts in the context of different higher taxa. This problem afflicts interpretation of many comparisons of the richness of different higher taxa, and the summation of richness values to give overall figures for areas. What does it mean to state that taxon A is more speciose than taxon B, if species in the two taxa are distinguished on the basis of entirely different criteria, or, as is more usual, when it is simply unclear how similar the criteria are?

Much that has been written about species concepts is somewhat divorced from the practical realities of taxonomic work, and distinction must be drawn between operational and theoretical species definitions (Otte, 1989). For the majority of higher taxa, species are discriminated on the basis of morphological features, on the pragmatic grounds that this is all that can be done given the resources available and the magnitude of the task. In general, using this approach, specialists studying the same taxon in the same period tend to recognize the same sets of species (although disputes naturally occur). The extent to which this implies that these species have any biological reality, or that the operational definition applied coincides with one or more theoretical species definitions, is doubtless variable. The likelihood that similar entities are being distinguished in different higher taxa is almost certainly broadly a function of the relatedness of those taxa. The more distantly related they are the less likely that the morphological criteria applied will distinguish equivalent entities (although there have been claims that the species concepts being applied to quite distantly related taxa, such as some insect orders, are very similar if not the same; e.g. Danks, 1993).

Concern over the consequences of differing species concepts for species richness estimates has been expressed on a number of occasions (Adis, 1990; May, 1990; Gaston, 1991; Vane-Wright, 1992). Ultimately, if we are to study species richness and use it as a measure of biodiversity, we must do so accepting that workers on different groups in effect use rather different species concepts and do not always apply them consistently. Heywood (1994) writes: 'It is a fact of systematic biology that the vast majority of the species recognized are equivalent by designation only, not through their degree of evolutionary, genetic, ecological or phenetic differentiation'.

4.3.2 Species discrimination

Although in practice the two are not always readily separated, a distinction can be made between species concepts and the process of species discrimination. Whatever the concept employed, a high degree of rigour, repeatability, and consistency in species discrimination is desirable in its practical application.

The above said, the adequacy of species discrimination in most studies of species richness is seldom open to the scrutiny of peers in the same way as the sampling and statistical methodology by which an estimate of richness is attained. This raises the possibility that serious errors in estimation can result because species have not been distinguished 'correctly'. The ease with which such errors can be avoided depends on the species concepts applied, real changes in species limits, the higher taxon or taxa under consideration, available literature, equipment and expertise. Errors are of two kinds: splitting, classifying one species as two or more; and lumping, classifying two or more species as one. The former inflates estimates of species richness; the latter makes them too small.

Comparatively little attention has been paid to determining empirically how serious these problems of splitting and lumping are in practice. With good taxonomy they should not pose a significant obstacle at the scale of small to moderate sized areas. However, recent study of patterns of synonymy has served to demonstrate that at the global and large regional scale, substantial errors may exist (Gaston & Mound, 1993; Mound & Gaston, 1993). For example, Gaston and Mound (1993) found that 20% or more of species-group names for several large insect taxa have been synonymized. That is, for every five names which have been designated, one has pertained to a species described previously under a different name. The numbers of synonyms associated with a putatively valid species name are not distributed equally amongst those names. Rather, most species names have no or few synonyms, whilst a few have many synonyms (Gaston *et al.*, 1995b). Little is known of the factors associated with high levels of synonymy, although species with large geographic ranges and high levels of morphological variability seem particularly prone to being described multiple times (Table 4.1; Gaston *et al.*, 1995a). One consequence is that synonymy becomes more of a problem as study areas become larger. It also suggests that the net effects of splitting and lumping of species are unlikely to be distributed evenly in space. Rather, they will vary with spatial patterns in species' geographic range sizes and morphological variability. In broad terms, the former on average become smaller towards low latitudes (Rapoport, 1982; Stevens, 1989).

Faced with a desire rapidly to determine spatial patterns of species richness, and a small and declining body of taxonomic expertise (Gaston & May, 1992), interest has been expressed in the use of a non-specialist workforce to generate data on species richness. This has raised in some quarters the spectre of a dramatic increase in errors in species discrimination. Assessment of the effectiveness of the approach is at an early stage (e.g. Cranston & Hillman, 1992; Oliver & Beattie, 1993, 1994). However, it seems most useful for groups where morphological differences between species are readily apparent and unambiguous, where relative rather than absolute numbers of species are

Table 4.1 The extents of the geographic ranges of *Enicospilus* (Ophioninae, Ichneumonidae, Hymenoptera) species in the Indo-Papuan region, and the numbers of synonyms associated with each valid specific name. Data from Gauld & Mitchell (1981).

Range (km)	Number of synonyms											
	0	1	2	3	4	5	6	7	8	9	10	11
1–999	134	–	–	–	–	–	–	–	–	–	–	–
1000–1999	32	2	–	–	–	–	–	–	–	–	–	–
2000–2999	18	–	1	–	–	–	–	–	–	–	–	–
3000–3999	12	5	1	–	–	–	–	–	–	–	–	–
4000–4999	11	–	2	–	–	–	–	–	–	–	–	–
5000+	16	15	9	8	3	1	–	1	2	–	1	1

required, where differences in the richness of areas are great, and at scales over which intra-specific morphological variation is small.

4.3.3 Scale dependence

In addition to the difficulties associated with the definition and the discrimination of species, the other major group of problems with the application of species richness as a measure of biodiversity concern its scale dependence. In general, increases in either the spatial extent of a study or its duration result in a greater tally of species for an area.

The interaction between numbers of species and area, the species–area relationship, has been the subject of extensive study (e.g. MacArthur & Wilson, 1967; Connor & McCoy, 1979; Williamson, 1988; Anderson & Marcus, 1993; Palmer & White, 1994). Emphasis has been laid particularly on the comparatively strong positive relationships which have been documented at meso- and micro-scales (Fig. 4.1). However, across a very wide span of sizes of areas (local to biogeographic or global), from a range of latitudes, relationships can be (on log–log axes) far weaker, with maximum richness tending to increase as areas become larger but with areas of a wide range of sizes having low species numbers.

Accounting for the dependence of species richness on spatial scale can be hampered because of the difficulty of appropriately quantifying the relevant areas. There are two principal problems. First, as with densities of individuals (e.g. Elton, 1932, 1933), distinction can be drawn between 'crude' and 'ecological' densities of species. The former are assessed over essentially arbitrary areas, whereas the latter are assessed over areas which attempt to discount habitats in which species of the taxon concerned do not happen to occur. Second, the size of an area will progressively increase as more account is taken of the details of its topography; measuring the size of an area is a fractal problem. In practice both problems usually remain difficult to resolve satisfactorily, and most studies treat the earth as though

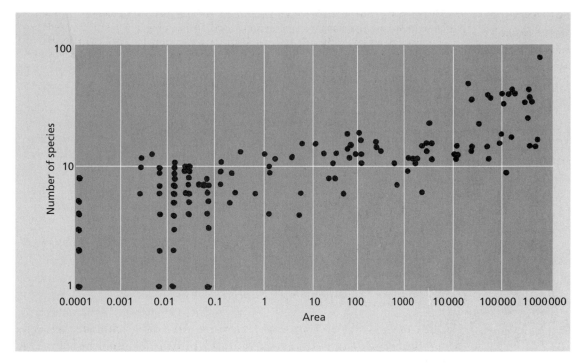

Fig. 4.1 Species–area relationship for European lumbricid earthworms, based on studies from throughout the region. Areas, expressed on graph in square kilometres, range from 100 m² to >500 000 km². From data in Judas (1988).

it were flat and as though the species of a given taxon used all the space available in a study area.

Standardization of the size of areas for the purpose of comparisons is further complicated, because few studies have documented within-region species–area relationships (Holt, 1993). This tends to result in corrections for differences in the size of areas being based on the between-region relationship, which may be wholly inappropriate. Species–area relationships may vary with taxon, latitude, and the size span of areas analysed, as well as the grain, extent and number of samples (e.g. Martin, 1981; Palmer & White, 1994).

The interaction between numbers of species and time, the species–time relationship, has been little explored over what can broadly be termed ecological rather than evolutionary or geological periods (for a review of the latter see Signor, 1990). Only over very brief time spans is there an absolute value to the species richness of almost any area, and there is little evidence for temporal equilibria in the numbers of species present in areas of moderate size (Boecklen & Nocedal, 1991). Over most time periods the numbers of species present changes. This change has essentially three components. The first is a result of regular movements of species, for example on daily, seasonal or perhaps annual bases. This is perhaps most apparent at smaller spatial scales, and may result in severe under-estimation of overall levels of species richness when sampling is carried out over insufficient periods of time. For a small area, the probability of an individual of a species being present at any one moment may be low, if it occurs at a low density and ranges more widely than the bounds of the area. This may be true even if the area is an important, or perhaps core, part of the individual's home range.

The second component to changes in the numbers of species present in an area is a consequence of colonization and extinction events. These events may result from stochastic processes or more deterministic changes in conditions. Gibbons *et al.* (1993) provide an example, in documenting turnover in the composition of bird species breeding in Britain and Ireland (Table 4.2).

The third component to temporal changes in species numbers are 'chance' occurrences of species in an area (see Section 4.3.4).

One consequence of turnover in species identities is growth in the cumulative number of species recorded with time. Based on the phenomenon, Grinnell (1922) estimated how long it would take before all the bird species recorded from North America had been seen in California:

> ... it is only a matter of time theoretically until the list of California birds will be identical with that for North America as a whole. On the basis of the rate for the last 35 years, $1^3/_5$ additions to the California list per year, this will happen in 410 years, namely in the year 2331, if the same intensity of observation now exercised can be

Table 4.2 The numbers of bird species which bred in 1968–72 and 1988–91, and did or did not breed in the other period, for (a) Britain and (b) Ireland. Figures in parentheses include species that bred in the wild but which were of introduced or reintroduced origin, or were feral. Data from Gibbons *et al.* (1993).

(a) Britain

	1968–72		
	+	−	Total
1988–91 +	197 (210)	7 (9)	204 (219)
−	4 (4)	[8 species bred in 1973–87 but in neither other period]	
Total	201 (214)		

(b) Ireland

	1968–72		
	+	−	Total
1988–91 +	130 (133)	7 (9)	137 (142)
−	5 (6)	[4 species bred in 1973–87 but in neither other period]	
Total	135 (139)		

maintained. If observers become still more numerous and alert, the
time will be shortened. (Grinnell, 1922, p. 375).

It remains to be seen whether his estimate will be correct. However, at the
present rate of discovery it is an over-estimate (Bock, 1987).

Differences in the duration of studies may cause differences in observed
patterns in species richness. Wolda (1987), for example, suggests that the
continuity of sampling may determine whether the local richness of insects is
seen to decline from low to high altitudes or to peak at intermediate altitudes.

4.3.4 Species status

The species occurring in an area are not all of an equivalent status. In
particular, many may be present for rather brief periods, may not breed, or
may not have self-sustaining populations (Table 4.3). Such species have
variously been termed accidentals, casuals, immigrants, incidentals, strays,
tourists, transients, vagrants and waifs; hereafter simply referred to as
vagrants. In sum, they serve to inflate species numbers in an area,

Table 4.3 The numbers of species of different status for (a) the British and Irish
macrolepidoptera (852 species; cases of possible or presumed status are included with
those for which information is more definite), and (b) the British and Irish Rhopalocera
(the butterflies) (110 species). The schemes used are rather different for the two
assemblages. Data from Skinner & Wilson (1984) and Emmet & Heath (1989).

(a)	Number of species
Extinct resident	11
Resident	683
Immigrant	67
Resident and immigrant	62
Extinct resident and immigrant	1
Importation	1
Importation and/or immigrant	9
Status uncertain	18

(b)	Number of species
Former native, now extinct	7*
Native	52
Native and common immigrant	2
Former native and immigrant	2
Common immigrant	3
Infrequent immigrant	9
Rare immigrant	11†
Deliberate introduction	1
Accidental introduction, adventive	23‡

* 4 of uncertain status, † 1 of uncertain status, ‡ 2 are also rare immigrants.

constituting what is known as a 'mass effect' (Shmida & Wilson, 1985). Whether it is desirable to include them in measuring the species richness of an area is debatable, and will depend largely on the objective of the exercise. Nonetheless, it is important to recognize that the relative contribution of vagrants to levels of species richness will depend crucially on the taxon, the habitat, the size of the study area, and the duration of the study. The proportion of species which are vagrants tends to increase with the mobility of the taxon, and the duration of the study (e.g. Hammond, 1992; Rich &

Box 4.1 The species richness of birds in Britain and Ireland.

The numbers of bird species recorded in Britain and Ireland tended to remain quite consistent from year to year through the 1980s, with a mean of 388, a low of 371 and a high of 408 (a). In total, 491 species were recorded between 1980 and 1989, 92% of the total British and Irish list. Of the overall list of 537 species, 258 are regarded as 'rare'. Most of these rare species were seen in only some of the 10 years between 1980 and 1989 (b) and the frequency

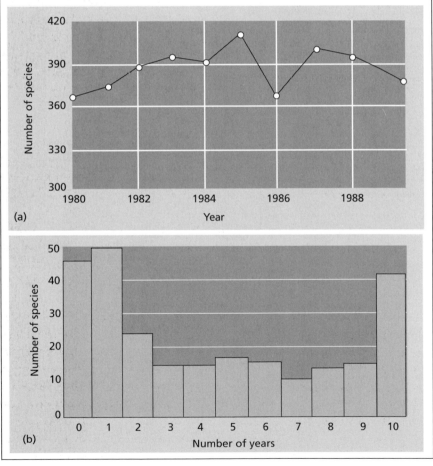

continued

Box 4.1 *contd*

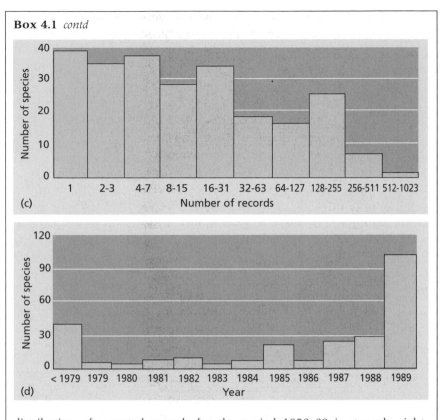

(c)

(d)

distribution of accepted records for the period 1958–89 is strongly right-skewed on logarithmic scales (c). The frequency distribution of the years of the most recent accepted sighting of these rare species is strongly left-skewed, peaking towards the present (d). From data in Whiteman & Millington (1991).

Woodruff, 1992). Moist tropical forest habitats are thought to comprise a higher proportion of vagrant species than are temperate forest habitats (Stevens, 1989), and the proportion of species which are vagrants is argued to peak at meso-scales (Shmida & Wilson, 1985).

The proportion of species in an area which are vagrants can be high. Pimentel and Wheeler (1973) classify as incidental to the system almost a half of the arthropod species encountered in their study of alfalfa. Gaston *et al.* (1993) find that a fifth of the beetle species sampled from oak trees by insecticide fogging are tourists to the system. Both these studies were of relatively brief duration, and the proportions can be expected to grow markedly over long periods. At a larger spatial scale, up to the end of 1989 the official British and Irish bird list stood at 537 species (there have been a few more recent alterations), of which 279 are 'common' and annual, and a further 258 are 'rare' and could reasonably be termed vagrants (Box 4.1; Whiteman & Millington, 1991).

The existence of temporal turnover in the identities of species in an area, and hence the coexistence at any one point in time of species of very different status (e.g. resident, vagrant, migrant) needs constantly to be borne in mind in considerations of species richness, and particularly when species numbers are being compared for different areas. Major discrepancies in different evaluations of the numbers of species documented for individual areas can often be explained on the basis of differences in the treatment of vagrants.

4.3.5 Recorder effort

The bulk of available information on levels of species richness is a product of the inventory process, the accumulation for an area of records of previously unrecorded species. Such lists are maintained for the species of many higher taxa in areas spanning in size from the global through the regional to the very local. Very often they are based on geopolitical units. In perhaps the majority of cases, there is no obvious means of quantitatively determining how close to completion such inventories are. For example, no record is kept of

Table 4.4 The number of sites at which different numbers of species of ophionines (Ophioninae, Ichneumonidae, Hymenoptera) have been collected in SE Asia. The asterisked sites are those regarded as having been intensively studied. Data from Gauld & Mitchell (1981).

Number of species per site	Number of sites	
1	383	
2	71	
3	41	
4	26	
5	34	
6	21	
7	13	
8	7	
9	14	
10	3	
11	6	
12	2	
13	2	
14	4	
15	3	
16	6	
19	2	*
26	1	*
27	1	*
34	1	*
35	1	*
46	1	*
	643	

precisely when new species records were acquired, of the amounts of effort (e.g. time elapsed, person-hours, number of individual organisms sampled) required to generate those additions, or of the status of the species. At best, some subjective, or semi-quantitative assessment can be made of the ease or difficulty encountered in making the most recent additions.

The paucity of hard quantitative information on the status of most inventories seriously undermines their value in studies of species richness. Comparisons of the richness of different areas often remain insecure, because there is no basis for demonstrating the relative completeness of different inventories.

Whilst estimates of actual effort are often difficult to obtain, it is evident that in many instances levels of effort have been inadequate to detect all of the species of a given higher taxon in an area (Table 4.4). The problem manifests itself in a number of ways. These include correlations between levels of effort and the numbers of species recorded from different areas (Fig. 4.2; e.g. Connor & Simberloff, 1978; Nelson *et al.*, 1990), between-recorder variation in observed levels of richness (e.g. Slater *et al.*, 1987), and pseudo-turnover (Lynch & Johnson, 1974; Nilsson & Nilsson, 1983, 1985) (note, species richness and some 'measures of effort' such as numbers of individuals may be correlated for biological reasons, not just sampling ones; e.g. Brown & Kurzius, 1987). With regard to the long-term inventory of areas, for most higher taxa and most regions the inadequacy of recorder effort is also evidenced by the collection of previously undescribed species. Various assessments have been made of the progress toward, and time to completion, of inventories of particular major taxa, of the steps which could be taken to

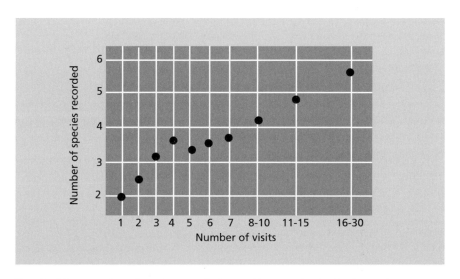

Fig. 4.2 Mean number of species recorded per grid square (each 4 ha in area) as a function of the number of visits, for landbirds on Silhouette island. Redrawn from Greig-Smith (1986).

89

speed up the process or make it more effective, and of the resources necessary to complete it (e.g. Wilson, 1985; Prance & Campbell, 1988; Soulé, 1990; Gaston, 1993, 1994; Hawksworth, 1993; Mound & Gaston, 1993; Parnell, 1993). For speciose groups, counts of numbers of described species that are regarded as valid may be very imprecise. Table 4.5 lists estimates of the proportions of the insect species which have been described in different, predominantly geopolitical, areas. The small magnitude of many of these figures is accentuated when one remembers that on the whole such estimates are only made for better known regions.

The problems in attaining the levels of effort necessary to document the numbers of species in an area over a given time frame have been reiterated repeatedly with regard to entire faunas, and to insects in particular. The point is equally amply illustrated with reference to comparatively well-known vertebrate higher taxa. New species continue to be discovered; a number of species thought to have become extinct have been rediscovered, and other species have been re-encountered after long periods of time (e.g. Terborgh & Weske, 1972; Raxworthy, 1988; Ferrari & Lopes, 1992; Glynn & Feingold, 1992; Heywood & Stuart, 1992; Queiroz, 1992; Vuilleumier *et al.*, 1992; Dung *et al.*, 1993; Shuker, 1993; Dinesen *et al.*, 1994;

Table 4.5 The numbers of species of insects described from and estimated to occur in different areas. Note, the accuracy of these figures, and how up to date they are varies enormously. Moreover, the definitions of 'insects' applied are not entirely consistent.

	Number of species		
	Described	Estimated	Described (%)
Arctic North America	1408	3100+	45 Danks (1986)
Queen Elizabeth Is.	237	600+	40 Danks (1986)
Bathurst Is.	65	130+	50 Danks (1986)
Canada	29976	54629+	55 Danks (1979)
Finland	18890	22500+	84 A. Jansson (pers. comm.)
Britain	22437	24000+	<93 Stubbs (1982)
Switzerland	16530	31400+	53 W. Sauter (pers. comm.)
Mediterranean basin	105000	150000+	70 Balletto & Casale (1991)
Canary Islands	4400	5000+	88 A. Machado (pers. comm.)
Southern Africa	40000	56–64000+	63–71 S. Endrody-Younga (pers. comm.)
	80000	250000+	32 Samways (1994)
Taiwan	13862	45–55000+	25–31 Maa (1956)
Japan	29000	70–100000+	29–41 Chu (1989)
Tasmania	6310	11140+	57 P.B. McQuillan (pers. comm.)
Micronesia	4000	10000+	40 Gressitt (1951)
Hawaii	5000	12000+	42 Gressitt (1951)
Samoa	1800	7000+	26 Gressitt (1951)
Marquesas	850	2000+	43 Gressitt (1951)
Society Islands	250	1000+	25 Gressitt (1951)
Fiji	3000	25000+	12 Gressitt (1951)

Ferrari & Queiroz, 1994; Humphries & Fisher, 1994; Pine, 1994; Short & Smith, 1994).

4.3.6 Absolute and relative measures

Particularly given the large amounts of effort (and hence costs) required to generate complete inventories of the species in areas, even where these are only moderately speciose, careful consideration needs to be given to whether studies can be based on estimates of relative species numbers, or whether they need estimates of overall (absolute) species richness. The former are obviously more readily obtained than the latter. Indeed, the rather poor attention which has been paid in the ecological literature to methods of estimating overall species richness is perhaps best explained as a consequence of the fact that it has been possible to address most issues of species richness using estimates of relative richness. Most frequently, this has been done using rarefraction techniques to determine the numbers of species which would have been found in samples taken from different areas, at different times, in different ways, and so on, if those samples were of equivalent size (e.g. Etter & Grassle, 1992; Grassle & Maciolek, 1992; Poore & Wilson, 1993; Rex *et al.*, 1993; Brey *et al.*, 1994; Haila *et al.*, 1994; Niemelä *et al.*, 1994). These techniques are typically based on individuals (Hurlbert, 1971; Simberloff, 1978; Smith *et al.*, 1979), but may equally be based on other measures of sampling effort such as number of visits to an area (Prendergast *et al.*, 1993), or the frequency of occurrence of species in sampling units (Kobayashi, 1974, 1979; Lewinsohn, 1991; Muller-Sharer *et al.*, 1991; Lawton *et al.*, 1993). These techniques have some significant limitations. These are associated particularly with assumptions of homogeneity and randomness, and the adequacy of differences in numbers of species at a given level of effort for making broader statements about relative richness (Simberloff, 1979; Gage & May, 1993; May, 1993).

A major complication with correcting for recorder effort is that it is not simply a function of the amount of time spent or the area traversed in conducting an inventory of an area. It may also be affected profoundly by such features as weather, altering species activity patterns. Moreover, in some instances there may be a positive feedback loop in species listing which is difficult to disentangle. Areas found to be high in richness are visited more often because they are known to be speciose and thus yet more species tend to be recorded, whilst areas which do not gain such a reputation are neglected (Bullock, 1991). The emergence of the concept of biodiversity has served to catalyse interest in absolute numbers of species, and in methods of estimating these numbers. However, some understanding of absolute numbers needs to be established for other reasons. Foremost, even if most studies can be performed using relative measures of species richness it is desirable to know how those measures relate to absolute species richness. As with abundances (e.g. Caughley, 1977), failure to understand relationships between relative and absolute estimates of species numbers can severely limit the interpretation of relative estimates.

Most estimates of species richness, even absolute ones, are perhaps in some sense relative, in that they are based not on measures or samples of the number of species actually present, but on measures or samples of the numbers detectable by the methods of sampling employed. The numbers of species recorded in any given assemblage may depend greatly on sampling methods, which can generate widely differing results (e.g. Disney *et al.*, 1982; Majer & Recher, 1988; Noyes, 1989; Gadagkar *et al.*, 1990; Hammond, 1990; Casson & Hodkinson, 1991; Martini, 1992). Indeed, upward revisions of estimates of global species numbers have often followed from the application of new sampling techniques (e.g. Erwin, 1983; Trüper, 1992; André *et al.*, 1994). Claims for the existence of new frontiers of previously unsuspected high species richness, have usually followed from the use of novel or previously untested sampling methods in a habitat. These are complexities for which statistical techniques for estimating species numbers (see below) cannot account.

4.3.7 Sampling strategies

Difficulties are encountered frequently in interpreting species richness figures because of the lack of information about levels and distributions of sampling effort. It is thus perhaps rather surprising how little attention has been directed toward establishing appropriate sampling strategies for best generating either relative or absolute estimates of species richness for an area. Indeed, discussion of methods of estimation has largely been devoid of consideration of this topic.

The most crucial issue would seem to concern the relative merits of random and stratified sampling strategies. Although ecology has a preoccupation with random survey designs, the advantages of stratified sampling on environmental gradients (gradsects–gradient-directed transect sampling) for obtaining representative samples of biological entities (e.g. vegetation types) and for making surveys more cost-effective have been demonstrated amply (Gillison & Brewer, 1985; Austin & Heyligers, 1989). The application of this approach as a basis for the measurement of species richness would seem entirely appropriate (but see Neave *et al.*, 1992).

One significant consequence of the poor attention paid to sampling strategies is that it remains a moot point how well different methods of estimating overall species numbers in an area will perform when based on samples generated using different strategies. Because different strategies are likely to produce sets of samples which differ in the patterns with which species occur amongst samples, presumably different methods of estimation may prove more or less reliable at estimating overall species numbers.

4.4 Methods of measurement

As already mentioned, the estimation of absolute species richness has not been a major preoccupation of ecologists. Rather little effort has therefore been placed either on the development or the testing of appropriate techniques for so doing. In the following section the principal methods presently available are outlined, with particular emphasis on broad approaches to estimation, rather than on statistical details (for which references are provided). Considerations centre on estimation of the numbers of species in an area, because such estimates essentially also provide the basis both for estimates of the numbers of species in a higher taxon (here the area is simply the planet, or whatever part contains all the species) and for estimates of changes in the numbers of species through time.

Two main groups of methods can be recognized, based on samples and on surrogacy. In the main, sample-based methods are more appropriate for estimation of species richness at relatively small than at relatively large spatial scales. Surrogacy-based methods are only appropriate where, at a given spatial scale, reliable estimates of absolute species richness are already available for some areas, and are probably most effective where species richness exhibits large variance between areas.

4.4.1 Sample-based methods

The estimation of the number of species in an area is one of a general set of problems in which an estimate of C is desired, where a population, finite or infinite, is partitioned into C classes (Bunge & Fitzpatrick, 1993). Other issues included in this set are estimation of the size of an author's vocabulary (Efron & Thisted, 1976), and the numbers of organic pollutants (Janardan & Schaeffer, 1981). Such problems remain difficult to solve. Various methods have been proposed for estimating C based on a sample of the total population, for which individuals can be sorted into classes (e.g. species). Bunge and Fitzpatrick (1993) divide these broadly into sampling-theoretic methods and data-analytic methods, and review them in some detail. In the context of estimating species richness from samples, most interest has centred on data-analytic methods (species accumulation curves and parametric models of relative abundance) and a subset of sampling-theoretic methods using non-parametric models.

Species accumulation curves

One common means of attempting to assess the completeness or otherwise of a sampling exercise is the species accumulation curve. A species accumulation curve is a plot of the cumulative numbers of recorded species as a function of effort (Fig. 4.3). As mentioned earlier, effort can be quantified in various

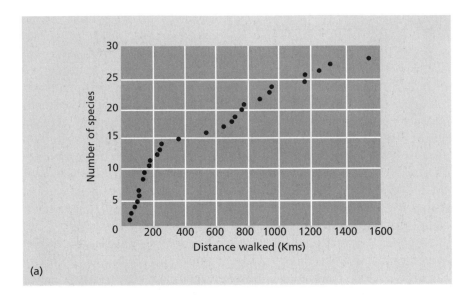

(a)

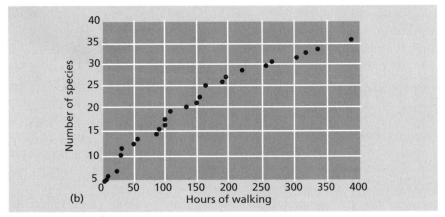

(b)

Fig. 4.3 Species accumulation curves for (a) cumulative number of snake species
encountered by one observer in the INPA–WWF reserves near Manaus (sampling effort
expressed in terms of number of kilometres walked; redrawn from Zimmerman &
Rodrigues, 1990); (b) cumulative number of frog species encountered by one observer in
the INPA–WWF reserves near Manaus (sampling effort expressed in terms of person-
hours of walking; from Zimmerman & Rodrigues, 1990); and (c) cumulative number of
species of hoverflies (Syrphidae) caught in a Malaise trap in 18 consecutive years in a
suburban garden in Leicester, England (sampling effort expressed in terms of the
cumulative number of individuals caught; from data in Owen & Owen, 1990).

ways. These include using the area sampled, such that species-area curves can
be viewed as species accumulation curves, where the areas are nested so that
small areas contribute to the large areas (they may or may not be physically
nested in space). Most measures of effort can be viewed as proxies for
numbers of individuals. However, they often necessitate rather careful
interpretation.

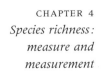

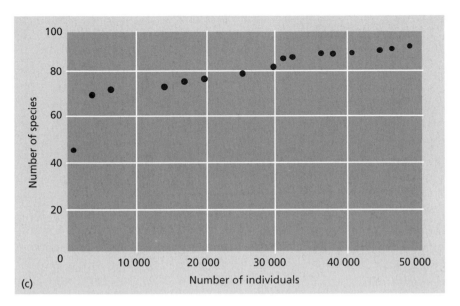

Fig. 4.3 *contd*

For well-known faunas and floras, species accumulation curves are seen to rise to an approximate asymptote, when the total number of species in the area has been recorded. With roughly constant effort being expended over time, new species initially are encountered at a comparatively rapid rate, which falls toward zero as the asymptote in cumulative numbers is reached. A true asymptote is seldom, however, strictly attained, because of temporal turnover in the identities of species present in the area (see Section 4.3.3).

Several mathematical functions have been suggested, used and discussed as descriptors of species accumulation curves (e.g. Stout & Vandermeer, 1975; Clench, 1979; Minshall *et al.*, 1985; Haynes, 1987; Miller & Wiegert, 1989; Palmer, 1990; Lamas *et al.*, 1991; Raguso & Llorente-Bousquets, 1991; Baltanás, 1992; Bunge & Fitzpatrick, 1993; Soberón & Llorente, 1993; Colwell & Coddington, 1994; Longino, 1994; Mawdsley, in press). These offer the possibility of estimating the overall numbers of species in an area by extrapolation. Models may also allow prediction of the numbers of additional species which would be encountered for a defined amount of effort, and conversely the amount of effort which would be required to generate a defined increase in species numbers.

A number of problems exist with using such functions to describe species-accumulation curves and estimate overall species richness. For example, data points are not independent and some functions lack estimators of variance. When little of the full accumulation curve has been documented a number of models may give good fits, but generate widely differing estimates of total species numbers (in the extreme, some functions generate estimates of over-all species richness which are lower than the observed number of species). The appropriateness of the models for the particular situation considered may help to narrow the possibilities but the fundamental problem remains, and most functions have no underlying biological interpretation.

The estimation of overall species richness by extrapolation of species-accumulation curves has been advocated principally at local to meso-scales. Its use has, however, also been debated for larger spatial scales, in the context of relationships between the cumulative numbers of species in a taxon which have been described, and the period (year, decade etc.) in which those descriptions took place (see Steyskal, 1965; Arnett, 1967; Frank & Curtis, 1979; O'Brien & Wibmer, 1979; White, 1979). Such relationships typically follow an approximately sigmoidal function, with low rates of description early and late in documenting a fauna, although for many groups an asymptote has yet to be reached (Arnett, 1967; Frank & Curtis, 1979; Hamilton, 1984; O'Brien & Wibmer, 1979; White, 1975, 1979; Illies, 1983; May, 1990; Gaston, 1991; Hammond, 1992; Gaston & Mound, 1993; Minelli, 1993). Although superficially similar, relationships between cumulative numbers of described species and time differ from many other species accumulation curves in some crucial respects. First, time may in this case be a particularly poor measure of effort, with levels of description often varying dramatically from one decade to the next (e.g. Gaston *et al.*, 1995b). This has the consequence that apparently unexpected increases in numbers of described species can occur (Hammond, 1992), causing substantial deviations from simple curvilinear functions. Second, these relationships are almost invariably retro-spective, in that they describe growth in the numbers of described species as presently recognized. They are not based simply on the numbers of species that were recognized at different points in the past, rather on the numbers of species which are presently regarded as valid and were described at those times (i.e. discounting the former numbers for subsequently recognized syn-onymy and other nomenclatural changes). Third, most interest in global species numbers pertains to hyper-rich groups for which a comparatively small proportion of species have been described. In these cases no asymptote is apparent to the growth of cumulative species numbers, and hence it can potentially be fit by a variety of functions. In sum, such difficulties severely limit the usefulness of attempting to generate extrapolations on the basis of relationships between cumulative numbers of described species and time.

An alternative method of attempting to estimate the total species richness of some higher taxa is based on using the ratio of described to undescribed species found in faunal or floral samples as a multiplier for the total number of species in the group which have already been described (e.g. May, 1990; Hawksworth, 1991, 1993; Hodkinson & Casson, 1991; Hodkinson, 1992; Hodkinson & Hodkinson, 1993). In some sense this can be viewed as related to methods of estimation employing a species accumulation curve, in that the ratio of described to undescribed species is derived from two points on such a curve.

Parametric models of relative abundance

Where data are available on the numbers of individuals of each species encountered in samples (and often they are not), potentially it becomes possible to use methods of estimating overall species numbers based on our understanding of species abundance distributions. Taking the lognormal distribution as the most general model of species abundances, especially at meso-scales (but see Nee *et al.*, 1991), an estimate of the overall number of species can be obtained by fitting a Gaussian curve to the log-transformed data, extrapolating the curve to the left, and integrating between − infinity and + infinity. This method and variants on it have been discussed and/or applied by a number of authors (e.g. Slocomb & Dickson, 1978; Krebs, 1989; Palmer, 1990; Coddington *et al.*, 1991; Baltanás, 1992; Bunge & Fitzpatrick, 1993; Colwell & Coddington, 1994; Longino, 1994). There are several difficulties in its application, particularly associated with the absence of a mode to species abundance curve for many samples (especially those for arthropod assemblages), curve-fitting, and the lack of an estimator for the confidence interval (Coddington *et al.*, 1991; Colwell & Coddington, 1994).

Non-parametric methods

Various non-parametric methods for obtaining estimates of species richness from samples have been proposed and applied. In the main these are based solely on species patterns of occurrence in sampling units, ignoring the numbers of individuals of a given species recorded for a given sampling unit. They include jackknife estimators (Heltshe & Forrester, 1983; Smith & van Belle, 1984; Krebs, 1989; Palmer, 1990, 1991; Coddington *et al.*, 1991; Baltanás, 1992; Neave *et al.*, 1992; Colwell & Coddington, 1994) and bootstrap estimators (Smith & van Belle, 1984; Krebs, 1989; Palmer, 1990), amongst others (see Bunge & Fitzpatrick, 1993; Colwell & Coddington, 1994). A major limitation on their application is that the maximum achievable estimate is often a low multiple of the number of observed species.

Not uncommonly, it is observed that sample-based estimators tend to under-estimate actual species richness (e.g. Palmer, 1990; Baltanás, 1992), particu-larly when sample sizes are small. The reliability of estimates generated by sample-based methods tends to increase markedly with the proportion of the overall number of species which are contained within the sample. However, a serious obstacle to evaluation of the various methods based on samples has been that rather few studies have been performed to compare the accuracy and precision of the estimates that in practice they generate. Work both on real assemblages (Palmer, 1990, 1991; Coddington *et al.*, 1991; Soberón & Llorente, 1993; Mawdsley, in press) and using computer simulations (Baltanás, 1992) has begun to rectify this situation. Nonetheless, many more such studies (embracing a wide range of taxa, habitats and spatial scales) are needed before it will be possible to judge what the different methods have to offer. Where methods are evaluated using data for real assemblages, it is necessary to have sound information on their actual species richness. Usually this has meant that such evaluations have tended to be performed at small spatial scales (e.g. Palmer, 1990). Because of the growing demand to apply such methods at comparatively large spatial scales, greater consideration will need to be given to studies of their efficacy using taxa where richness is well documented at regional levels (see Mawdesley in press, for an example of this approach).

Judgement of the usefulness of different methods needs to be based on a number of considerations. These include: (i) the performance of the estimator relative to the total number of species observed (Palmer, 1990); (ii) the availability of estimators for variances; (iii) whether estimates have consistent biases; (iv) whether methods tend reliably to estimate lower bounds to species richness; (v) whether there is a correlation between true species richness and that estimated (i.e. how good an estimate of relative richness does a method provide?); and (vi) the susceptibility of methods to levels of equability (evenness) of species abundances and patterns of aggregation. All these methods suffer from the usual, and severe, limitations associated with attempting to extrapolate beyond the limits of available data.

4.4.2 Surrogacy methods

The methods of estimation discussed this far necessitate, at the least, samples for the higher taxon or taxa of interest from the area of investigation. An alternative set of approaches has been proposed which circumvents this demand. These rely on the establishment of relationships between the species richness of the higher taxon of interest and another variable, such that in additional areas, estimation of the species richness of this higher taxon can be derived solely from the other variable. The general disadvantage of this methodology is that it necessitates the rigorous empirical establishment of the overall relationship. This presupposes that robust estimates of the actual numbers of species of the higher taxon can be established for at least some areas independently. For a few higher taxa in some, predominantly temperate, regions this may be possible.

Three main surrogacy methods have been suggested, based respectively on environmental variables, indicator groups, and numbers of higher taxa.

Environmental variables

There have been numerous studies of the relationships between the numbers of species of a higher taxon in an area and environmental parameters (e.g. Pianka, 1967; Harman, 1972; Kohn & Levitan, 1976; Brown & Davidson, 1977; Schall & Pianka, 1978; Moran, 1980; Boomsma & van Loon, 1982; Tonn & Magnuson, 1982; Abramsky & Rosenzweig, 1983; Wright, 1983; Miller, 1986; Currie & Paquin, 1987; Miller *et al.*, 1987, 1989; Turner *et al.*, 1987, 1988; Owen, 1988, 1989; Adams & Woodward, 1989; Currie, 1991; Etter & Grassle, 1992; Hill & Keddy, 1992; Wright *et al.*, 1993; Wylie & Currie, 1993a, 1993b; Dzwonko & Kornas, 1994; Gough *et al.*, 1994; McIntyre & Lavorel, 1994). These have been motivated largely by desires to understand the determinants of species richness, but increasingly to predict patterns of species richness for the purposes of conservation. The primary attraction of using such relationships to predict species richness is that environmental parameters are often substantially easier to measure than species richness. The environmental parameters explored in these studies have been varied, and include evapotranspiration, habitat heterogeneity and structure, productivity, temperature, and topographic diversity.

It is difficult to generalize about this work and the usefulness of relationships between species richness and environmental parameters for predicting the former. Documented relationships vary widely in levels of explained variance, and the ease with which simple descriptive models can be fitted. Using wide ranges of values of environmental parameters, relationships have often been found to be non-linear, the 'hump-shaped' relationship between richness and productivity being perhaps the best known example (e.g. Rosenzweig & Abramsky, 1993). Relationships are known in some instances to differ for different subsets of species, such as those which are common and those which are rare (e.g. White & Miller, 1988; Hill & Keddy, 1992; but see McIntyre & Lavorel, 1994), and the extent to which they reflect causal interactions, and how these operate, is usually poorly understood. The latter is particularly the case because there are frequently strong correlations between different environmental parameters. The best models usually involve multiple environmental parameters, which may interact in complex ways (e.g. Margules *et al.*, 1987).

The dream of many studying biodiversity would seem to be a robust relationship between species richness and one or a set of environmental parameters with strong predictive value, where these environmental parameters could be measured remotely (e.g. using satellites). This remains something of a dream. However, it is encouraging that there have been rapid developments in techniques for predicting the spatial distributions and abundances of individual species (e.g. Longmore, 1986; Margules & Stein, 1989; Osborne & Tigar, 1992; Buckland & Elston, 1993; Gates *et al.*, 1994) and of species richness in particular groups on the basis of habitat and environmental data which can, in part at least, be collected in such a fashion.

Indicator groups

The most widely cited method of estimating species richness is probably that based on 'predictor sets' (Kitching, 1993) or 'indicator' groups (this use of the term 'indicator' needs to be differentiated carefully from its numerous other applications). This method uses relationships between the species richness of different higher taxa in an area to enable that of a group whose richness is relatively hard to determine to be estimated from that of a group for which richness can be more readily determined. The method is commonly simply assumed to work, and statements about patterns in the species richness of one higher taxon are regularly made on the basis of patterns in the richness of another, with no explicit demonstration that the two are appropriately and adequately related.

As with relationships between species richness and environmental parameters, explained variances and the complexity of interactions can differ dramatically for relationships between the species richness of two higher taxa (Gaston, 1995). They are also liable to differ with region and spatial scale. In other words, indicator relationships exist in some situations but not in others. The apparent variation in relationships between the species richness of different higher taxa may explain the rather divergent opinions which have been expressed as to the utility of this method of estimating species richness.

To date, most proposed indicator groups have been constrained on the basis of taxonomy. They could, however, be constrained on some other basis. A functional one is particularly attractive. For example, correlations and other regularities have been observed between the numbers of species in different guilds, as have strong similarities in the proportions of species in different guilds (Table 4.6; Warren & Gaston, 1992), raising the possibility of estimating overall species richness on the basis of enumeration of the richness of one or a very few functional groups. The utility of the method would, however, depend greatly on the ease with which functional groups could in practice be recognized.

Relationships between the local or regional numbers of species in different higher taxa (often simple ratios of species numbers) have been used as the basis of extrapolations for estimating the global numbers of species in some groups (e.g. Erwin, 1982; Hawksworth, 1991, 1993; Hodkinson & Casson, 1991; Gaston, 1992; Gaston & Hudson, 1994; Gaston *et al.*, in press). Such extrapolations are by definition almost invariably extreme, and little reliance can be placed on the conclusions generated from individual calculations.

100

Table 4.6 The percentage of species classified into different trophic groups for the known insect faunas of various areas of North America. Phy, phytophagous; Sap, saprophagous; Har, harpactophagous; Par, parasitic; and Pol, pollen feeders etc. From Weiss (1939).

	Species	Phy	Sap	Har	Par	Pol
Western Arctic Coast	400	47	27	14	10	2
New Jersey	10 500	49	19	16	12	4
Connecticut	6 781	52	19	16	10	3
North Carolina	9 249	46	17	22	11	4
Mount Desert Is., Maine	5 177	52	17	14	15	2

Wheeler (1988) provides an interesting variant on the indicator group approach to species richness estimation, in work on rich-fen vegetation. He develops a richness indicator score, based on summing the median species richness scores for several richness-indicator species, where the median species richness score is the median number of species with which an indicator species is associated at sites where it occurs.

Higher taxon richness

A third surrogacy-based method of estimating species richness is founded on the relationship between the numbers of species and the numbers of supra-specific taxa (e.g. genera, tribes, families) in an area (Gaston & Williams, 1993; Williams & Gaston, 1994; Williams et al., 1994). Although its potential remains little explored with regard to extant faunas and floras, such an approach has substantial roots in the palaeontological literature (e.g. Sepkoski, 1991, 1992). Moreover, positive relationships between the numbers of species and higher taxa in different areas have been documented for a variety of extant groups (e.g. Gotelli & Abele, 1982; Jablonski & Flessa, 1986; Smith & Theberge, 1986; Gaston & Williams, 1993; Jones, 1993; Williams, 1993; Williams & Gaston, 1994; McAllister et al., 1994). Obviously for taxonomic units of very high rank (e.g. orders), such relationships tend rapidly to reach asymptotes with increasing species richness, however, this need not be so for units of lower rank. Here, approximately linear or curvilinear non-asymptotic relationships may be observed.

The principal advantage of an approach to estimating species richness based on higher taxon richness is that once an overall relationship between species richness and higher taxon richness has been established it is only necessary directly to determine the numbers of higher taxa in subsequent areas. The number of higher taxa in areas can be documented more rapidly than the number of species simply because there are fewer of them and they can more readily be enumerated (Table 4.7). The use of numbers of higher taxa is particularly attractive because species are not distributed evenly

Table 4.7 The growth of knowledge of the insect fauna of the Galapagos Archipelago. The figures in parentheses indicate the proportion of the 1977–1990 numbers discovered in each of the two preceding periods. From Peck (1991).

Year	Orders	Families	Genera	Species
1835–1966	19 (0.79)	129 (0.58)	395 (0.50)	618 (0.46)
1966–1977	22 (0.92)	164 (0.74)	531 (0.67)	883 (0.66)
1977–1990	24	221	790	1339

amongst higher taxa. The distribution is strongly right-skewed (Willis, 1922; Williams, 1964; Clayton, 1972, 1974; Anderson, 1974; Bock & Farrand, 1980; Dial & Marzluff, 1989). Those higher taxa with large numbers of species (e.g. Curculionidae, Pyralidae, Orchidaceae, Asteraceae) are frequently regarded as taxonomically the more complex, although the ultimate reasons are not always well understood. They absorb disproportionately large amounts of taxonomic effort. Use of the higher taxon method essentially avoids the need for such work.

The primary limitations to this method concern potential inconsistencies in the definition and discrimination of higher taxa, and genuine differences in the taxonomic structure (e.g. distribution of species amongst higher taxa) of assemblages in different regions. Higher taxa suffer from many of the difficulties of definition and recognition discussed with regard to species, and between-region differences in taxonomic structure have been documented (e.g. Ricklefs, 1987, 1989; Latham & Ricklefs, 1993; Prance, 1994). Both problems are likely to be most acute for comparisons of spatially disparate areas.

Studies of the utility of surrogacy-based approaches to estimating species richness have concentrated on establishing the appropriate relationships and have seldom explored how useful these are in predicting the richness of additional areas. All three surrogacy-based methods have some shared limitations. First, in each instance, relationships tend to be rather case-specific, usually dependent at a minimum on spatial scale, region, and taxa. Second, relationships generated by all three methods are subject to spatial autocorrelation. Data points are not independent, being derived from different points in space. The closer together they are, the more similar they are in richness of species or higher taxa and in environmental conditions. Techniques are becoming available to analyse bivariate relationships where both variables exhibit spatial autocorrelation, but these have not as yet been used in the exploration of the relationships of the surrogacy-based methods discussed (e.g. Clifford & Richardson, 1985; Clifford *et al.*, 1989; Legendre & Fortin, 1989; Legendre, 1993). Third, these methods are most useful for estimating the species richness of areas whose richness lies within the range of those areas used to generate the relationship being used. Extrapolation beyond these limits suffers all the usual problems of such actions.

4.5 Conclusion

By way of conclusion, it is perhaps useful to revisit the four possible reasons suggested at the outset of this chapter for species richness being the most frequently and widely applied measure of biodiversity.

• Species richness does indeed capture something of the essence of biodiversity. However, in no way does species richness equate to biodiversity, it is simply one potentially useful measure of biodiversity.

• The meaning of species richness is generally understood, and there is no need to derive complex indices to express it.

• Species richness is not as readily measurable a parameter as one might be led to believe. Most discussion centres on relative rather than absolute measures, and both are severely complicated by issues of spatial and temporal scale, and recorder effort. Estimators for absolute richness remain poorly explored.

• Whilst many data on species richness are available in the literature, they should be treated circumspectly. Details are seldom available as to the completeness or otherwise of inventories, of the effort used to generate richness estimates, or of the status of different species included.

Acknowledgments

I am grateful to Tim Blackburn, Nick Nicholls, Malcolm Scoble and Phil Warren for discussion and comments on the manuscript, and to Ian Gauld for assistance in data compilation.

References

Abramsky Z. & Rosenzweig M.L. (1983) Tilman's predicted productivity–diversity relationship shown by desert rodents. *Nature* **309**, 150–151. [4.4.2]

Adams J.M. & Woodward F.I. (1989) Patterns in tree species richness as a test of the glacial extinction hypothesis. *Nature* **339**, 699–701. [4.4.2]

Adis J. (1990) Thirty million arthropod species – too many or too few? *J. Trop. Ecol.* **6**, 115–118. [4.3.1]

Andersen R.A. (1992) Diversity of eukaryotic algae. *Biodiv. Conserv.* **1**, 267–292. [4.3]

Anderson S. (1974) Patterns of faunal evolution. *Q. Rev. Biol.* **49**, 311–332. [4.4.2]

Anderson S. & Marcus, L.F. (1993) Effect of quadrat size on measurements of species density. *J. Biogeog.* **20**, 421–428. [4.3.1] [4.3.3]

André H.M., Noti M.I. & Lebrun P. (1994) The soil fauna: the other last biotic frontier. *Biodiv. Conserv.* **3**, 45–56. [4.3.6]

Arnett R. H. (1967) Present and future systematics of the Coleoptera in North America. *Ann. Entomol. Soc. Am.* **60**, 162–170. [4.4.1]

Austin M.P. & Heyligers P.S. (1989) Vegetation survey design for conservation: gradsect sampling of forests in north-eastern New South Wales. *Biol. Conserv.* **50**, 13–32. [4.3.7]

Balletto E. & Casale A. (1991) Mediterranean insect conservation. In: *The Conservation of Insects and their Habitats* (eds N.M. Collins & J.A. Thomas), pp. 121–142. Academic Press, London. [4.3.5]

Baltanás A. (1992) On the use of some methods for the estimation of species richness. *Oikos* **65**, 484–492. [4.4.1]

Bengtsson J. (1994) Confounding variables and independent observations in comparative analyses of food webs. *Ecology* **75**, 1282–1288. [4.2]

Bock C.E. (1987) Distribution–abundance relationships of some Arizona landbirds: a matter of scale? *Ecology* **68**, 124–129. [4.3.3]

Bock W.J. & Farrand Jr J. (1980) The number of species and genera of Recent birds: a contribution to comparative systematics. *Am. Mus. Nov.* **2703**, 1–29. [4.4.2]

Boecklen W.J. & Nocedal J. (1991) Are species trajectories bounded or not? *J. Biogeog.* **18**, 647–652. [4.3.3]

Boomsma J.J. & van Loon A.J. (1982) Structure and diversity of ant communities in successive coastal dune valleys. *J. Anim. Ecol.* **51**, 957–974. [4.4.2]

Brewer A. & Williamson M. (1994) A new relationship for rarefraction. *Biodiv. Conserv.* **3**, 373–379. [4.2]

Brey T., Klages M., Dahm C., Gorny M., Gutt J., Haln S., Stiller M., Arntz W.E., Wägele J-W. & Zimmermann A. (1994) Antarctic benthic diversity. *Nature* **368**, 297. [4.3.6]

Brown J.H. & Davidson D.W. (1977) Competition between seed-eating rodents and ants in desert ecosystems. *Science* **196**, 880–882. [4.4.2]

Brown J.H. & Kurzius M.A. (1987) Composition of desert rodent faunas: combinations of coexisting species. *Ann. Zool. Fennici* **24**, 227–237. [4.3.5]

Buckland S.T. & Elston D.A. (1993) Empirical models for the spatial distribution of wildlife. *J. Appl. Ecol.* **30**, 478–495. [4.4.2]

Bullock J.A. (1991) The distribution of a taxon is that of its students and the diversity of a site is a matter of serendipity? *Antenna* **15**, 6–7. [4.3.6]

Bunge J. & Fitzpatrick M. (1993) Estimating the number of species: a review. *J. Amer. Stat. Assoc.* **88**, 364–373. [4.4.1]

Casson D.S. & Hodkinson I.D. (1991) The Hemiptera (Insecta) communities of tropical rain forest in Sulawesi. *Zool. J. Linn. Soc.* **102**, 253–275. [4.3.6]

Caughley G. (1977) *Analysis of Vertebrate Populations.* Wiley, New York. [4.3.6]

Chu Y.I. (1989) *A Check List of Japanese Insects.* Entomological Laboratory, Faculty of Agriculture, Kyushu University. [4.3.5]

Clayton W.D. (1972) Some aspects of the genus concept. *Kew Bull.* **27**, 281–287. [4.4.2]

Clayton W.D. (1974) The logarithmic distribution of angiosperm families. *Kew Bull.* **29**, 271–279. [4.4.2]

Clench H. (1979) How to make regional lists of butterflies: some thoughts. *J. Lepid. Soc.* **33**, 216–231. [4.4.1]

Clifford P. & Richardson S. (1985) Testing the association between two spatial processes. *Statistics and Decisions* Suppl. Issue **2**, 155–160. [4.4.2]

Clifford P., Richardson S. & Hémon D. (1989) Assessing the significance of the correlation between two spatial processes. *Biometrics* **45**, 123–134. [4.4.2]

Coddington J.A., Griswold C.E., Dávila D.S., Peñaranda E. & Larcher S.F. (1991) Designing and testing sampling protocols to estimate biodiversity in tropical ecosystems. In: *The Unity of Evolutionary Biology: proceedings of the fourth international congress of systematic and evolutionary biology* (2 vols) (ed. E.C. Dudley), pp. 44–60. Dioscorides Press, Portland. [4.4.1]

Colwell R.K. & Coddington J.A. (1994) Estimating terrestrial biodiversity through extrapolation. *Phil. Trans. R. Soc. Lond. B* **345**, 101–118. [4.4.1]

Connor E.F. & McCoy E.D. (1979) The statistics and biology of the species–area relationship. *Am. Nat.* **113**, 791–833. [4.3.3]

Connor E.F. & Simberloff D. (1978) Species number and compositional similarity of the Galapagos flora and avifauna. *Ecol. Monogr.* **48**, 219–248. [4.3.5]

Cousins S. (1977) Sample size and edge effect on community measures of farm bird populations. *Polish Ecol. Stud.* **3**, 27–35. [4.2]

Cousins S.H. (1994) Taxonomy and functional biotic measurement, or, will the ark work?

In: *Systematics and Conservation Evaluation* (eds P.L. Forey, C.J. Humphries & R.I. Vane-Wright), pp. 397–419. Oxford University Press. [4.2]

Cracraft J. (1992) The species of the birds-of-paradise (Paradisaeidae): applying the phylogenetic species concept to a complex pattern of diversification. *Cladistics* **8**, 1–43. [4.3.1]

Cranston P. & Hillman T. (1992) Rapid assessment of biodiversity using 'biological diversity technicians'. *Aust. Biol.* **5**, 144–154. [4.3.2]

Currie D.J. (1991) Energy and large-scale patterns of animal and plant species richness. *Am. Nat.* **137**, 27–49. [4.4.2]

Currie D.J. & Paquin V. (1987) Large-scale biogeographical patterns of species richness of trees. *Nature* **329**, 326–327. [4.4.2]

Danks H.V. (1979) Summary of the diversity of Canadian terrestrial arthropods. In: *Canada and its Insect Fauna* (ed. H.V. Danks), pp. 240–244. *Mem. Entomol. Soc. Can.* **108**. [4.3.5]

Danks H.V. (1986) Insect–plant interactions in arctic regions. *Revue Entomol. Quebec* **31**, 52–75. [4.3.5]

Danks H.V. (1993) Patterns of diversity in the Canadian insect fauna. *Mem. Entomol. Soc. Can.* **165**, 51–74. [4.3.1]

Dial K.P. & Marzluff J.M. (1989) Nonrandom diversification within taxonomic assemblages. *Syst. Zool.* **38**, 26–37. [4.4.2]

Dinesen L., Lehmberg T., Svendsen J.O., Hansen L.A. & Fjeldså J. (1994) A new genus and species of perdicine bird (Phasianidae, Perdicini) from Tanzania; a relict form with Indo-Malayan affinities. *Ibis* **136**, 2–11. [4.3.5]

Disney R.H.L., Erizinclioglu Y.Z., Henshaw D.J. de C., Howse D., Unwin D.M., Withers P. & Woods A. (1982) Collecting methods and the adequacy of attempted fauna surveys, with reference to the Diptera. *Field Studies* **5**, 607–621. [4.3.6]

Dung V.V., Giao P.M., Chinh N.N., Tuoc D., Arctander P. & MacKinnon J. (1993) A new species of living bovid from Vietnam. *Nature* **363**, 443–445. [4.3.5]

Dzwonko Z. & Kornas J. (1994) Patterns of species richness and distribution of pteridophytes in Rwanda (Central Africa): a numerical approach. *J. Biogeog.* **21**, 491–501. [4.4.2]

Efron B. & Thisted R. (1976) Estimating the number of unseen species: how many words did Shakespeare know? *Biometrika* **63**, 435–447. [4.4.1]

Elton C. (1932) Territory among wood ants (*Formica rufa* L.) at Picket Hill. *J. Anim. Ecol.* **1**, 69–76. [4.3.3]

Elton C. (1933) *The Ecology of Animals.* Methuen, London. [4.3.3]

Emmet A.M. & Heath J. (eds) (1989) *The Moths and Butterflies of Great Britain and Ireland. Volume 7, Part 1. Hesperiidae-Nymphalidae: The Butterflies.* Harley, Colchester. [4.3.4]

Erwin T.L. (1982) Tropical forests: their richness in Coleoptera and other arthropod species. *Coleopt. Bull.* **36**, 74–75. [4.4.2]

Erwin T.L. (1983) Tropical forest canopies: the last biotic frontier. *Bull. Entomol. Soc. Am.* **30**, 14–19. [4.3.6]

Etter R.J. & Grassle J.F. (1992) Patterns of species diversity in the deep sea as a function of sediment particle size diversity. *Nature* **360**, 576–578. [4.3.6] [4.4.2]

Ferrari S.F. & Lopes M.A. (1992) A new species of marmoset, genus *Callithrix* Erxleben 1777 (Callitrichidae, Primates), from western Brazilian Amazonia. *Goeldiana Zool.* **12**, 1–13. [4.3.5]

Ferrari S.F. & Queiroz H.L. (1994) Two new Brazilian primates discovered, endangered. *Oryx* **28**, 31–36. [4.3.5]

Frank J.H. & Curtis G.A. (1979) Trend lines and the number of species of Staphylinidae. *Coleopt. Bull.* **33**, 133–149. [4.4.1]

Gadagkar R., Chandrashekara K. & Nair P. (1990) Insect species diversity in the tropics: sampling methods and a case study. *J. Bombay Nat. Hist. Soc.* **87**, 337–353. [4.3.6]

Gage J.D. & May R.M. (1993) A dip into the deep seas. *Nature* **365**, 609–610. [4.3.6]

Gaston K.J. (1991) The magnitude of global insect species richness. *Conserv. Biol.* **5**, 283–296. [4.3.1] [4.4.1]

Gaston K.J. (1992) Regional numbers of insect and plant species. *Functional Ecol.* **6**, 243–247. [4.4.2]

Gaston K.J. (1993) Spatial patterns in the description and richness of the Hymenoptera. In: *Hymenoptera and Biodiversity* (eds J. LaSalle & L.D. Gauld), pp. 277–293. CAB International, Wallingford. [4.3.5]

Gaston K.J. (1994) Spatial patterns of species description: how is our knowledge of the global insect fauna growing? *Biol. Conserv.* **67**, 37–40. [4.3.5]

Gaston K.J. Spatial covariance in the species richness of higher taxa. In: *The Genesis and Maintenance of Biological Diversity* (eds M. Hochberg, M.E. Clobert & R. Barbault). Oxford University Press, in press. [4.4.2]

Gaston K.J. & Hudson E. (1994) Regional patterns of diversity and estimates of global insect species richness. *Biodiv. Conserv.* **3**, 493–500. [4.4.2]

Gaston K.J. & May R.M. (1992) Taxonomy of taxonomists. *Nature* **356**, 281–282. [4.3] [4.3.2]

Gaston K.J. & Mound L.A. (1993) Taxonomy, hypothesis testing and the biodiversity crisis. *Proc. R. Soc. Lond. B* **251**, 139–142. [4.3.2] [4.4.1]

Gaston K.J. & Williams P.H. (1993) Mapping the world's species – the higher taxon approach. *Biodiv. Lett.* **1**, 2–8. [4.2] [4.3] [4.3.4] [4.4.2]

Gaston K.J., Blackburn T.M., Hammond P.M. & Stork N.E. (1993) Relationships between abundance and body size – where do tourists fit? *Ecol. Entomol.* **18**, 310–314. [4.3.4]

Gaston K.J., Blackburn T.M. & Loder N. (1995a) Which species are described first? The case of North American butterfly species. *Biodiv. Conserv.* **4**, 119–127. [4.3.2]

Gaston K.J., Scoble M.J. & Crook A. (1995b) Patterns in species description: a case study using the Geometridae. *Biol. J. Linn. Soc.*, **55**, 225–237. [4.3.2] [4.4.1]

Gaston K.J., Gauld I.D. & Hanson P. The size and composition of the hymenopteran fauna of Costa Rica. *J. Biogeog.*, in press. [4.4.2]

Gates S., Gibbons D.W., Lack P.C. & Fuller R.I. (1994) Declining farmland bird species: modelling geographical patterns of abundance in Britain. In: *Large-scale Ecology and Conservation Biology* (eds P.J. Edwards, R.M. May & N.R. Webb), pp. 153–177. Blackwell Scientific, Oxford. [4.4.2]

Gauld I.D. & Mitchell P.A. (1981) *The Taxonomy, Distribution and Host Preferences of Indo-Papuan Parasitic Wasps of the Subfamily Ophioninae (Hymenoptera: Ichneumonidae).* Commonwealth Agricultural Bureaux, Slough. [4.3.2] [4.3.5]

Gibbons D.W., Reid J.B. & Chapman R.A. (1993) *The New Atlas of Breeding Birds in Britain and Ireland: 1988–1991.* T. & A.D. Poyser, London. [4.3.3]

Gillison A.N. & Brewer K.R.W. (1985) The use of gradient directed transects or gradsects in natural resource surveys. *J. Environ. Manage.* **20**, 103–127. [4.3.7]

Glynn P.W. & Feingold J.S. (1992) Hydrocoral species not extinct. *Science* **257**, 1845. [4.3.5]

Gotelli N.J. & Abele L.G. (1982) Statistical distributions of West Indian land bird families. *J. Biogeog.* **9**, 421–435. [4.4.2]

Gough L., Grace J.B. & Taylor K.L. (1994) The relationship between species richness and community biomass: the importance of environmental variables. *Oikos* **70**, 271–279. [4.4.2]

Gressitt J.L. (1951) Introduction. *Insects of Micronesia*, Vol. 1. Bishop Museum, Honolulu. [4.3.5]

Grinnell J. (1922) The role of the 'accidental'. *The Auk* **39**, 373–380. [4.3.3]

Grassle J.F. & Maciolek N.J. (1992) Deep-sea species richness: regional and local diversity estimates from quantitative bottom samples. *Am. Nat.* **139**, 313–341. [4.3.6]

Greig-Smith P.W. (1986) The distribution of native and introduced landbirds on Silhouette Island, Seychelles, Indian Ocean. *Biol. Conserv.* **38**, 35–54. [4.3.5]

Haila Y., Hanski I.P., Niemelä J., Punttila P., Raivio S. & Tukia H. (1994) Forestry and the boreal fauna: matching management with natural forest dynamics. *Ann. Zool. Fennici* **31**, 187–202. [4.3.6]

Hall S.J. & Raffaelli D.G. (1993) Food webs: theory and reality. *Adv. Ecol. Res.* **24**, 187–239. [4.2]

Hamilton K.G.A. (1984) The tenth largest family? *Tymbal* **3**, 4–5. [4.4.1]

Hammond P.M. (1990) Insect abundance and diversity in the Dumoga-Bone National Park, N. Sulawesi, with special reference to the beetle fauna of lowland rain forest in the Toraut region. In: *Insects and Rain Forests of South East Asia (Wallacea)* (eds W.J. Knight & J.D. Holloway), pp. 197–254. Royal Entomological Society of London, London. [4.3.6]

Hammond P.M. (1992) Species inventory. In: *Global Biodiversity: status of the earth's living resources* (ed. B. Groombridge), pp. 17–39. Chapman & Hall, London. [4.3.4] [4.4.1]

Harman W. (1972) Benthic substrates: their effect on freshwater molluscs. *Ecology* **53**, 271–272. [4.4.2]

Havens K. (1992) Scale and structure in natural food webs. *Science* **257**, 1107–1109. [4.2]

Havens K.E. (1993a) Predator–prey relationships in natural community food webs. *Oikos* **68**, 117–124. [4.2]

Havens K.E. (1993b) Effect of scale on food web structure. *Science* **260**, 243. [4.2]

Hawksworth D.L. (1991) The fungal dimension of biodiversity: magnitude, significance, and conservation. *Mycol. Res.* **95**, 441–456. [4.4.1] [4.4.2]

Hawksworth D.L. (1993) The tropical fungal biota: census, pertinence, prophylaxis, and prognosis. In: *Aspects of Tropical Mycology* (eds S. Issac, J.C. Frankland, R. Watling & A.J.S. Whalley), pp. 265–293. Cambridge University Press. [4.3.5] [4.4.1] [4.4.2]

Haynes A. (1987) Species richness, abundance and biomass of benthic invertebrates in a lowland tropical stream on the island of Viti Levu, Fiji. *Arch. Hydrobiol.* **110**, 451–459. [4.4.1]

Heltshe J.F. & Forrester N.E. (1983) Estimating species richness using the jackknife procedure. *Biometrics* **39**, 1–11. [4.4.1]

Heywood V.H. (1994) The measurement of biodiversity and the politics of implementation. In: *Systematics and Conservation Evaluation* (eds P.L. Forey, C.J. Humphries & R.I. Vane-Wright), pp. 15–22. Oxford University Press. [4.3]

Heywood V.H. & Stuart S.N. (1992) Deforestation and species extinction in tropical moist forests. In: *Tropical Deforestation and Species Extinction* (eds T.C. Whitmore & J.A. Sayer), pp. 91–117. Chapman & Hall, London. [4.3.5]

Hill N.M. & Keddy P.A. (1992) Prediction of rarities from habitat variables: coastal plain plants on Nova Scotian lakeshores. *Ecology* **73**, 1852–1859. [4.4.2]

Hodkinson I.D. (1992) Global insect diversity revisited. *J. Trop. Ecol.* **8**, 505–508. [4.4.1]

Hodkinson I.D. & Casson D. (1991) A lesser predilection for bugs: Hemiptera (Insecta) diversity in tropical rain forests. *Biol. J. Linn. Soc.* **43**, 101–109. [4.4.1] [4.4.2]

Hodkinson I.D. & Hodkinson E. (1993) Pondering the imponderable: a probability-based approach to estimating insect diversity from repeat faunal samples. *Ecol. Entomol.* **18**, 91–92. [4.4.1]

Holland P.G. (1978) An evolutionary biogeography of the genus. *Aloe. J. Biogeog.* **5**, 213–226. [4.2]

Holt R.D. (1993) Ecology at the mesoscale: the influence of regional processes on local communities. In: *Species Diversity in Ecological Communities: historical and geographical perspectives* (eds R.E. Ricklefs & D. Schluter), pp. 77–88. Chicago University Press. [4.3.3]

Humphries C.J. & Fisher C.T. (1994) The loss of Banks's legacy. *Phil. Trans. Roy. Soc., Lond. B.* **344**, 3–9. [4.2] [4.3.5]

Hurlbert S.H. (1971) The non-concept of species diversity: a critique and alternative parameters. *Ecology* **52**, 577–586. [4.3.6]

Illies J. (1983) Changing concepts in biogeography. *Annu. Rev. Entomol.* **28**, 391–406. [4.4.1]

Jablonski D. & Flessa K.W. (1986) The taxonomic structure of shallow-water marine faunas: implications for Phanerozoic extinctions. *Malacologia* **27**, 43–66. [4.4.2]

Janardan K.G. & Schaeffer D.J. (1981) Methods for estimating the number of identifiable organic pollutants in the aquatic environment. *Water Resources Res.* **17**, 243–249. [4.4.1]

Jones A.R. (1993) Horses for courses: pragmatic measures of marine benthic invertebrate biodiversity in response to capacity and need. In: *Rapid Biodiversity Assessment* (ed. A.J. Beattie), pp. 69–74. Research Unit for Biodiversity and Bioresources, Macquarie University. [4.4.2]

Judas M. (1988) The species–area relationship of European Lumbricidae (Annelida, Oligochaeta). *Oecologia* **76**, 579–587. [4.3.3]

Kaesler R.L., Herricks E.E. & Crossman J.S. (1978) Use of indices of diversity and hierarchical diversity in stream surveys. In: *Biological Data in Water Pollution Assessment: quantitative and statistical analysis* (eds K.L. Dickson, J. Cairns Jr & R.J. Livingston), pp. 92–112. American Society for Testing and Materials, STP 652. [4.2]

Kershaw M., Williams P.H. & Mace G.M. (1994) Conservation of Afrotropical antelopes: consequences and efficiency of using different site selection methods and diversity criteria. *Biodiv. Conserv.* **3**, 354–372. [4.2]

Kitching R.L. (1993) Towards rapid biodiversity assessment—lessons following studies of arthropods of rainforest canopies. In: *Rapid Biodiversity Assessment* (ed. A.J. Beattie), pp. 26–30. Research Unit for Biodiversity and Bioresources, Macquarie University. [4.4.2]

Kobayashi S. (1974) The species–area relation. I. A model for discrete sampling. *Res. Popul. Ecol.* **15**, 223–237. [4.3.6]

Kobayashi S. (1979) Species–area curves. In: *Statistical Distributions in Ecological Work* (eds J.K. Ord, G.P. Patil & C. Taillie), pp. 349–368. International Co-operative Publishing House, Fairland, Maryland. [4.3.6]

Kohn A.J. & Levitan P.J. (1976) Effect of habitat complexity on population density and species richness in tropical intertidal predatory gastropod assemblages. *Oecologia* **25**, 199–210. [4.4.2]

Krebs C.J. (1989) *Ecological Methodology*. Harper & Row, New York. [4.4.1]

Lamas G., Robbins R.K. & Harvey D.J. (1991) A preliminary survey of the butterfly fauna of Pakitza, Parque Nacional del Manu, Peru, with an estimate of its species richness. *Publicaciones del Museo de Historia Natural Universidad Nacional Mayor de San Marcos Serie A Zoologia* **40**, 1–19. [4.4.1]

Latham R.E. & Ricklefs R.E. (1993) Continental comparisons of temperate-zone tree species diversity. In: *Species Diversity in Ecological Communities: historical and geographical perspectives* (eds R.E. Ricklefs & D. Schluter), pp. 294–314. University of Chicago Press. [4.4.2]

Lawton J.H., Lewinsohn T.W. & Compton S.G. (1993) Patterns of diversity for the insect herbivores on bracken. In: *Species Diversity in Ecological Communities: historical and geographical perspectives* (eds R.E. Ricklefs & D. Schluter), pp. 178–184. University of Chicago Press. [4.3.6]

Legendre P. (1993) Spatial autocorrelation: trouble or new paradigm? *Ecology* **74**, 1659–1673. [4.4.2]

Legendre P. & Fortin M-J. (1989) Spatial pattern and ecological analysis. *Vegetatio* **80**, 107–138. [4.4.2]

Lewinsohn T.M. (1991) Insects in flower heads of Asteraceae in southeast Brazil: a case study on tropical species richness. In: *Plant–animal Interactions: evolutionary ecology in tropical and temperate regions* (eds P.W. Price, T.M. Lewinsohn, G.W. Fernandes & W.W. Benson), pp. 525–559. Wiley, New York. [4.3.6]

Longino J. (1994) How to measure arthropod diversity in a tropical rainforest. *Biol. Int.* **28**, 3–13. [4.4.1]

Longmore R. (ed.) (1986) *Snakes: atlas of elapid snakes of Australia*. Australian Government Publishing Service, Canberra. [4.4.2]

Lynch J.F. & Johnson N.K. (1974) Turnover and equilibria in insular avifaunas, with special reference to the California Channel islands. *Condor* **76**, 370–384. [4.3.5]

Maa T.C. (1956) Notes on the insect fauna of Taiwan. *Scientific Agriculture* **4**, 228–237. [4.3.5]

MacArthur R.H. & Wilson E.O. (1967) *The Theory of Island Biogeography*. Princeton University Press, New Jersey. [4.3.3]

Magurran A.E. (1988) *Ecological Diversity and its Measurement*. Croom Helm, London. [4.2]

Majer J.D. & Recher H.F. (1988) Invertebrate communities on Western Australian euca-lypts: a comparison of branch clipping and chemical knockdown procedures. *Aust. J. Ecol.* **13**, 269–278. [4.3.6]

Margules C.R. & Stein J.L. (1989) Patterns in the distribution of species and the selection of nature reserves: an example from *Eucalyptus* forests in south-eastern New South Wales. *Biol. Conserv.* **50**, 219–238. [4.4.2]

Margules C.R., Nicholls A.O. & Austin M.P. (1987) Diversity of *Eucalyptus* species predicted by a multi-variable environmental gradient. *Oecologia* **71**, 229–232. [4.4.2]

Martin T.E. (1981) Species–area slopes and coefficients: a caution on their interpretation. *Am. Nat.* **118**, 823–837. [4.3.3]

Martin J. & Gurrea P. (1990) The peninsular effect in Iberian butterflies (Lepidoptera: Papilionoidea and Hesperioidea). *J. Biogeog.* **17**, 85–96. [4.2]

Martinez N.D. (1991) Artifacts or attributes? Effects of resolution on the Little Rock lake food web. *Ecol. Monogr.* **61**, 367–392. [4.2]

Martinez N.D. (1993) Effect of scale on food web structure. *Science* **260**, 242–243. [4.2]

Martini A. (1992) Biodiversity and conservation of yeasts. *Biodiv. Conserv.* **1**, 324–333. [4.3.6]

Mawdsley N. (1995) The theory and practice of estimating regional species richness from local samples, in press. [4.4.1]

May R.M. (1990) How many species? *Phil. Trans. Roy. Soc., Lond. B.* **330**, 293–304. [4.3.1][4.4.1]

May R.M. (1993) Marine species richness. *Nature* **361**, 598. [4.3.6]

McAllister D.E., Schueler F.W., Roberts C.M. & Hawkins J.P. (1994) Mapping and GIS analysis of the global distribution of coral reef fishes on an equal-area grid. In: *Mapping the Diversity of Nature* (ed. R.I. Miller), pp. 155–175. Chapman & Hall, London. [4.4.2]

McIntyre S. & Lavorel S. (1994) Predicting richness of native, rare, and exotic plants in response to habitat and disturbance variables across a variegated landscape. *Conserv. Biol.* **8**, 521–531. [4.4.2]

Miller R.I. (1986) Predicting rare plant distribution patterns in the southern Appalachians of the southeastern USA. *J. Biogeog.* **13**, 293–311. [4.2] [4.4.2]

Miller R.I. & Wiegert R.G. (1989) Documenting completeness, species–area relations, and the species-abundance distribution of a regional flora. *Ecology* **70**, 16–22. [4.4.1]

Miller R.I., Bratton S.P. & White P.S. (1987) A regional strategy for reserve design and placement based on an analysis of rare and endangered species' distribution patterns. *Biol. Conserv.* **39**, 255–268. [4.2] [4.4.2]

Miller R.I., Stuart S.N. & Howell K.M. (1989) A methodology for analyzing rare species distribution patterns utilizing GIS technology: the rare birds of Tanzania. *Landscape Ecol.* **2**, 173–189. [4.4.2]

Minelli A. (1993) *Biological Systematics: the state of the art*. Chapman & Hall, London. [4.4.1]

Minshall G.W., Petersen R.C. Jr & Nimz C.F. (1985) Species richness in streams of different size from the same drainage basin. *Am. Nat.* **125**, 16–38. [4.4.1]

Moran V.C. (1980) Interactions between phytophagous insects and their *Opuntia* hosts. *Ecol. Entomol.* **5**, 153–164. [4.4.2]

Mound L.A. & Gaston K.J. (1993) Conservation and systematics — the agony and the ecstasy. In: *Perspectives on Insect Conservation* (eds K.J. Gaston, T.R. New & M.J. Samways), pp. 185–195. Intercept, Andover. [4.3.2] [4.3.5]

Muller-Sharer H., Lewinsohn T.M. & Lawton J.H. (1991) Searching for weed biocontrol agents — when to move on? *Biocontrol Sci. Technol.* **1**, 271–280. [4.3.6]

Neave H.M., Cunningham R.B., Norton T.W. & Nix H.A. (1992) Evaluation of field sampling strategies for estimating species richness by Monte Carlo methods. *Math. Comp. Simul.* **33**, 391–396. [4.3.7] [4.4.1]

Nee S., Harvey P.H. & May R.M. (1991) Lifting the veil on abundance patterns. *Proc. Roy. Soc., Lond. B* **243**, 161–163. [4.4.1]

Nelson B.W., Ferreira C.A.C., da Silva M.F. & Kawasaki M.L. (1990) Endemism centres,

109

refugia and botanical collection density in Brazilian Amazonia. *Nature* **345**, 714–716. [4.3.5]

Niemelä J., Tukia H. & Halme E. (1994) Patterns of carabid diversity in Finnish mature taiga. *Ann. Zool. Fennici* **31**, 123–129. [4.3.6]

Nilsson I.N. & Nilsson S.G. (1985) Experimental estimates of census efficiency and pseudoturnover on islands: error trend and between-observer variation when recording vascular plants. *J. Ecol.* **73**, 65–70. [4.3.5]

Nilsson S.G. & Nilsson I.N. (1983) Are estimated species turnover rates on islands largely sampling errors? *Am. Nat.* **121**, 595–597. [4.3.5]

Noyes J.S. (1989) A study of five methods of sampling Hymenoptera (Insecta) in a tropical rainforest, with special reference to the Parasitica. *J. Nat. Hist.* **23**, 285–298. [4.3.6]

O'Brien C.W. & Wibmer G.J. (1979) The use of trend curves of rates of species descriptions: examples from the Curculionidae (Coleoptera). *Coleopt. Bull.* **33**, 151–166. [4.4.1]

Oliver I. & Beattie A.J. (1993) A possible method for the rapid assessment of biodiversity. *Conserv. Biol.* **7**, 562–568. [4.3.2]

Oliver I. & Beattie A.J. (1994) A possible method for the rapid assessment of biodiversity. In: *Systematics and Conservation Evaluation* (eds P.L. Forey, C.J. Humphries & R.I. Vane-Wright), pp. 133–136. Oxford University Press. [4.3.2]

Osborne P.E. & Tigar B.J. (1992) Interpreting bird atlas data using logistic models: an example from Lesotho, Southern Africa. *J. Appl. Ecol.* **29**, 55–62. [4.4.2]

Otte D. (1989) Speciation in Hawaiian crickets. In: *Speciation and its Consequences* (eds D. Otte & J.A. Endler), pp. 482–526. Sinauer, Sunderland, Massachusetts. [4.3.1]

Otte D. & Endler J.A. (eds) (1989) *Speciation and its Consequences*. Sinauer, Sunderland, Massachusetts. [4.3.1]

Owen D.F. & Owen J. (1990) Assessing insect species-richness at a single site. *Environ. Conserv.* **17**, 362–364. [4.4.1] [4.4.2]

Owen J.G. (1988) On productivity as a predictor of rodent and carnivore diversity. *Ecology* **69**, 1161–1165. [4.4.2]

Owen J.G. (1989) Patterns on herpetofaunal species richness: relation to temperature, precipitation, and variance in elevation. *J. Biogeog.* **16**, 141–150. [4.4.2]

Paasivirta L. (1976) Species, biomass and production of macrozoobenthos in Lake Suomunjarvi (Lieksa). *Univ. Joensuu, Karelian Institute Publ.* **18**, 1–17. [4.2]

Palmer M.W. (1990) The estimation of species richness by extrapolation. *Ecology* **71**, 1195–1198. [4.4.1]

Palmer M.W. (1991) Estimating species richness: the second-order jackknife reconsidered. *Ecology* **72**, 1512–1513. [4.4.1]

Palmer M.W. & White P.S. (1994) Scale dependence and the species–area relationship. *Am. Nat.* **144**, 717–740. [4.3.3]

Parnell J. (1993) Plant taxonomic research, with special reference to the tropics: problems and potential solutions. *Conserv. Biol.* **7**, 809–814. [4.3.5]

Peck S.B. (1991) The Galapagos Archipelago, Ecuador: with an emphasis on terrestrial invertebrates, especially insects; and an outline for research. In: *The Unity of Evolutionary Biology: Proceedings of the Fourth International Congress of Systematic and Evolutionary Biology* (2 vols) (ed. E.C. Dudley), pp. 319–336. Dioscorides Press, Portland, Oregon. [4.4.2]

Pianka E.R. (1967) On lizard species diversity: North American flatland deserts. *Ecology* **48**, 333–351. [4.4.2]

Pimentel D. & Wheeler A.G. (1973) Species and diversity of arthropods in the Alfalfa community. *Environ. Entomol.* **2**, 659–668. [4.3.4]

Pine R.H. (1994) New mammals not so seldom. *Nature* **368**, 593. [4.3.5]

Poore G.C.B. & Wilson G.D.F. (1993) Marine species richness. *Nature* **361**, 597–598. [4.3.6]

Prance G.T. (1994) A comparison of the efficacy of higher taxa and species numbers in the assessment of biodiversity in the neotropics. *Phil. Trans. Roy. Soc., Lond. B* **345**, 89–99. [4.4.2]

Prance G.T. & Campbell G.G. (1988) The present state of tropical floristics. *Taxon* **37**, 519–548. [4.3.5]

Prendergast J.R., Wood S.N., Lawton J.H. & Eversham B.C. (1993) Correcting for variation in recording effort in analyses of diversity hotspots. *Biodiv. Lett.* **1**, 39–53. [4.3.6]

Queiroz H.L. (1992) A new species of capuchin monkey, genus *Cebus* Erxleben 1777 (Cebidae: Primates) from eastern Brazilian Amazonia. *Goeldiana Zool.* **15**, 1–13. [4.3.5]

Raguso R.A. & Llorente-Bousquets J. (1991) The butterflies (Lepidoptera) of the Tuxtlas Mts., Veracruz, Mexico, revisited: species-richness and habitat disturbance. *J. Res. Lepid.* **29**, 105–133. [4.4.1]

Rapoport E.H. (1982) *Areography: geographical strategies of species.* Pergamon, Oxford. [4.3.2]

Raxworthy C.J. (1988) Reptiles, rainforest and conservation in Madagascar. *Biol. Conserv.* **43**, 181–211. [4.3.5]

Rex M.A., Stuart C.T., Hessler R.R., Allen J.A., Sanders H.L. & Wilson G.D.F. (1993) Global-scale latitudinal patterns of species diversity in the deep-sea benthos. *Nature* **365**, 636–639. [4.3.6]

Rich T.C.G. & Woodruff E.R. (1992) Recording bias in botanical surveys. *Watsonia* **19**, 73–95. [4.3.4]

Ricklefs R.E. (1987) Community diversity: relative roles of local and regional processes. *Science* **235**, 167–171. [4.4.2]

Ricklefs R.E. (1989) Speciation and diversity: the integration of local and regional processes. In: *Speciation and its Consequences* (eds D. Otte & J.A. Endler), pp. 599–622. Sinauer, Sunderland, Massachusetts. [4.4.2]

Rojas M. (1992) The species problem and conservation: what are we protecting? *Conserv. Biol.* **6**, 170–178. [4.3.1]

Rosenzweig M.L. & Abramsky Z. (1993) How are diversity and productivity related? In: *Species Diversity in Ecological Communities: historical and geographical perspectives* (eds R.E. Ricklefs & D. Schluter), pp. 52–65. University of Chicago Press. [4.4.2]

Samways M.J. (1994) *Insect Conservation Biology.* Chapman & Hall, London. [4.3.5]

Schall J.J. & Pianka E.R. (1978) Geographical trends in numbers of species. *Science* **201**, 679–686. [4.2] [4.4.2]

Schoener T.W. (1989) Food webs from the small to the large. *Ecology* **70**, 1559–1589. [4.2]

Sepkoski J.J. Jr (1991) Diversity in the Phanerozoic oceans: a partisan review. In: *The Unity of Evolutionary Biology: Proceedings of the Fourth International Congress of Systematic and Evolutionary Biology* (2 vols) (ed. E.C. Dudley), pp. 210–236. Dioscorides Press, Portland, Oregon. [4.4.2]

Sepkoski J.J. Jr (1992) Phylogenetic and ecologic patterns in the Phanerozoic history of marine biodiversity. In: *Systematics, Ecology, and the Biodiversity Crisis* (ed. N. Eldredge), pp. 77–100. Columbia University Press, New York. [4.4.2]

Shmida A. & Wilson M.V. (1985) Biological determinants of species diversity. *J. Biogeog.* **12**, 1–20. [4.3.4]

Short J. & Smith A. (1994) Mammal decline and recovery in Australia. *J. Mammal.* **75**, 288–297. [4.3.5]

Shuker, K. (1993) *The Lost Ark: new and rediscovered animals of the twentieth century.* Harper-Collins, London. [4.3.5]

Signor P.W. (1990) The geologic history of diversity. *Annu. Rev. Ecol. Syst.* **21**, 509–539. [4.3.3]

Simberloff D. (1979) Rarefraction as a distribution-free method of expressing and estimating diversity. In: *Ecological Diversity in Theory and Practice* (eds J.F. Grassle, G.P. Patil, W. Smith & C. Taillie), pp. 159–176. International Co-operative Publishing House, Fairland, Maryland. [4.3.6]

Simberloff D.S. (1978) Use of rarefraction and related methods in ecology. In: *Biological Data in Water Pollution Assessment: quantitative and statistical analysis* (eds K.L. Dickson, J. Cairns Jr & R.J. Livingston), pp. 150–165. American Society for Testing and Materials, STP 652. [4.3.6]

Skinner B. & Wilson D. (1984) *Colour Identification Guide to Moths of the British Isles (Macrolepidoptera)*. Viking Penguin, Harmondsworth, Middlesex. [4.3.4]

Slater F.M., Curry P. & Chadwell C. (1987) A practical approach to the evaluation of the conservation status of vegetation in river corridors in Wales. *Biol. Conserv.* **40**, 53–68. [4.3.5]

Slocomb J. & Dickson K.L. (1978) Estimating the total number of species in a biological community. In: *Biological Data in Water Pollution Assessment: quantitative and statistical analyses* (eds K.L. Dickson, J. Cairns Jr. & R.J. Livingston), pp. 38–52. American Society for Testing and Materials. STP 652. [4.4.1]

Smith E.P. & van Belle G. (1984) Nonparametric estimation of species richness. *Biometrics* **40**, 119–129. [4.4.1]

Smith P.G.R. & Theberge J.B. (1986) Evaluating biotic diversity in environmentally significant areas in the Northwest Territories of Canada. *Biol. Conserv.* **36**, 1–18. [4.4.2]

Smith W., Grassle J.F. & Kravitz D. (1979) Measures of diversity with unbiased estimates. In: *Ecological Diversity in Theory and Practice* (eds J.F. Grassle, G.P. Patil, W. Smith & C. Taillie), pp. 177–191. International Co-operative Publishing House, Fairland, Maryland. [4.2] [4.3.6]

Soberón M.J. & Llorente B.J. (1993) The use of species accumulation functions for the prediction of species richness. *Conserv. Biol.* **7**, 480–488. [4.4.1]

Soulé M.E. (1990) The real work of systematics. *Ann. Missouri Bot. Gard.* **77**, 4–12. [4.3.5]

Stevens G.C. (1989) The latitudinal gradient in geographical range: how so many species coexist in the tropics. *Am. Nat.* **133**, 240–256. [4.3.2] [4.3.4]

Steyskal G.C. (1965) Trend curves of the rate of species description in zoology. *Science* **149**, 880–882. [4.4.1]

Stout J. & Vandermeer J. (1975) Comparison of species richness for stream-inhabiting insects in tropical and mid-latitude streams. *Am. Nat.* **109**, 263–280. [4.4.1]

Stubbs A.E. (1982) Conservation and the future for the field entomologist. *Proc. Trans. Brit. Entomol. Nat. Hist. Soc.* **15**, 55–67. [4.3.5]

Terborgh J. & Weske J.S. (1972) Rediscovery of the imperial snipe in Peru. *Auk* **89**, 497–505. [4.3.5]

Tonn W.M. & Magnuson J.J. (1982) Patterns in the species composition and richness of fish assemblages in N. Wisconsin lakes. *Ecology* **63**, 1149–1166. [4.4.2]

Trüper H.G. (1992) Prokaryotes: an overview with respect to biodiversity and environmental importance. *Biodiv. Conserv.* **1**, 227–236. [4.3.6]

Turner J.R.G., Gatehouse C.M. & Corey C.A. (1987) Does solar energy control organic diversity? Butterflies, moths and the British climate. *Oikos* **48**, 195–205. [4.4.2]

Turner J.R.G., Lennon J.J. & Lawrenson J.A. (1988) British bird species distributions and the energy theory. *Nature* **335**, 539–541. [4.4.2]

Vane-Wright R.I. (1992) Species concepts. In: *Global Biodiversity: status of the Earth's living resources* (ed. B. Groombridge), pp. 13–16. Chapman & Hall, London. [4.2] [4.3]

Vane-Wright R.I. & Rahardja D.P. (1993) An evaluation of the diversity of subspecies, species and genera of Hesperiidae within The Philippines, using the WORLDMAP computer program. *Zoologische Verhandelingen* **228**, 116–121. [4.3.1]

Vuilleumier F., LeCroy M. & Mayr E. (1992) New species of birds described from 1981 to 1990. *Bull. B.O.C. Centenary Suppl.* **112A**, 267–309. [4.3.5]

Warren P.H. (1990) Variation in food web structure: the determinants of connectance. *Am. Nat.* **136**, 689–700. [4.2]

Warren P.H. & Gaston K.J. (1992) Predator–prey ratios: a special case of a general pattern? *Phil. Trans. Roy. Soc., Lond. B* **338**, 113–130. [4.4.2]

Weiss H.B. (1939) Insect food habit ratios of North Carolina, and Mount Desert Island, Maine. *J. New York Entom. Soc.* **97**, 155–157. [4.4.2]

Wheeler B.D. (1988) Species richness, species rarity and conservation evaluation of rich-fen vegetation in lowland England and Wales. *J. Appl. Ecol.* **25**, 331–353. [4.4.2]

White P.S. & Miller R.I. (1988) Topographic models of vascular plant richness in the southern Appalachian high peaks. *J. Ecol.* **76**, 192–199. [4.4.2]

White R.E. (1975) Trend curves of the rate of species description for certain North American Coleoptera. *Coleopt. Bull.* **29**, 281–295. [4.4.1]

White R.E. (1979) Response to the use of trend curves by Erwin, Frank and Curtis, and O'Brien and Wibmer. *Coleopt. Bull.* **33**, 167–168. [4.4.1]

Whiteman P. & Millington R. (1991) The British list and rare birds in the eighties. *Birding World* **3**, 429–434. [4.3.4]

Williams C.B. (1964) *Patterns in the Balance of Nature*. Academic Press, London. [4.4.2]

Williams P.H. (1993) Measuring more of biodiversity for choosing conservation areas, using taxonomic relatedness. In: *International Symposium on Biodiversity and Conservation* (ed. T-Y. Moon), pp. 194–227. Manus. Col. ISBC KEI. Korean Entomological Institute, Seoul. [4.2] [4.4.2]

Williams P.H. & Gaston K.J. (1994) Measuring more of biodiversity: can higher-taxon richness predict wholesale species richness? *Biol. Conserv.* **67**, 211–217. [4.2] [4.4.1] [4.4.2]

Williams P.H. & Humphries C.J. (1994) Biodiversity, taxonomic relatedness, and endemism in conservation. In: *Systematics and Conservation Evaluation* (eds P.L. Forey, C.J. Humphries & R.I. Vane-Wright), pp. 269–287. Oxford University Press. [4.2]

Williams P.H., Humphries C.J. & Gaston K.J. (1994) Centres of seed-plant diversity: the family way. *Proc. Roy. Soc. Lond. B* **256**, 67–70. [4.2] [4.4.2]

Williamson M. (1988) Relationship of species number to area, distance and other variables. In: *Analytical Biogeography: an integrated approach to the study of animal and plant distributions* (eds A.A. Myers & P.S. Giller), pp. 91–115. Chapman & Hall, London. [4.3.3]

Willis J.C. (1992) *Age and Area: a study in geographical distribution and origin of species*. Cambridge University Press. [4.4.2]

Wilson E.O. (1985) The biological diversity crisis: a challenge to science. *Issues Sci. Technol.* **2**, 20–29. [4.3.5]

Wolda H. (1987) Altitude, habitat and tropical insect diversity. *Biol. J. Linn. Soc.* **30**, 313–323. [4.3.3]

Wright D.H. (1983) Species–energy theory: an extension of species–area theory. *Oikos* **41**, 496–506. [4.4.2]

Wright D.H., Currie D.J. & Maurer B.A. (1993) Energy supply and patterns of species richness on local and regional scales. In: *Species Diversity in Ecological Communities: historical and geographical perspectives* (eds R.E. Ricklefs & D. Schluter), pp. 66–74. University of Chicago Press. [4.4.2]

Wylie J.L. & Currie D.J. (1993a) Species energy theory and patterns of species richness. I. Patterns of bird, angiosperm, and mammal species richness on islands. *Biol. Conserv.* **63**, 137–144. [4.4.2]

Wylie J.L. & Currie D.J. (1993b) Species energy theory and patterns of species richness. II. Predicting mammal species richness on isolated nature reserves. *Biol. Conserv.* **63**, 145–148. [4.4.2]

Yodzis P. (1993) Environment and trophodiversity. In: *Species Diversity in Ecological Communities: historical and geographical perspectives* (eds R.E. Ricklefs & D. Schluter), pp. 26–38. University of Chicago Press. [4.2]

Zimmerman B.L. & Rodrigues M.T. (1990) Frogs, snakes and lizards of the INPA-WWF reserves near Manaus, Brazil. In: *Four Neotropical Forests* (ed AH Gentry), pp. 426–454. Yale University Press, New Haven. [4.4.1]

5: Defining and measuring functional aspects of biodiversity

NEO D. MARTINEZ

5.1 Introduction

Concern with biodiversity from genes to ecosystems focuses primarily on what these entities are and where they are located. A more recently emphasized research agenda concerns what these living entities do or what their 'function' is (e.g. Grassle *et al.*, 1991; Solbrig, 1991; Younès, 1992). Research in this area is barely a half decade old and there is only a small and immature body of literature that explicitly addresses relationships between biodiversity and function. Furthermore, this literature lacks generally accepted definitions of central and frequently reiterated terms. For example, practically no one defines 'function' in the context of biodiversity. Also, only one brave author explicitly defines 'functional diversity' (Steele, 1991) and the level of acceptance of this definition is unclear. As Lawton and Brown (1993, p. 261) comment, 'We lack even a preliminary theoretical framework to guide our observations and experiments in this area and to avoid a morass of special cases.' A 'careful reading' of Schulze and Mooney's (1993) edited volume of eminently authored papers on biodiversity and ecosystem function 'makes it clear that in fact we know embarrassingly little' about this subject (Huenneke, 1994).

Despite this lack of clarity, there are several consistent concepts explicitly and implicitly embraced by investigators of relations between function and biodiversity. For example, there is nearly universal acceptance that function has something to do with ecological processes. The first section of this chapter (Section 5.2) elaborates the meaning of ecological and ecosystem function because 'much confusion arises from our imprecise use of those terms' (Huenneke, 1994). The section continues by developing a preliminary conceptual framework that uses several prominent themes in the literature that explicitly focus on functional aspects of biodiversity. Two aspects in particular are discussed: ecological function addresses the relationships between living entities and ecological processes; and functional diversity addresses the variety of these relationships. After these concepts are distinguished, they are more fully developed by reviewing relevant research.

While several hypotheses are alluded to in the first section, the second section (Section 5.3) more systematically examines a wide range of hypotheses that relate biodiversity and function. The purpose of investigating these relationships is taken to be the generation and testing of such hypotheses. Conceptual difficulties with prominent hypotheses employing redundancy and keystone concepts are identified and ways to ameliorate the difficulties are examined. Further discussion identifies other hypotheses that relate biodiversity to ecological function which have been largely ignored in previous research. The third section (Section 5.4) provides a detailed example of the definition and measurement of certain indices of functional diversity using current food-web research. This example also demonstrates how such measurements are used to test hypotheses such as those that relate diversity and stability. The fourth section (Section 5.5) revisits functional redundancy and replaces the concept with more precise and scientifically useful terms. The conclusion (Section 5.6) considers the future of research in this area and the types of generality that may be achieved by employing more rigorous approaches to functional aspects of biodiversity.

5.2 What is functional about biodiversity?

5.2.1 Defining 'functional' within the context of biodiversity

Following Franklin (1988), Noss (1990, p. 357) was one of the earliest authors to explicitly distinguish functional from structural aspects of biodiversity: 'Structure is the physical organisation or pattern of a system, from habitat complexity as measure within communities to the pattern of patches and other elements at a landscape scale. Function involves ecological and evolutionary processes, including gene flow, disturbance, and nutrient cycling.' Structure typically refers to an amount or arrangement of substance within a particular space. Function typically refers to how such substances change over time. Biodiversity has been evaluated primarily in terms of criteria (e.g. taxonomic, phylogenetic) that qualify the morphological and genetic structure of organisms and also implicitly, or explicitly, address evolutionary processes (e.g. Chapters 3 & 4). Still, there are many other processes that organisms interact with. Interactions between biodiversity and ecosystem processes such as material and energetic stocks and flows have received the most attention (e.g. Schulze & Mooney, 1993; Naeem *et al.*, 1994). Other processes of interest include ecological stability (MacArthur, 1955; May, 1973; Ehrlich & Ehrlich, 1981; Tilman & Downing, 1994), behaviour (Stone *et al.*, 1994), economic activities (Mikluski, 1994; Solow, 1993), and ecosystem services (Westman, 1977; Ehrlich & Ehrlich, 1981).

After familiarizing ourselves with the multitude of concepts concerning biodiversity or the variability of life from genes to ecosystems, we might hope that qualifying 'biodiversity' with 'functional' would restrict it to a more tractable concept. However disappointing, the assertion here is that the functional aspects of biodiversity are a broad and vague concept that

115

needs substantial added specification in order to become scientifically more useful. This vagueness follows from the many uses of 'function' in the English language. Webster's Dictionary (Woolf, 1979) ascribes a variety of meanings to function from a person's occupation to a social gathering, not to mention a mathematical operation. Perhaps the most relevant definition with regard to biodiversity is: 'the acts or operations expected of a person or thing, referable to anything living, material, or constructed, implies a definite end or purpose that the one in question serves or a particular kind of work it is intended to perform.' 'Functional,' in turn, is defined as 'used to contribute to the development or maintenance of a larger whole.'

The teleological implications of definitions involving *purpose* pervade discussions of function in biology and ecology. Several theories considered in this chapter that relate biodiversity to function also hinge on purpose. For example, consider the redundancy hypothesis that many species needlessly perform functions identical to other species in ecological systems. Walker (1992) compares redundant species to passengers in an automobile and juxtaposes the passengers to non-redundant drivers. Lawton and Brown (1993) compare functionally redundant species to words of a sentence that can be deleted without changing the meaning of the sentence. Eliminating the passengers and redundant words does not interfere with the purposeful activity of travelling from point A to point B or communicating an idea. However, other activities, such as a driver talking with someone or a sentence flowing in a particular way, *are* interfered with. The teleological implications are expressed by the unstated but tacitly accepted hierarchy where one activity (going from point A to point B) is generally considered more important or purposeful than another (talking with someone). An activity lower in this hierarchy is redundant if elimination of the entity engaged in the activity does not interfere with the more important activity, also called its 'proper function' (Solbrig, 1992). In an alternative theory to redundancy, Ehrlich and Ehrlich (1981) compare species to rivets in an airplane where each is functionally important. Though loss of a few may not be missed, the loss of many leads to catastrophic failure. Each of these examples employs entities designed by humans for widely accepted purposeful activities or functions. Theories that make use of such anthropocentric analogies are encumbered by scientifically indefensible notions of purpose and proper function.

Teleology has a long and problematic history in biology (Mayr, 1988). In recognition of this, Lawton and Brown (1993) eschew teleological purposes of ecological function and instead assert that ecosystems simply process materials and energy. Following Lawton and Brown, this chapter replaces the notion of purpose with process (an activity by which some effect is obtained). Doing so avoids the distracting discussion of which activities are most important and instead encourages explicit specification of the activity of interest. The replacement of purpose with process retains the intuition that

116

if something is functional, it must do something. Here, only entities that interact with ecological processes are considered to be functional. Ecological processes include more than just material and energetic or ecosystem processes. Ecological processes also encompass information exchanges represented by gene flows and communication represented by certain bird songs. Wider and more explicit consideration of these and other ecological processes such as mutualism, competition, pollination, and population dynamics broadens the framework for addressing ecological function beyond one that only admits ecosystem processes.

5.2.2 A framework for investigating functional aspects of biodiversity

Perhaps the most useful role of a scientific framework is to facilitate consistent determination of which hypotheses are corroborated or refuted by certain observations. This role is achieved through the definition of terms and development of concepts in a manner that can be widely accepted. Redefining function to imply ecological processes instead of purpose is an important precursor to describing a framework for investigating relationships between biodiversity and ecological function.

Further explication of function might not be necessary if we accept Lawton and Brown's (1993, p. 255) suggestion that ' "ecological processes" and "ecosystem function" [are] synonymous.' However, this synonymy assumes that interest in a process (e.g. primary productivity) is restricted to entities whose function is to engage directly in the process (primary producers). This assumption is warranted in many, but not all, cases. Interest in primary productivity could also be focused on nitrogen-fixing bacteria that are not primary producers but do interact with primary productivity by increasing the supply of nitrogen to primary producers (Vitousek & Hooper, 1993). In other words, there are different ways that living entities affect a process besides engaging in it. These include increasing or decreasing the rate of a process. Such associations and relationships seem well described as interactions with processes. Therefore, a function in this context may be defined as an interaction with a process. This definition allows entities that do not interact with a process to be accurately characterized as non-functional relative to that process. More examples below demonstrate that in order to specify a function, both process and interaction should be described.

Biodiversity interacts with ecological processes in many ways. Perhaps most basically, living entities enable or disable the occurrence of ecological processes. For example, communities with high species diversity may contain pollinators for a certain plant (pollination occurs) whereas the pollinators may be absent in communities with lower species richness (pollination does not occur). One of the more commonly examined ways biodiversity interacts with ecological processes is its influence on the rates of processes. Increasing some aspect of biodiversity may increase, decrease, or leave unaffected the rate (Naeem *et al.*, 1994) or stabilize that rate (Tilman & Downing, 1994). A different measure could focus on the product of a process such as stocks of minerals or nutrients (Vitousek & Hooper, 1993; Tilman & Downing, 1994). For example, the stock of vegetative biomass could be measured instead of primary productivity. An increase in this stock could result from an increase in primary productivity or allocthonous inputs or from a decrease in decomposition. Besides affecting rates, changes in biodiversity can also alter the temporal dynamics and spatial patterns of processes.

This categorization of interactions leads to the following definition of *ecological functions: interactions with ecological processes.* Ecological processes are those activities that result from interactions among organisms and between organisms and their environment. Frost *et al.* (1995, p. 229) 'define ecological function in terms of the effects of a given species on the population growth rates of all species in the community.' Interactions with processes generalize Frost *et al.*'s definition to entities other than species and functions other than those that involve population processes. Hypotheses that relate biodiversity to ecological function can be tested by measuring both biodiversity and interactions with processes in the same system. Application of this framework to other examples relating biodiversity to ecological function is discussed below.

A particular aspect of measuring a function is measuring the diversity of the function, i.e. functional diversity. Measurements of relationships between biodiversity and functional diversity require more than measuring a particular interaction with a process (e.g. the magnitude of herbivory, the occurrence of pollination). It also involves distinguishing between different types of interactions with processes. For example, Cornell and Hawkins (1993) describe four functional groups of insect herbivores: well-concealed endophytics (gall formers and stem, flower, fruit, and seed borers); poorly concealed endophytics (casebearers and leaf miners), exophytics (leaf chewers); and mixed exophytics/endophytics. One could also add intracellular herbivores (microbial parasites and diseases) to expand the included taxa to protists and viruses. Each of these groups has a different and somewhat distinct type of herbivorous interaction with the process of primary productivity. Here, *functional diversity is defined as the variety of interactions with ecological processes.* The number of types of herbivorous interactions within a community is a measure of its functional diversity.

Another important issue involves specification of the scale of interest (Allen & Starr, 1982; O'Neill *et al.*, 1986). Scale is specified by range or extent of phenomena of interest as well as the grain or resolution of the observations within the range. For example, the range of concern with the functional diversity of herbivorous insects may be defined as that between the present and the last ice age. Furthermore, the resolution of interest may be 1000 year intervals. Variations of diversity over a few tens of years and before the last ice age may not be of interest. Specification of the scale of interest is critical to an understanding of what theories and data may be reasonably employed to address phenomena. Explicit scales of interest are also very useful for clarifying the level of generalization that theories and observations warrant. For example, theories developed within the scale specified above may have little to do with why a small plot of land may have a functionally more diverse set of herbivores this year than were observed last year.

5.2.3 Biodiversity and ecological function

Specification of function in terms of processes and interactions lays a foundation for defining, observing and comparing relationships between biodiversity and ecological function. For example, consider the process of herbivory, the consumption and subsequent respiration of live primary production. Ecological entities can interact with herbivory as do many insects, by engaging in it or, as do many plants, by producing the material consumed. Entities such as parasitoids can also interact with herbivory by consuming herbivores. The role each of these entities plays with respect to herbivory can be defined as one of their functions. The effects of biodiversity on herbivory could be investigated by observing how the species richness of plants, herbivores, and consumers of herbivores vary with rates of herbivory. All else equal, increasing plant species richness may decrease herbivory due to increased heterogeneity of the herbivores' trophic resources (Gilbert, 1980). Increasing parasitoid species richness may also reduce herbivory by increasing predation pressure on the herbivores (Hawkins *et al.*, 1993). Of these functional groups, perhaps only increasing species richness of herbivores increases the rate of herbivory. The effects of varying diversity within each of these functional groups on herbivory could be compared within a community. Among communities, it would be more convenient to choose only one of these functional groups and compare its diversity with rates of herbivory.

As previous chapters have demonstrated, defining and measuring biodiversity is not a simple matter. A definition of biodiversity that explicitly uses the structure-function dichotomy embraced here is, 'the spatial and temporal variability of the structure and function of living systems'. We could measure biodiversity in terms of genetic polymorphism or in terms of species diversity that takes into account both the presence and relative abundance of species (Magurran, 1988). As is commonly done, Tilman and Downing (1994) measure plant biodiversity simply in terms of species richness. They measure function in terms of the rate of primary productivity. Only the species that engage in primary productivity were examined and the researchers found a positive correlation between species richness and stability of primary productivity during drought, which led them to accept the rivet hypothesis and reject the redundancy hypothesis. Vitousek and Hooper

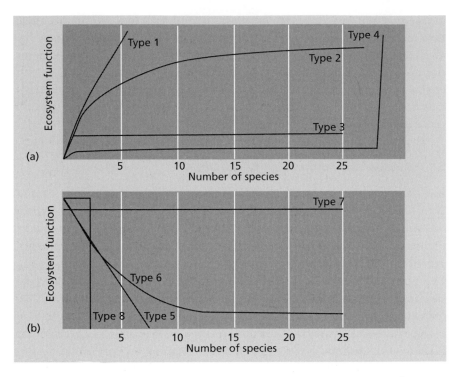

Fig. 5.1 Eight possible simple relationships between biodiversity and ecological function (after Vitousek & Hooper, 1993). (a) Describe mostly positive relationships between diversity and function. (b) Describe mostly negative relationship between diversity and function.

(1993) describe several frequently reiterated potential relationships between species richness and ecosystem function (Fig. 5.1a). They define ecosystem function in terms of the stocks of soil organic matter, soil nitrogen, extractable calcium and soil acid saturation and suggest that several studies are consistent with a modified rivet hypothesis (Fig. 5.1, type 2), where increasing species richness asymptotically increases ecosystem function.

Beyond these details of relating biodiversity to ecological function is the more fundamental choice of which processes to focus on. Are biologists interested in material, energetic, or population processes? What about behavioural or genetic processes? Is there any limit to the number of processes that occur in ecological systems? Even if there is such a limit, would the different ways biodiversity can interact with ecological processes result in an infinite amount of different functions of biodiversity? No precise answers to these questions are proposed here. However, a reading of the literature demonstrates that there are very many ecological processes that interact with biodiversity in many different ways. A large set of functions that has received

much attention has been termed ecosystem services (e.g. Westman, 197?
Ehrlich & Ehrlich, 1992), which may be defined as ecological functions th
are beneficial to humans. Westman (1977) recommended quantifying man
of nature's services (Table 5.1) represented by ecosystem functions 'th
enable humans to obtain the food, fibre, energy, and other material needs f
survival.' Ehrlich and Ehrlich (1992) related a similar and overlapping set oi
ecosystem services of biodiversity (Table 5.2).

These comments about the relationship of biodiversity to ecological function focus on function. The problems of measuring biodiversity will not be dwelt upon. There are many measures of biodiversity (e.g. alpha-diversity, beta-diversity) and likewise function (e.g. rate of flows, stability of processes) that do not necessarily correlate with each other. While authors emphasize different measures, none are generally accepted as preferred or better than the others. They are simply different. Without clear and generally accepted preferences, it is unlikely that the disposition of hypotheses (redundancy, rivet, etc.) will be clearly articulated and generally accepted. Instead it appears that their disposition will depend on the processes, interactions and the scales invoked during examination and possibly the system investigated.

The scientific challenge to investigators of biodiversity and ecological function is to discover hypotheses that apply to the broadest range of systems experiencing the widest variety of conditions. The first paragraph in this subsection suggests that the rate of a consumptive process is positively correlated with the diversity of the entities engaged in the process and negatively correlated with the diversity of the entities consumed. Another example is Vitousek and Hooper's (1993) proposal that a type 2 relationship generally applies to interactions between species diversity and biogeochemical functions (Fig. 5.1a). Such discoveries lay the groundwork for delineating general mechanisms that determine relationships between biodiversity and function in all, or at least many, ecological systems. However, even Vitousek and Hooper's unusually well articulated example does not clarify why stock-based measures are used as opposed to flow-based measures (e.g. Naeem *et al.*,

Table 5.1 List of 'nature's services' provided by ecosystem functioning discussed by Westman (1977).

Sulphate reduction	Air pollutant (e.g. carbon monoxide)
Carbon dioxide fixation	absorption
Oxygen release	Radiation balance
Waterfowl support	Regulation of global climate
Ground-water storage	Water vapour release
Soil binding	Nutrient storage
Water purification	Nitrogen fixation
Streamside fertilization	

Table 5.2 Ecosystem services discussed by Ehrlich and Ehrlich (1992).

Climate and water

Maintenance of the gaseous composition of the atmosphere
Climate control
Maintenance of biotically necessary moisture levels
Water recycling
Hydrological regulation

Soils, nutrients, and wastes
Generation and maintenance of soils
Conversion of nitrogen, phosphorus, and sulphur into forms usable by higher plants
Weathering of underlying parent rock
Holding soils in place
Disposal of wastes

Maintaining biogeochemical cycles
Carbon
Nitrogen
Phosphorus
Sulphur

Pest control and pollination
Control of pest and disease that attack crops or domestic animals
Pollination of 90 different crops in the United States alone

1994). Also, it would seem that there are other biogeochemical functions in the same soils examined that are not consistent with type 2 relationships. More specific description of the processes and interactions referred to as biogeochemical functions would help clarify the generality and testability of their type 2 hypothesis.

5.2.4 Functional diversity, similarity, groups, and guilds

Functional diversity is very closely related to ecological guilds, functional groups, and functional similarity. Steele (1991, p. 470) defines 'functional diversity' as 'the variety of different responses to environmental change'. He explicitly includes biotic change as environmental change, suggesting a high degree of synonymy between 'responses to environmental change' and 'interactions with ecological processes'. If a living entity responds to biotic change, then that change could easily be seen as an ecological process. Processes themselves are changes over time and ecological processes are biotic and abiotic changes. For example, soil nutrient cycles could be described as changes in the stocks and flows of nutrients in the biotic and abiotic components of the soil over time. However, this redescription is rather clumsy in light of the wide acceptance of nutrient cycles as ecological processes (e.g. Noss, 1990; Lawton & Brown, 1993; Vitousek & Hooper, 1993).

122

The 'responses to' part of Steele's definition connotes a one-way inter-action—the reaction of an entity to a process or environmental change. 'Interactions with', as used here, more explicitly embraces two-way interactions (reciprocal reactions). This appears to more accurately describe relations between living entities and ecological processes. For example, a nitrogen-fixing bacteria may react to an increase in molecular nitrogen by converting more of that nitrogen into nitrate. This not only increases the nitrate available to the bacteria but it also decreases the nitrogen in the environment and thus, the nitrogen available to other organisms. If we accept that interactions include responses and that ecological processes include biotically consequential environmental changes, then the definition advanced here can be seen as including the special case of interactions and processes that Steele describes.

The most problematic aspect of Steele's definition is his explicit exclusion of ecological diversity from the notion of functional diversity. Functional diversity is considered a fourth category of biodiversity distinct from genetic, species, and ecological diversity (Steele, 1991, p. 470). This distinction is not elaborated further and seems difficult to sustain. For example, vegetation dynamics in a landscape is both a change in the biotic environment and an ecological process. There seems to be no major distinction between Steele's biotic and abiotic environmental changes that living entities respond to and the ecological processes invoked by many investigators of functional diversity. Zak *et al.* (1994) define microbial functional diversity in terms of the utilization of substrates by a bacterial community. Such utilization is perhaps more often thought of as an ecological process than an environmental change. Cowling *et al.* (1994) measure plant functional diversity in terms of growth forms that 'attempt to categorize species on the basis of known differences in ecophysiological processes' (p. 145). Following these authors, no significant distinction between ecological and functional diversity is claimed here.

Functional diversity can be further clarified by comparing it with functional similarity. This similarity is inherent in the notion of functional groups. That is, identification of a functional group means that its members are functionally similar and also that entities excluded from the group are functionally different. 'Functionally similar' means that entities have similar interactions with the same process(es) (Chapin *et al.*, 1992). The interactions may be specified qualitatively (e.g. consumers that prey on the same species) or quantitatively (e.g. consumers that consume similar amounts of the prey). Functional diversity refers to differences among functional groups. The most widely accepted, albeit tautological, measure of functional diversity may be the number of functional groups in an ecological system. Also, the most widely accepted measure of functional redundancy may be the number of functionally similar entities within a functional group.

Vitousek and Hooper (1993) make a corresponding distinction between

123

'diversity among functional groups and [diversity] within them.' The diversity of herbivorous interactions represents diversity *among* functional groups. In a previous example, diversity among functional groups refers to exophytic, well-concealed endophytic, poorly concealed endophytic and mixed exo-/endophytic herbivores. Systems with more functional diversity have more of these functional groups represented than systems with less functional diversity. Diversity *within* these groups could be described in terms of the variable types of primary production consumed by organisms or in terms of their taxonomic diversity. An example that employs variability of consumption could distinguish between pollination related exophytic consumers (e.g. hummingbirds, butterflies, etc.) and non-pollination related exophytic consumers (e.g. grasshoppers, aphids, etc.). While this quantitative distinction maintains an emphasis on the function of herbivory, taxonomic distinctions broaden the emphasis to include evolutionary processes. This example brings up two important points. First, functional diversity within functional groups is often based on more finely resolved distinctions among interactions with a process than the distinctions that determined group membership. Second, distinctions within functional groups may, and often are, based on interactions or processes different from those used to group members.

Like Vitousek and Hooper, several other researchers examining biodiversity and function describe 'one approach [they] find particularly useful [that] subdivides the species in an ecosystem into two levels on the basis of function: (1) different functional types (e.g. feeding guilds, plant growth forms) and (2) functionally similar taxa within a functional type' (Solbrig, 1991, p. 38). A description of functional types presents a typological scheme that enables classification of living entities into functional groups. As suggested by the functional groups exemplified above, 'any typological scheme applied to the functional diversity of species will be, to some degree, arbitrary' (Solbrig, 1991, p. 38). It is important to recognize that the number of functional groups is highly contingent upon the processes and interactions chosen. Still, once these choices are made, the number of functional groups can be very useful.

Perhaps the most useful functional typology in ecology has been based on the notion of guilds (Simberloff & Dayan, 1991). Guilds are functional groups of organisms whose members exploit environmental resources in a similar way (Root, 1967). Like functional groups and 'as with the genus in taxonomy, the limits that circumscribe the membership of any guild must be somewhat arbitrary' (Root, 1967, p. 335—quoted in Simberloff & Dayan, 1991). The development of the guild concept over the years provides many lessons for research on functional groups such as the importance of clear definitions and delineations (Simberloff & Dayan, 1991). Simberloff and Dayan (1991, p. 137) explain the preference for the term 'guild' over 'functional group' as almost a haphazard coincidence: ' "Functional group" has evolved almost simultaneously to carry the same connotation, but the meta-phor of the "guild" must have seemed more elegant than this term.' This blurring of the distinction between guilds and functional groups is largely due to the departure from the original emphasis of guild on exploitation of food resources towards inclusion of other ecological processes (habitat exploi-tation, taxonomic relatedness, etc.).

Before moving on to the quantification of functional diversity, a few general comments on the entities being measured is warranted. The distinc-tion between structure and function leads to questioning the functional relevance of biogeochemical stocks (Vitousek & Hooper, 1993), body size (Cousins, 1991) and growth forms (Cowling *et al.*, 1994). Strictly speaking, these entities appear to be distinguished on structural criteria such as physical arrangements rather than process-oriented criteria. As is common in biology, structural differences often imply functional differences but such implications are best made explicit if function is the main focus of research. For example, it would be helpful to know whether energetic (Harvey & Godfray, 1987), hydrologic (Le Maitre *et al.*, 1995), or competitive (Simberloff & Boecklen, 1981) processes are being addressed by investigators of body-size and growth-form distributions. Without such specification, the purported functional rel-evance of more structurally defined (e.g. physical appearance) functional groups is difficult to assess.

5.2.5 Quantifying functional diversity

Functional diversity can be quantified by determining which and how many functional groups are represented in an ecological system. First, a typology of functional groups should be developed based on a variety of interactions with an ecological process. Once this typology is arrived at, representation of a functional group is expressed by the occurrence of interactions with the process that defines the group. Membership of living entities in functional groups is evaluated quantitatively (e.g. quantities of prey consumed) by characterizing the interaction with a process both qualitatively (which prey were consumed) and quantitatively (how much was consumed). Such measurements can be analysed for the number of functional groups in standard statistical procedures that quantify the similarity and difference between entities according to the qualified and quantified functions. Since a guild is a type of functional group, Simberloff and Dayan's (1991, p. 125) review of quantitative analyses of guild membership applies quite well to assigning functional group membership:

> Quantitative methods used for guild assignment include nearest
> neighbour statistics (Inger & Colwell, 1977; Winemiller & Pianka,
> 1990), cluster analysis (Crome, 1978; Landres & MacMahon, 1980),
> principal component analysis (Holmes *et al.*, 1979; Short &
> Burnham, 1982; Toda, 1984), canonical correlation (Folse, 1981)

and Monte Carlo techniques (Joern & Lawlor, 1981). All of these approaches, though explicit, do not unambiguously determine guilds because the investigator sets arbitrary levels for clustering. Various Monte Carlo methods (e.g. Strauss, 1982) can allow tests of hypotheses such as whether potential guild associates have diets more similar than would be expected given specified randomisation's of the data (e.g. Jaksic & Medel, 1990), but the level of nonrandomness required to qualify for membership in the same guild is still arbitrary. Further, so long as the basic data for these analyses consist of relative amounts of some resources used by each species, the classification depends on which resources are selected for analysis (Terborgh & Robinson, 1986).

Such analyses arrive at a delineation and enumeration of functional groups. The number of functional groups is comparable among systems in as much as the criteria for measurement are consistent among systems.

Zak *et al.* (1994) provide a cogent example of measuring functional diversity and comparing it among systems. They measure functional diversity of soil bacteria in six plant communities in the Chihuahuan Desert in the southwestern United States in terms of 128 specific carbon sources used. A functional portrait of each community is based on the activity of the soil bacteria allowed to grow on each of the sources. Three communities (mesquite-playa fringe, herbaceous bajada, black grama grassland) had the highest functional diversity measured both in terms of the number of carbon sources utilized and in terms of substrate diversity. The number of sources utilized is analogous to the species richness measure of a community. Substrate diversity is analogous to community diversity measured in terms of both species richness and the evenness of abundance among species. Indeed, Zak *et al.* used a classic community diversity index, the Shannon index, to measure substrate diversity. Substrate diversity takes into account both the number and evenness of substrates used. It is lower when substrates are used equitably than when the same number of substrates are used more variably.

Zak *et al.* (1994) employed all the basic components of the framework developed here. They identify the process of carbon utilization. The variety of interactions with the process are identified in terms of the variety of substrates on which bacteria are active. Functional diversity is quantified in terms of which different substrates are utilized by the communities. Scale is given explicit attention and is standardized by the consistent soil sampling regime among six plant communities during early December 1993.

As with comparing communities based on measuring biodiversity in terms of species richness, measuring functional diversity in terms of the number of functional groups does not completely quantify how different ecological systems are from one another. In a taxonomic example, consider a system with three closely related species of grasses. Is this system as diverse as another three-species system with one grass species, one forb species and one tree species? In terms of species richness, the answer is yes. However, sometimes it may be desirable to measure biodiversity in a manner that weighs more distantly related species greater than closely related species. Indeed, the title of this volume emphasizes this desire to understand not only the quantities of different living entities, but also how different the entities are from one another. Zak *et al.* (1994, p. 1104) make this point about functional diversity: 'Measures of substrate richness, evenness and diversity do not provide information about the types of substrate that are utilized by the bacterial community. Two sites could exhibit identical substrate richness, evenness or diversity but still catabolize totally different substrates.'

The same methods for measuring differences among species are directly

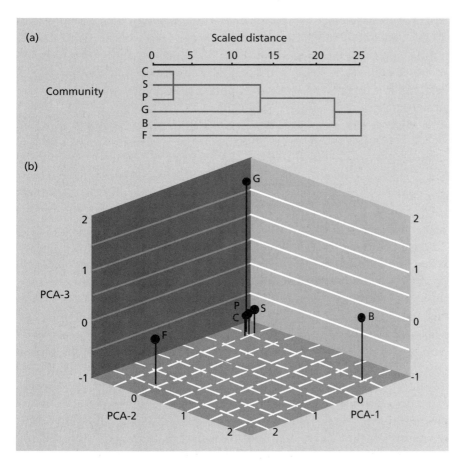

Fig. 5.2 Measures of functional diversity of soil bacteria from several plant communities investigated by Zak *et al.* (1994). The abbreviations correspond to plant communities: F, mesquite-playa fringe; B, herbaceous bajada; G, black grama grassland; C, creosotebush bajada; P, playa grassland; S, *Sporobolus* grassland. Functional diversity is based on overall substrate utilization using (a) cluster analysis based on presence or absence of activity and (b) principal components analysis based on levels of activity.

applicable to functional groups (Gaston, 1994). Measures of the difference between species are typically based on phylogenetic or evolutionary distance (Cousins, 1991). To measure the difference between functional groups, species' evolutionary relatedness is simply replaced by functional similarity. Zak *et al.* (1994) measure functional similarity using cluster and principal component analyses based on Jaccard's similarity index. These analyses show that the mesquite-playa fringe, herbaceous bajada, and black grama grassland communities are functionally different from one another while the other three communities were found to be functionally similar to each other (Fig. 5.2).

While ecosystem services typically focus on whether or not a function is performed and, if so, to what degree, the diversity of ecosystem service functions has received some recent interest. For example, Solow (1993) describes a measure of functional diversity that is derived from genetic diversity by 'assuming that genetically similar species tend to share more characteristics, including medicinal ones, than do dissimilar species.' By associating an economic option value with medical possibilities, Solow transforms a measure of taxonomic diversity into a measure of the functional diversity of ecosystem services.

A more detailed example of quantifying functional diversity is discussed below. First, the purpose of measuring functional aspects of biodiversity is discussed.

5.3 Hypothesis specification and testing

While no purpose or goal of biodiversity is assumed here, scientists do have goals which should be explicitly stated since this facilitates evaluation of their accomplishments. One would also hope that goal specification would help in accomplishing goals more efficiently but this is not always the case (Feyerabend, 1993). As is implied throughout this chapter, the purpose of measuring functional aspects of biodiversity is to test hypotheses that relate biodiversity to ecological function. The purpose of a scientific framework is to enable reproducible conclusions as to whether or not certain observations are consistent with explicit hypotheses (e.g. Table 5.3). Hypothesis-driven research agendas have been elaborately embraced in calls to study the relationship between biodiversity and ecological function (Solbrig, 1991; Grassle *et al.*, 1991; Younès, 1992), although these references lack general definitions that would provide a framework to examine the hypotheses. We now turn to general hypotheses that relate biodiversity to ecological function. Then, hypotheses that relate biodiversity to functional diversity are discussed.

5.3.1 Prominent hypotheses relating biodiversity and function

Some of the earliest theories relating biodiversity to ecological function focused on population processes. MacArthur (1955) asserted that populations of species would be more stable in communities with more species because these communities provide a greater variety of trophic resources. Increasing this variety reduces the dependence of species on the population fluctuations of their individual resources. However, this theory is challenged by the assertion that the number of trophic resources per species is independent of the number of species in the community (Pimm, 1982; Cohen & Briand, 1984). May (1973) argued that increasing species diversity could decrease the stability of populations because greater numbers of interactions in large communities increase the likelihood of destabilizing positive feedback loops whereby populations explode to infinity or go extinct. According to May's mathematics, these instabilities could be avoided if more diverse communities did not have more interactions on a per species basis. The assertion that the

Table 5.3 Examples of hypotheses that address functional aspects of biodiversity (from Solbrig, 1991).

HYPOTHESIS B.1
'Keystone' species are essential for maintaining species richness in communities under all environmental conditions

Alternative Hypothesis B.1a
The effect of a keystone predator on the species richness of a community is dependent on the abiotic environmental disturbance regime

Alternative Hypothesis B.1b
The effect of a keystone resource species on species richness of dependent species is dependent on the temporal variability in total resource availability

Alternative Hypothesis B.1c
Keystone species play a more important role in marine than in terrestrial ecosystems and this role is more important in the lower latitudes

HYPOTHESIS B.2
Both the richness of functional types and species richness of functional analogues within functional types will respond the same way to a change in resource availability

Alternative Hypothesis B.2a
The number of different functional types in a community is unaffected by the total resource availability (possibly indexed by the net productivity of the community) or by the number of different types and forms of the resources

Alternative Hypothesis B.2b
The number of potentially competing species within a given functional type is unaffected by the total amount of resources available to that functional type

HYPOTHESIS B.3
The number of functionally analogous species within functional types has no effect on ecosystem function or stability

---

continued on p. 130

Table 5.3 (*continued*)

HYPOTHESIS B.4
Local species diversity is determined by local environmental properties and processes, not by the number, dispersal ability, life history adaptations, or reproductive strategies of species in the regional pool. Communities with the same degree of spatial heterogeneity, and the same dynamic equilibrium between the opposing processes of competition and mortality-causing disturbances should exhibit the same level of species diversity

Alternative Hypothesis B.4a
The representation of the regional pool of species (within a functional type) in assemblages of locally coexisting species of that functional type is unaffected by the local disturbance regime or potential level of productivity

Alternative Hypothesis B.4b
In local communities, the proportion of a regional species pool that is due to species within a functional type or taxon is independent of the dispersal ability of the organisms involved

Alternative Hypothesis B.4c
In local communities, taxonomic richness within a specific functional type or higher taxon is independent of the reproductive biology and genetic structure of the taxa

HYPOTHESIS B.5
Spatial heterogeneity of the regional landscape has no effect on the number of functional types or coexisting species in a local community

Alternative Hypothesis B.5a
The rate of increase of the species/area curve is independent of the disturbance regime or average productivity of a 'mosaic' landscape

---

number of trophic interactions per species was independent of species diversity (Pimm, 1982; Cohen & Briand, 1984) was seen as consistent with May's stability criteria (May, 1983). However, the assumptions necessary for this result such as linear deterministic population interactions and population equilibria may not be biologically reasonable (Law & Blackford, 1992; Haydon, 1994).

More recent theories relating biodiversity to ecological function have drifted towards anthropocentric analogies exemplified by the rivet and redundancy hypotheses. These hypotheses have the strength of being applicable in principle to any function. However, as discussed previously, the analogies assume that an unstated and relatively narrowly defined function is widely accepted among investigators. The rivet hypothesis is quite similar to MacArthur's theory that diversity contributes stability to ecological systems, while the redundancy hypothesis is similar to May's theory. Species richness does not affect stability as long as May's criteria are adhered to. His theory has also been interpreted as going beyond redundancy by asserting that diversity could be destabilizing and therefore have a negative rather than a neutral relationship to ecological function.

Given that function may refer to any ecological process, there seems to be no *a priori* restrictions on hypotheses that relate biodiversity to ecological function. We could assume that an increase in biodiversity correlates with an increase in the quantity of ecological function, but exceptions easily come to mind. For example, highly oligotrophic lakes such as Lake Tahoe have hundreds of species of producers (phytoplankton) in each square metre. Still, these systems produce much less net primary production than a monocultural sugar cane field. We could guess that, due to the great variety of life, more biotic diversity leads to more functional diversity (e.g. Cody, 1991), but this is not necessarily the case. Cowling *et al.* (1994) found that arid and semi-arid sites in southern Africa relatively rich in species had less functional diversity in terms of 21 above-ground growth forms. However, this result could be criticized by asserting that the growth forms represent structural diversity instead of functional diversity.

The rivet and redundancy hypotheses previously described have received much attention in the literature. Another frequently considered hypothesis concerns the functional importance of 'keystone species' that by definition are not redundant (Chapin *et al.*, 1992): 'A keystone species is a species whose impacts on its community or ecosystem are large, and much larger than would be expected from its abundance' (Power & Mills, 1995). It is these taxa to which policy makers and scientists are asked to pay particular attention because their presence is relatively inconspicuous compared to the effects of their extirpation. Keystone species are often described as strong interactors as compared to weakly interacting redundant species. The litmus test for a keystone species is to experimentally remove its population from an enclosed ecological system and observe the effects (Paine, 1980). If unexpectedly large changes in abundance or composition of the community ensue, the species is considered a strong interactor and therefore a keystone. If the removal does not result in such changes, it is a weak interactor and may be considered redundant. The number of strongly interacting populations in a community is thought to be small compared to the large number of weakly interacting populations (Paine, 1992).

Since keystone species or strong interactors ['drivers' in Walker's (1992) automobile analogy] are many fewer than the weak interactors ('passengers'), diversity *per se* may not be critical to ecological function (Paine, 1992). Instead, a small number of strong interactors is all that is needed. In this context, ecological function is closely related to stability and typically refers to the maintenance of species composition and relative abundance within a community. However, similar to the teleological aspects of the rivet and redundancy hypotheses, the keystone hypothesis suffers from unstated and inaccurate assumptions (Mills *et al.*, 1993). Perhaps most basically, the keystone species designation applies to populations, not whole species. It is a population that is removed from an enclosure, not the larger group of organisms asserted by Power and Mill's definition and implied by 'species removal' experiments. This leads to confusion, such as whether the classic keystone seastar *Pisaster ochraceus* (Paine, 1980) is a keystone in all the intertidal systems in which it is found (Foster, 1991). This difficulty may be avoided by explicit acknowledgment and critical evaluation of the disparity between the scales at which experiments are conducted and the scales at which the results are presumed valid.

Another difficult aspect of the keystone hypothesis is the clause, 'much larger than would be expected from its abundance.' What one expects from abundance depends on how abundance is measured. Since abundance, both in terms of population numbers and in terms of biomass, is used for the keystone designation, there is considerable flexibility in one's expectations. Consider a removal of Douglas fir *Pseudotsuga menziesii* trees from a forest monoculture that leads to large changes in the forest community. Based on the large abundance of fir biomass, such large changes would be expected. However, based on the small number of trees in comparison with orders of magnitude more microbes and insects, the large community changes could be unexpected which would lead the fir to be considered a keystone. This flexibility combined with the extremely subjective and therefore problematic notion of human expectation casts serious doubts as to the utility of the keystone concept.

Other uncertainties have prompted abandonment of the keystone/weak-interactor dichotomy in favour of more formally measuring the interaction strength of the group of organisms in question (Mills *et al.*, 1993). As with explicit measurements of function, measurements of interaction-strength eliminate many uncertainties. For example, interaction-strength measurements do not include the problematic judgement of whether the strength is disproportionate to expectations.

5.3.2 Less prominent hypotheses relating biodiversity and function

Chapin *et al.* (1992) and Lawton and Brown (1993), among many others, consider the rivet and redundancy hypotheses to be extremes of a continuum. Still, there are other hypotheses that do not lie within these extremes. One may be called the uniqueness hypothesis, which asserts that whenever a species is lost, a particular function of an ecological system is largely eliminated (Fig. 5.1a, type 4). In other words, a function can be found for practically every species which leads to its designation as a keystone species. For example, Daily *et al.* (1993) observed that a single species of woodpecker provides the only habitat available for several other species in an ecological system in the Rocky Mountains. If this warbler is extirpated, all of this ecological engineering function (Jones *et al.*, 1994) would apparently be lost. Beaver (*Castor canadensis*) are similar in that their elimination leads to the local extinction of many other species (Pollock *et al.*, 1995). While these may be considered merely examples of relatively uncommon keystone species, it has been claimed that each species of herbivorous insect has several parasitoid species that specialize and therefore critically depend on it (Price, 1977). It is also plausible that all species have host-specific endosymbiotic or exosymbiotic viral, microbial and metazoan endemics whose survival depends on the host.

The redundancy, rivet, and uniqueness hypotheses appear to cover a comprehensive range of simple hypotheses that assert positive relationships between ecological processes and biodiversity. A different set of hypotheses that have not received as much attention assert negative relationships between ecological processes and biodiversity (Fig. 5.1b). One such hypothesis may be called the inverse rivet hypothesis (Fig. 5.1b, type 5). Consider the hydrologic processes in a catchment basin and how vegetation may affect the yield of water emerging from the catchment. Le Maitre *et al.* (1995) show that invasions of alien wooden plants into a catchment reduces the water yield from that catchment. The invasions increase the species richness of a system and decrease an ecosystem service function. As invasions progress, water yield progressively decreases.

In other cases, the negative effects of biodiversity on ecological function may be more immediate, leading to an almost complete loss of ecological function with a slight increase of biodiversity from an already very low value (Fig. 5.1b, type 8). This may be most obvious in hydroponic plant growing systems where additional species are often pathogens whose presence reduces or eliminates plant growth. Other examples include crop systems whose low diversity must be maintained by insecticides, fungicides, herbicides, cultivation, and other methods to prevent consumer and competitor species from establishing and reducing the growth of crop species available for harvest. This example may follow the type 5 or type 6 relationship in Fig. 5.1b.

Perhaps the most heretical hypothesis to consider is that an ecological function could be maintained at a high level during the progressive decrease and ultimate elimination of species diversity (Fig. 5.1b, type 7). An example could be the amount of water that filtrates down to the water table below a sand dune ecosystem. Le Maitre *et al.* (1995) show that species diversity interacts with hydrological processes. It is plausible that plants on a sand dune filter particulates and pollutants out of precipitation that percolates down to a water table. However, as the diversity and the biomass of species decrease, the sand below the biota may begin performing the filtration function, resulting in a negligible change in filtration despite the eventual elimination of the biota. In other words, the sand may be functionally equivalent to the biota in narrow terms of water-filtration; and the filtration capacity of the biota-sand system may be well above that necessary for maintaining a certain water quality in the water table.

Each of these eight hypotheses (Fig. 5.1, types 1–8) address the relationship of ecological function to taxonomic or species diversity within functional groups, whether it be in terms of positive, neutral, or negative interactions with ecological processes. The framework has been useful in developing hypotheses not yet seriously considered in the literature. However, it should be apparent from the previous hypotheses and examples that biodiversity may have all or perhaps none of the hypothesized relationships to ecological function, depending on which processes, interactions, and measures of biodiversity are emphasized.

5.3.3 General notes on hypotheses relating biodiversity to ecological function

In the search for general relationships between biodiversity and ecological function, it should be recognized that almost any function can be specified in an inverse form. For example, rather than maintaining a habitat hospitable to a particular species, an ecological function could be defined as maintaining a habitat inhospitable to it (or hospitable to its consumers and competitors). Merely respecifying the function like this may reverse which hypotheses are rejected or corroborated. For example, Vitousek and Hooper (1993) consider the relation between species richness and biogeochemical functions. They fit a type 2 response (Fig. 5.1a) to function in terms of percent soil organic matter. A type 6 response (Fig. 5.1b) would logically fit the same data if function was evaluated in terms of percent inorganic matter. This is why the framework developed here is based on *a priori* identification of function in order to reproducibly evaluate specific hypotheses.

A hypothetical alternative to *a priori* identification of a function of interest is the *a priori* identification of a species of interest. In this case, one or many species may be identified and the functional effects of their addition or loss to the system can be investigated. Still, the practically infinite number of functions cannot all be investigated. Therefore, even this purported alternative demands *a priori* identification of functions to be observed.

Another difficulty with these general hypotheses is that species are not all the same and no consistent ranking of species along an X-axis of Fig. 5.1 is apparent. Species have different effects which could greatly affect which hypotheses are corroborated or rejected (Hughes & Noss, 1992). For example, whether an asymptote is reached or not may depend on the order of species along the species-richness axis in Fig. 5.1. If species that are almost entirely responsible for a function are on the right side of the species-richness axis, an asymptote may not be apparent. An asymptote would be apparent if those same species were arranged on the left side of the species-richness axis. This difficulty can be somewhat ameliorated by establishing a generally acceptable convention for arranging species along the X-axis. Hypotheses concerning aspects of function not easily described by the Y-axes in Fig. 5.1 (e.g. temporal dynamics) and other measures of biodiversity (e.g. landscape diversity) also deserve consideration.

5.4 Measuring functional diversity using food webs

5.4.1 Specifying the functions of interest and related hypotheses

Strategies and methods for measuring functional diversity will be illustrated using food webs. Food webs depict trophic processes between classes of organisms that consume and then respire the biomass of other or the same (cannibalism) classes of organisms (see Lawton, 1989; Pimm *et al.*, 1991 and Hall & Raffaelli, 1993 for reviews). The purpose of conducting such measurements is to test hypotheses that predict certain patterns of functional diversity among and within ecological systems (e.g. Pimm *et al.*, 1991). These statements specify the process being studied and the purpose of investigation. Additionally, the manners in which organisms interact with this process are simply either as consumers or resources. More elaborate specification of trophic interactions involves quantifying associated interaction strengths (Paine, 1992) and flows of energy and materials such as carbon (Odum, 1973; Strayer & Likens, 1986; Baird & Ulanowicz, 1989; Hairston & Hairston, 1993).

Initial documentation of food webs is often based on taxonomic classes of organisms such as Linnean species, due to their general use by ecologists that study trophic interactions of organisms. These taxa may be aggregated into functional groups that employ criteria closely related to the process being investigated and pragmatically reduce the number of classes to be considered. Trophic species are the functional group of choice in food web studies (e.g. Cohen, 1989; Pimm *et al.*, 1991 and many others). Trophic species are aggregations of organisms within an ecological system that appear to share the same set of trophic consumers and resources (Briand & Cohen, 1984). 'Trophic consumers' and 'trophic resources' are general terms that encompass several more specific ones denoting the type of trophic interaction such as predation, parasitism, pathogenicity and herbivory. As defined, trophic species are functional groups of organisms closely related to guilds. Indeed, the notion of guild motivates the delineation of trophic species but trophic species take the guild concept farther than it is usually employed. This is because organisms are not only lumped together due to shared resources but also due to shared consumers.

A difficulty with trophic species that extends to other functional groups is in deciding whether to include ontogenetic functional shifts in the delineation of trophic species. Immature and mature organisms of the same species often function differently. Problems with including this temporal resolution in food webs involve definition, consistency and pragmatism. How much ontogenetic change warrants classification into a different potential trophic species? Can such changes, once defined, be applied consistently among highly diverse taxa? Would such an application result in an intractably large number of entities to investigate? Space does not allow explanation of why ontogenetic trophic species may not be consistently defined in a pragmatic manner. However, it is important to realize that this class of problems has already been addressed by evolutionary biologists in debates about species concepts.

Mishler and Brandon (1987) describe such problems as an issue of specifying grouping and ranking criteria: 'All species concepts must have two components: one to provide criteria for placing organisms together into a taxon ("grouping") and another to decide the cut-off point at which the taxon is designated a species ("ranking")' (p. 404). These two components apply to functional groups as well as taxonomic species. The grouping criterion for trophic species is the similarity in a taxon's sets of trophic consumers and resources. The ranking criterion is complete similarity. If a group of organisms has a single different consumer or resource different from that of another group, these groups belong to different trophic species. The challenge for distinguishing ontogenetic trophic species is to develop ranking criteria that allow consistent determination of ontogenetic classes among taxa as diverse as microbes, fungi, plants and animals.

To measure functional diversity, the framework is specifically applied to the function of interest at a particular scale which specifies the extent and resolution of observations. Let us choose a lake volume over a year as the spatial and temporal bounds of our system of interest. Furthermore, we will target trophic species as aggregates of taxonomic species and trophic interactions as either present or absent at the desired level of resolution. As an example we will address Little Rock Lake, Wisconsin (Martinez, 1991).

In the interest of scientific advancement, one of the initial considerations is whether there are any specific hypotheses regarding food webs that may be tested by observing the trophic structure in Little Rock Lake. Pimm *et al.* (1991) specify a suite of patterns that apply to food web structure that may be tested. The only hypothesis we consider here, has been called the 'link-species scaling law' that asserts that the number of trophic links in a food web is roughly twice the number of trophic species (Cohen & Briand, 1984; Cohen *et al.*, 1990). Another initial step is identification of which taxa participate in trophic interactions. Virtually all taxa participate in trophic interactions, so the first measurement may include an all-taxa survey of the organisms within the lake. An initial, albeit limited, survey documented 182 taxa in Little Rock Lake (Martinez, 1991).

5.4.2 Conducting the measurement

Ideal documentation of the food web would include testing the hypothesis that each taxon consumes each other taxon against the hypothesis that no such trophic interaction occurs. This would involve over 33 000 (182^2) tests. However, basic knowledge of natural history can greatly reduce this number. For example, since 60 of the taxa are algae, phytoplankton, or otherwise producer taxa that are apparently physiologically unable to trophically consume other taxa, we can eliminate the 10 920 (60 multiplied by 182) hypotheses that these producers consume any of the taxa within the web. Further specification of basic natural history, such as which taxa are known solely to consume primary producers or secondary producers, further reduces the hypotheses in need of further investigation. Several ecologists who were studying the organisms in Little Rock Lake were asked to provide their educated guesses as to the trophic interactions involving the organisms of their speciality that were likely to occur over a typical year. This resulted in roughly 2500 links being designated. These links were checked against available data such as gut contents of fishes and also cross-checked with available literature and other specialists studying identical taxa in other ecological systems.

Contrary to the discussion in the previous section [5.4.1], five of the 182 taxa are ontogenetically immature stages of fishes and three more of the taxa are immature zooplankton. This represents a classic bias of ecologists toward more highly resolving vertebrates, which make up only six Linnean species, compared to the invertebrates that constitute over one hundred Linnean species in Little Rock Lake. Such bias in taxonomic resolution should be avoided. Fortunately, the ontogenetic species are a small fraction of all the species in the Little Rock Lake web and should not interfere with the heuristic purposes for which the web is used here.

Aggregation of the 182 taxa results in 93 distinct trophic species. The trophic species web has 1036 links. Both of these quantities measure among group functional diversity of the web. That is, there are 93 different functional groups of taxa in the web as well as 1036 functional groups of trophic links among these groups. An increasingly popular way of comparing the functional diversity of webs with different numbers of species is dividing the number of links by the square of species richness to calculate a measure of ecological complexity called directed connectance (the fraction of all possible trophic links including cannibalism and mutual predation that are realized; Martinez, 1991; Warren, 1994). Directed connectance equals 0.12 in the 93 trophic species web.

Besides these among-group measures of functional diversity, we can also calculate within-group functional diversity given the functional portrait that is a food web. For example, we can define new functional groups called specialists and generalists. Generalists are trophic species that consume a higher than average fraction of species. This average is equal to directed connectance or 0.12 in Little Rock Lake. Within the specialist group, there are 64 trophic species that consume less than 0.12 of the species in the web, while the remaining 29 trophic species consume more than 0.12 of the species. Another such measure calculates the fraction of trophic species within functional groups at basal, intermediate and top trophic positions (Briand & Cohen, 1984). The basal group includes species with consumers but no resource species (e.g. many plant species). The intermediate group includes

137

species that have consumers and also consume. Top species consume but have no consumers. The variation of these measures with overall species richness is actively debated (e.g. Martinez, 1994). In Little Rock Lake, 13% of the trophic species are basal, 86% are intermediate and only 1% are top. Interestingly, the variation of these fractions with the number of trophic species addresses the relationship between within- and among-functional group diversity.

Besides these sorts of within-group measures based primarily on trophic processes, one can also expand the measures to invoke evolutionary processes by examining taxonomic diversity within functional groups. The most taxonomically diverse trophic species are among phytoplankton whose taxa are largely lumped according to size class and taxonomic lineage (e.g. large diatoms). These groups are highly diverse because they appear to consume no other organisms and therefore, unlike intermediate and top species, basal species cannot be distinguished by differences among their organismal trophic resources. Also, because these organisms are small, there is little information about their consumers other than consumers who engulf their resources. There may be a rich group of highly specialized bacterial consumers of phytoplankton that could be used to distinguish the trophic roles of phytoplankton. However, such bacteria were not investigated and there is little, if any, locally based information that can inform this issue. Certainly, the hypothesis that phytoplankton within a trophic species have identical sets of trophic interactions would be important to test if the functional diversity of phytoplankton was an issue of special interest.

At a different level, one could also measure the within-functional group diversity by counting among the 182 taxa to discover how many are within top, intermediate and basal functional groups. The count is 1, 121, and 60 richness. Variation of the number of taxa in these groups with species diversity is also actively debated (e.g. Havens, 1992; Martinez, 1993a; Murtaugh, 1994). This variation addresses the relationship between within-functional group diversity in terms of number of taxa in functional groups and biodiversity in terms of species richness.

Again, this sort of descriptive reorganization of functional data such as a food web, could go on forever. We could look for the most trophically unique taxa (e.g. those who share the fewest predators or prey with all other taxa). We could examine vulnerability (the fraction of species that consume each taxa or trophic species) in the same way as we examined generality. Such explorations can be very useful for hypothesis generation. However, we now turn to the previously discussed hypotheses that could be tested by measuring the functional diversity of Little Rock Lake.

5.4.3 Testing hypotheses with the measurements

Among the objectives for measuring functional diversity was the testing of the link-species scaling law that predicts that the number of trophic links in a food web is roughly twice the number of trophic species (Cohen & Briand, 1984). Analysis of the Little Rock Lake food web demonstrated that there are over 1000 trophic links among the 93 trophic species. This is highly inconsistent with the link-species scaling law that estimates less than 200 links for this web. This discrepancy suggests that the 'law' should be rejected as inapplicable to Little Rock Lake.

One possible reason for the discrepancy is that the number of links in Little Rock Lake are highly over-estimated due to the inclusion of extremely rare links that are typically excluded from the webs that the law was based on. This possibility was examined by reducing the detail or resolution of the web to be more similar to less speciose and less resolved webs. These sensitivity analyses were conducted via a hierarchical cluster analysis that varied the ranking criterion from complete similarity to progressively less similarity (Martinez, 1991, 1993b). The analysis progressively aggregates the most trophically similar taxa in order to represent the web with progressively fewer species (Fig. 5.3). Once the web was aggregated to include 30 or less 'trophospecies,' webs began to look much more like that which would be estimated by the link-species scaling law which was based on webs with 30 or fewer species (Cohen & Briand, 1984) even though all original links were included in the aggregated webs. This consistency suggests that no over-estimation is necessary to observe the large number of links in Little Rock

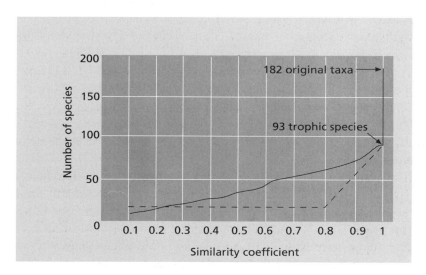

Fig. 5.3 Number of trophospecies or clusters of taxa as a function of the minimum trophic similarity among taxa within clusters, based on an analysis of the food web of Little Rock Lake, Wisconsin. As less trophic overlap is required for taxa to be within a cluster, fewer clusters or trophospecies are present in the web. Dashed line represents a hypothetical result discussed in the text. Trophically equivalent taxa are aggregated into trophic species in the vertical portion of the curve at the 1.0 similarity coefficient. From Martinez (1991).

Lake. If the links were over-estimated in the original web, the more aggregated webs should also have relatively many links.

This aggregation addresses the arbitrariness of functional group delineation. The decision to base trophic species on complete similarity of trophic consumers and resources is an arbitrary but pragmatic and easily applicable criterion. One clear result of this application is the immediate reduction of the number of 182 taxonomic species to 93 trophic species. Changing the ranking criterion from 100% to, for example, 90% similarity reduces the 93 trophic species to fewer numbers of trophospecies (Fig. 5.3). As Simberloff and Dayan's (1991) discussion of guilds argues, choosing 100% instead of 90% is an arbitrary decision in the definition of trophic species as functional groups.

It is interesting to imagine what less arbitrary functional groups might look like in Fig. 5.3. For example, imagine that there were 20 relatively distinct functional groups of taxa in Little Rock Lake and that each taxon shares 80% of its trophic consumers and resources with other taxa in its group. Also imagine that the taxon within each group shares less than 10% of their consumers and resources with taxa in the other functional groups. While there could still be 93 trophic species at 100% similarity, there would be a steep drop to 20 trophospecies at 80% similarity (see dashed line, Fig. 5.3). There would be no decline to fewer trophospecies until at least 10% similarity. That no such step functions at lower than 100% similarity was observed and suggests that no such relatively distinct and less arbitrary functional groups are present in Little Rock Lake.

A hypothesis that rectifies the discrepancy between the numbers of links in small webs and that of Little Rock Lake is the constant connectance hypothesis that links increase as the square of the number of species (Warren, 1990; Martinez, 1992). This hypothesis is consistent with both the Little Rock Lake web and the much less speciose webs (Martinez, 1992). Many recent data have corroborated the predictions resulting from this hypothesis (Martinez, 1993a, 1995; Warren, 1994).

An important scientific component of testing hypotheses is bringing the results of the test back to bear on theory that originally motivated the test in a way that leads to new hypotheses. The rejection of the link–species scaling law in favour of constant connectance affects the key variables employed in the stability–diversity debate that relates population processes to biodiversity. Loosely translated, May's (1973) stability criterion holds that the product of average interaction strength (i) and square root of species richness (S) multiplied by connectance (C) must stay below one ($i(SC)^{1/2} < 1$) for the population processes of a multi-species system to be stable. Assuming constant i with increasing S means that C must decrease to maintain stability if this criterion constrains food-web structure. Maintaining a constant ratio of trophic links (L) to species (L/S) maintains this stability because C equals L/S^2 which means that SC equals L/S. In other words, as long as i and L/S are constant, stability is maintained and C varies inversely with S. However, the constant C hypothesis suggests that i should decrease with increasing S to maintain stability (Warren, 1994; Martinez, 1995). This new hypothesis can be studied by conducting interaction strength experiments among communities with variable species diversity. Such experiments could empirically illuminate whether i varies with S in the manner suggested by the constant C hypothesis. These experiments would also be an exciting contribution to the study of the relation between ecological function and biodiversity.

This food web example demonstrates the utility of measuring functional aspects of biodiversity within a framework that tests hypotheses that relate biodiversity, ecological function and functional diversity. The rejection of earlier hypotheses, their replacement with more empirically consistent and successfully predictive alternatives, along with the generation of new testable hypotheses, represents basic scientific progress.

5.5 Functional redundance revisited and replaced

Walker's (1992) automobile passengers and Lawton and Brown's (1993) unnecessary words in a sentence are quite consistent with Webster's (Woolf, 1979) definition of redundancy as superfluous or needless repetition. It is, again, the teleological and subjective notions of superfluousness and needlessness that degrade the scientific utility of redundance. Taxa within a trophic species are functionally similar or repetitive because they interact with the same process in a similar manner. These taxa may even be called trophically equivalent (Cohen, 1989) because there are no topologically apparent differences among them. Is such similarity, repetition or equivalence (along with people's blood cells and second kidneys) superfluous or unnecessary?

'Redundancy', 'superfluous', and 'needless' are value-laden terms that have pejorative and anthropocentric connotations that seem out of place in a science that purports to be an objective activity. An economic organization would be widely presumed to be flawed if it retained many redundant workers. It seems revealing that, whereas the dichotomy between structure and function has been useful throughout biology, ecologists alone are drawn towards characterizing functional similarity, repetition or equivalence as redundance. For example, even though red blood cells in a human may be functionally repetitive and equivalent to the point of no measurable functional response to removal of several cells, 'redundant' is rarely if ever used to describe red blood cells. Why should ecologists much more readily consider redundance in an analogous situation with species? At a time when conserving biodiversity and other social activities (e.g. politics, economics) appear in conflict, scientists should be wary of advancing terms that serve particular ends that irreversibly disturb and destroy the objects of their study. The finding that many species are functionally redundant would seem to contribute a scientific patina to efforts to weaken conservation efforts. Therefore redundance warrants heightened scrutiny.

141

While the motivation behind the term 'functional redundance' can be questioned, the scientific utility of the term should be decided on whether it accurately describes the current scientific understanding of living entities' functions in ecological systems. Consider the type 3 and 7 responses in Fig. 5.1 that are purported examples of functional redundance. There are at least two distinct situations that lead to these responses, neither of which are accurately described as redundance. The first is that the species removed or added are simply non-functional. They do not interact with the process measured. Such species do not repeat the function and therefore cannot be redundant. For example, a species of grasshopper could be excluded from a pollination experiment that results in no change in pollination rates. The grasshopper is not redundant if it does not interact with pollination even though type 3 or 7 responses would be observed. The second is a situation that Frost *et al.* (1995, p. 224) describe as functional complementarity, which occurs when 'processes are maintained at constant levels despite stresses that induce shifts in the [entities] driving those processes.' Here, we have bonafide repetition by functionally similar and compensatory (Frost *et al.*, 1995) entities, but the classification of 'superfluous', 'needless' and hence, 'redundance' appears to be a scientifically illegitimate determination.

Functionally inconsequential repetition seems to retrieve the most scientifically defensible connotations of functional redundance and would accurately describe a situation where addition or removal of the repetitive entity has no functional effect. In order to defend the assertion of functional inconsequence, a scientist must be able to assert that all the functions of the redundant entity have been accounted for. This appears to be a theoretical and practical impossibility. The most a scientist can defend is a description of a specified set of functions and an investigation of observable differences among the manners in which entities interact with the processes responsible for the specified functions over the scales investigated. Not only are all processes impossible to investigate, but it is impossible to investigate any one process at every scale. Given these limitations, functionally inconsequential repetition can only be demonstrated for certain functions over certain scales. However, such highly qualified inconsequence appears more accurately described by functional similarity left unqualified.

Functional redundance and inconsequence implies great certainty regarding the future of ecological systems. Billions of years ago, a single mutation in a single organism may have been largely responsible for the conversion of Earth's reducing atmosphere to an oxidizing atmosphere which resulted in the vast majority of species going extinct. Scientists do not appear to have any certainty that similar mutation or recombination of genetic material will not result in another global calamity or prevention thereof. Describing a living entity as redundant or inconsequential claims that that entity has no such potential. This claim, however likely to be correct, appears to be beyond scientists' ability to determine with absolute certainty. Hobbs and Mooney (1995) more specifically make this point about vegetation dynamics:

> Clearly, changing conditions or particular episodic events can lead to hitherto relatively unimportant species suddenly becoming very important. This applies equally to minor native components of the vegetation and to non-native species which have previously been present only at low abundances. Such results suggest that care has to be taken when the issue of ecological redundancy is raised (Solbrig, 1991; Walker, 1992). The change in *Microseris douglasii* from a minor component representing 1–2% of the vegetation cover at the start of the study to a dominant with over 70% cover at the end (11 year) illustrates that redundancy depends on the time scale involved.

Another scale-related difficulty with redundance and inconsequential repetition can be illustrated using an extreme example concerning leaves on a tree. Removing one leaf may be compensated by the tree growing more leaf area that eventually replaces all known and perhaps all unknown functions of the removed leaf. Still, during the time when the leaf area was reduced, the photosynthetically active area of the tree is reduced along with particulate adsorbing capacity and all other functions associated with the lost leaf surface area. This demonstrates that even loss of a leaf of a tree is not functionally inconsequential. Loss of the leaf results in loss of function even though the loss may be small in effect and consequence. Indeed, leaves of trees are functionally similar in many ways but even in this extreme case, redundant and inconsequential appear to be inaccurate characterizations of the biophysical situation. Investigators could come up with even more extreme cases such as a single bacterium in a soil system but this does not escape the assertion that the loss of any living entity results in the loss of some amount of function, however small in scale.

These arguments demonstrate that functional redundance has anthropocentric and pejorative connotations, that it implies an ability of ecologists to see into the future with absolute certainty, and that it misrepresents scientists' understanding of living entities' interactions with ecological processes. Furthermore, it appears that other terms such as functional similarity, equivalence and compensation do not suffer from these problems and more accurately describe what scientists can defend in terms of living entities' known interactions with ecological processes. For example, type 3 and 7 responses are more accurately described as functional inconsequence than as functional redundance. Trophically equivalent taxa within trophic species may be called functionally inconsequential because the taxa represent topologically inconsequential repetition. It is inaccurate to call such taxa functionally repetitive and inconsequential repetition without qualifications such as, 'with respect to binary trophic interactions and the topology of trophic-species webs observed over a certain spatial and temporal scale.'

143

Without such qualifications, such taxa are accurately described as function-ally similar.

5.6 Conclusion: prospects for general theory relating biodiversity to function

Research on functional aspects of biodiversity runs the risk of losing its scientific utility by becoming so variably defined that determination of whether hypotheses are accepted or rejected will depend more on whimsical semantic interpretations than on reproducible determinations of whether or not certain observations are consistent with hypothesized predictions. Such a loss would limit both scientific achievement and the ability of society to rationally assess the consequences of the rapid and widespread loss of biodiversity at all scales. In contrast, the development of a successfully predictive and general theoretical core that relates biodiversity to function could do much to enhance scientific achievement and increase human society's abilities to rationally address our current biodiversity crisis. This chapter aims to accomplish this more positive end by embracing more precise, accurate and explicit terminology and avoiding teleologic, anthropocentric, and politically-charged terms used to relate biodiversity to function.

The framework developed in this review leads to the suggestion that ecologists do 'more of the same' regarding function and biodiversity, and continue to specify their consideration of function by more rigorously articu-lating and restricting the processes that they consider. The focus on ecosystem functions restricts consideration to energetic and material processes that are somewhat separate from population and evolutionary processes. Still, as the discussion of Vitousek and Hooper's (1993) work demonstrates, such restric-tion needs to be more extreme and precise in order to articulate reproducibly testable theory. Instead of specifying biogeochemical processes, it may be more useful to specify 'the consumption of live primary producers by meta-zoan primary consumers.' Even leaving out 'live' from this phrase which describes the process of interest could result in confusing herbivory with decomposition. Leaving out 'metazoan' could confuse herbivory with patho-genic consumption by disease organisms such as bacteria and viruses.

Such extreme restriction may be thought to limit the generality of hypotheses in a scientifically counter-productive manner. However, a frame-work based on more precisely defined processes and interactions provides a means to organize research beyond the current 'morass of special cases' (Lawton & Brown, 1993). Instead of a 'morass' of research specialized upon each ecological function, research on many different functions can be use-fully integrated if the functions represent different interactions with the same processes. Such integrated research could lead to a scientific understanding of the relations between biodiversity and particular processes. Such understand-ing would be demonstrated by successfully predictive hypotheses that (Peters,

1991) relate biodiversity and a particular process. After this understanding is achieved for several processes such as primary productivity and nitrogen cycling, similarities in the relation between biodiversity and both processes could be discovered. Further work may achieve greater generality by illuminating patterns common to larger classes of processes such as the cycling of plant nutrients and eventually back to ecosystem processes or, even more ambitiously, ecological processes in general. This chapter does not dismiss these more grandiose and somewhat ill-defined conceptual leaps but merely recommends achieving the same goal via smaller and more well-defined steps.

Acknowledgments

Many discussions with colleagues were critical to the development of this essay. Jennifer Dunne, John Harte, Terry Chapin, Jim Clegg, Curt Daehler and Tom Parker provided especially useful comments. Jennifer Dunne's editing was extremely helpful, particularly during the manuscript's earlier and less comprehensible stages. Similarly, John Lawton is greatly appreciated for a very effective and early review. Support was provided by the National Science Foundation Grant BIR-9207426.

References

Allen T.F.H. & Starr T.B. (1982) *Hierarchy: perspectives for ecological complexity*. University of Chicago Press. [5.2.2]

Baird D. & Ulanowicz R.E. (1989) The seasonal dynamics of the Chesapeake Bay ecosystem. *Ecol. Monogr.* **59**, 326–364. [5.4.1]

Briand F. & Cohen J.E. (1984) Community food webs have scale-invariant structure. *Nature* **307**, 264–266. [5.3.1] [5.4.1] [5.4.2]

Chapin F.S., Schulze E-D. & Mooney H.A. (1992) Biodiversity and ecosystem processes. *Trends Ecol. Evol.* **7**, 107–108. [5.2.4] [5.3.1] [5.3.2]

Cody M.L. (1991) Niche theory and plant growth form. *Vegetatio* **97**, 39–55. [5.3.1]

Cohen J.E. (1989) Food webs and community structure. In: *Perspectives in Ecological Theory* (eds J. Roughgarden, R.M. May & S.A. Levin), pp. 181–202. Princeton University Press, New Jersey. [5.4.1] [5.5]

Cohen J.E. & Briand F. (1984) Trophic links of community food webs. *Proc. Nat. Acad. Sci., USA* **81**, 4105–4109. [5.4.3]

Cohen J.E., Briand F. & Newman C.M. (1990) Community food webs: data and theory. In *Biomathematics*, Vol. 20. Springer-Verlag, Berlin. [5.4.1]

Cornell H.V. & Hawkins B.A.(1993) Accumulation of native parasitoid species on introduced herbivores: a comparison of hosts as natives and host as invaders. *Am. Nat.* **141**, 847–865. [5.2.2]

Cousins S.H. (1991) Species diversity measurement: choosing the right index. *Trends Ecol. Evol.* **6**, 190–192. [5.2.4] [5.2.5]

Cowling R.M., Esler K.J., Midgley G.F. & Honig M.A. (1994) Plant functional diversity, species diversity and climate in arid and semi-arid southern Africa. *J. Arid Environ.* **27**, 141–158. [5.2.4] [5.3.1]

Crome F.H.J. (1978) Foraging ecology of an assemblage of birds in lowland rainforest in northern Queensland. *Aust. J. Ecol.* **3**, 195–212. [5.2.5]

Daily G.C., Ehrlich P.R. & Haddad N.M. (1993) Double keystone bird in a keystone species complex. *Proc. Nat. Acad. Sci., USA* **90**, 592–594. [5.3.2]

Ehrlich P.R. & Ehrlich A.H. (1981) *Extinction: the causes and consequences of the disappearance of species.* Ballantine, New York. [5.2.1]

Ehrlich P.R. & Ehrlich A.H. (1992) The value of biodiversity. *Ambio* **21**, 219–226. [5.2.3]

Feyerabend P. (1993) *Against Method,* 3rd edn. Verso, London. [5.3]

Folse L.T. Jr (1981) Ecological relationships of grassland birds to habitat and food supply in east Africa. In: *The Use of Multivariate Statistics in Studies of Wildlife Habitat.* USDA Forest Service, General Technical Report RM-87 (ed. D.E. Capen), pp. 160–166. Rocky Mountain Forest Range Experimental Station, Fort Collins, Colorado. [5.2.5]

Foster M.S. (1991) Rammed by the Exxon Valdez: a reply to Paine. *Oikos* **62**, 93–96. [5.3.1]

Franklin J.F. (1988) Structural and functional diversity in temperate forests. In: *Biodiversity* (eds E.O. Wilson & F.M. Peter), pp. 166–175. National Academy Press, Washington DC. [5.2.1]

Frost T.M., Carpenter S.R., Ives A.R. & Kratz T.K. (1995) Species compensation and complementarity in ecosystem function. In: *Linking Species and Ecosystems* (eds C.G. Jones & J.H. Lawton), pp. 224–339. Chapman & Hall, London. [5.2.2] [5.5]

Gaston K.J. (1994) Biodiversity—measurement. *Progr. Phys. Geog.* **18**, 565–574. [5.2.5]

Gilbert L.E. (1980) Food web organization and the conservation of neotropical diversity. In: *Conservation Biology: an evolutionary-ecological perspective* (eds M.E. Soulé & B.A. Wilcox), pp. 11–33. Sinauer, Sunderland, Massachusetts. [5.2.3]

Grassle J.F., Lasserre P., McIntyre A.D. & Ray G.C. (1991) Marine biodiversity and ecosystem function. *Biol. Int. Special Issue* **23**, i–iv, 1–19. [5.1] [5.3]

Hairston N.G. Jr & Hairston N.G. Sr (1993) Cause-effect relationships in energy flow, trophic structure and interspecific interactions. *Am. Nat.* **142**, 379–411. [5.4.1]

Hall S.J. & Raffaelli D. (1993) Food webs: theory and reality. *Adv. Ecol. Res.* **24**, 187–239. [5.3.3] [5.4.1]

Harvey P.H. & Godfray H.C.J. (1987) How species divide resources. *Am. Nat.* **129**, 318–320. [5.2.4]

Havens K. (1992) Scale and structure in natural food webs. *Science* **257**, 1107–1109. [5.4.2]

Hawkins B.A., Thomas M.B. & Hochberg M.E. (1993) Refuge theory and biological control. *Science* **262**, 1429–1432. [5.2.3]

Haydon D. (1994) Pivotal assumptions determining the relationship between stability and complexity: an analytical synthesis of the stability–complexity debate. *Am. Nat.* **144**, 14–30. [5.3.1]

Hobbs R. & Mooney H.A. (1995) Spatial and temporal variability in California annual grasslands: results from a long-term study. *J. Veg. Sci.* **6**, 43–56. [5.5]

Holmes R.T., Bonney R.E. & Pacala S.W. (1979) Guild structure of the Hubbard Brook bird community: a multivariate approach. *Ecology* **60**, 512–520. [5.2.5]

Huenneke L.F. (1994) Redundancy in natural systems. *Trends Ecol. Evol.* **9**, 76. [5.1]

Hughes R.M. & Noss R.F. (1992) Biological diversity and biological integrity: current concerns for lakes and streams. *Fisheries* **17**, 11–19. [5.3.3]

Inger R.F. & Colwell R.K. (1977) Organization of contiguous communities of amphibians and reptiles in Thailand. *Ecol. Monogr.* **47**, 229–253. [5.2.5]

Jaksic F.M. & Medel R.G. (1990) Objective recognition of guilds: testing for statistically significant species cluster. *Oecologia* **82**, 87–92. [5.2.5]

Joern A. & Lawlor L.R. (1981) Guild structure in grasshopper assemblages based on food and microhabitat resources. *Oikos* **37**, 93–104. [5.2.5]

Jones C.G., Lawton J.H. & Shachak M. (1994) Organisms as ecosystem engineers. *Oikos* **69**, 373–386. [5.3.2]

Landres P.B. & MacMahon J.A. (1980) Guilds and community organization: analysis of an oak woodland avifauna in Sonora, Mexico. *Auk* **97**, 351–365. [5.2.5]

Law R. & Blackford J.C. (1992) Self-assembling food webs: a global viewpoint of coexistence of species in Lotka–Volterra communities. *Ecology* **73**, 567–579. [5.3.1]

Lawton J.H. (1989) Food Webs. In: *Ecological Concepts* (ed. J.M. Cherrett), pp. 43–78.

Blackwell Scientific Publications, Oxford. [5.4.1]

Lawton J.H. & Brown V.K. (1993) Functional redundancy. In: *Biodiversity and Ecosystem Function* (eds E-D. Schulze & H.A. Mooney), pp. 255–270. Springer-Verlag, Berlin. [5.1] [5.2.1] [5.2.2] [5.2.4] [5.3.2] [5.5]

Le Maitre D.C., Van Wilgen B.W., Chapman R.A. & McKelley D.H. (1995) Invasive plants and water resources in the western Cape Province, South Africa: modelling the consequences of a lack of management. *J. Appl. Ecol.*, in press. [5.2.4] [5.3.2]

MacArthur R.H. (1955) Fluctuations of animal populations as a measure of community stability. *Ecology* **36**, 533–536. [5.2.1] [5.3.1]

Magurran A.E. (1988) *Ecological Diversity and its Measurement*. Princeton University Press, New Jersey. [5.2.3]

Martinez N.D. (1991) Artifacts or Attributes? Effects of resolution on the Little Rock Lake food web. *Ecol. Monogr.* **61**, 367–392. [5.4.1] [5.4.2] [5.4.3]

Martinez N.D. (1992) Constant connectance in community food webs. *Am. Nat.* **139**, 1208–1218. [5.4.3]

Martinez N.D. (1993a) Effect of scale on food web structures. *Science* **260**, 242–243. [5.4.2] [5.4.3]

Martinez N.D. (1993b) Effects of resolution on food web structure. *Oikos* **66**, 403–412. [5.4.2]

Martinez N.D. (1994) Scale-dependent constraints on food-web structure. *Am. Nat.* **144**, 935–953. [5.4.2]

Martinez N.D. (1995) Unifying ecological subdisciplines with ecosystem food webs. In: *Linking Species and Ecosystems* (eds C. Jones & J.H. Lawton), pp. 166–175. Chapman & Hall, London. [5.4.3]

May R.M. (1973) *Stability and Complexity in Model Ecosystems*. Princeton University Press, New Jersey. [5.2.1] [5.3.1]

May R.M. (1983) The structure of food webs. *Nature* **301**, 566–568. [5.3.1]

Mayr E. (1988) *Toward a New Philosophy of Biology: observations of an evolutionist*. Belknap Press of Harvard University Press, Cambridge, Massachusetts. [5.2.1]

Mikulski B.A. (1994) Science in the national interest. *Science* **264**, 221–222. [5.2.1]

Mills L.S., Soulé M.E. & Doak D.F. (1993) The keystone-species concept in ecology and conservation. *BioScience* **43**, 219–224. [5.3.1]

Mishler B.D. & Brandon R.N. (1987) Phylogenetic species concept. *Biol. Philos.* **2**, 397–414. [5.4.1]

Murtaugh P.A. (1994) Statistical analysis of food webs. *Biometrics* **50**, 1199–1202. [5.4.2]

Naeem S., Thompson L.J., Lawler S.P., Lawton J.H. & Woodfin R.M. (1994) Declining biodiversity can alter the performance of ecosystems. *Nature* **368**, 734–738. [5.2.1] [5.2.2] [5.2.3]

Noss R.F. (1990) Indicators for monitoring biodiversity: a hierarchical approach. *Conserv. Biol.* **4**, 355–364. [5.2.1] [5.2.4]

Odum E.P. (1973) *Fundamentals of Ecology*. W.B. Saunders, Philadelphia. [5.4.1]

O'Neill R.V., DeAngelis D.L., Waide J.B. & Allen T.F.H. (1986) *A Hierarchical Concept of the Ecosystem*. Princeton University Press, New Jersey. [5.2.2]

Paine R.T. (1980) Food webs: linkage, interaction strength and community infrastructure. *J. Anim. Ecol.* **49**, 667–685. [5.3.1]

Paine R.T. (1992) Food web analysis through field measurement of per capita interaction strength. *Nature* **355**, 73–75. [5.3.1] [5.4.1]

Peters R.H. (1991) *A Critique for Ecology*. Cambridge University Press, New York. [5.6]

Pimm S.L. (1982) *Food Webs*. Chapman & Hall, London. [5.3.1]

Pimm S.L., Lawton J.H. & Cohen J.E. (1991) Food web patterns and their consequences. *Nature* **350**, 669–674. [5.4.1]

Pollock M.M., Naiman R.J., Erickson H.E., Johnston C.A., Pastor J. & Pinay G. (1995) Beaver as engineers: influences on biotic and abiotic characteristics of drainage basins. In: *Linking Species and Ecosystems* (eds C. Jones & J.H. Lawton), pp. 117–126. Chapman & Hall, London. [5.3.2]

Power M.E. & Mills L.S. (1995) The Keystone cops meet in Hilo. *Trends Ecol. Evol.* **10**, 182–184. [5.3.1]

Price P.W. (1977) General concepts on the evolutionary biology of parasites. *Evolution* **31**, 405–420. [5.3.2]

Root R.B. (1967) The niche exploitation pattern of the blue–gray gnatcatcher. *Ecol. Monogr.* **37**, 317–350. [5.2.4]

Schulze E-D. & Mooney H.A. (eds) (1993) *Biodiversity and Ecosystem Function.* Springer-Verlag, Berlin. [5.1] [5.2.1]

Short H.L. & Burnham K.P. (1982) Techniques for structuring wildlife guilds to evaluate impacts on wildlife communities. *USDI Fish and Wildlife Service, Special Science Report—Wildlife* 244. [5.2.5]

Simberloff D. & Boecklen W. (1981) Santa Rosalia reconsidered: size ratios and competition. *Evolution* **35**, 1206–1228. [5.2.5]

Simberloff D. & Dayan T. (1991) The guild concept and the structure of ecological communities. *Annu. Rev. Ecol. Syst.* **22**, 115–143. [5.1] [5.2.4] [5.2.5] [5.4.3]

Solbrig O.T. (ed.) (1991) *From Genes to Ecosystems: a research agenda for biodiversity.* International Union of Biological Sciences, Paris. [5.2.4] [5.3] [5.5]

Solbrig O.T. (1992) The IUBS–SCOPE–UNESCO program of research in biodiversity. *Ecol. Appl.* **2**, 131–138. [5.2.1]

Solow A.R. (1993) Measuring biological diversity. *Environ. Sci. Technol.* **27**, 25–26. [5.2.1] [5.2.5]

Steele J.H. (1991) Marine functional diversity: ocean and land ecosystems may have different time scales for their responses to change. *BioScience* **41**, 470–474. [5.1] [5.2.4]

Stone P.A., Snell H.L. & Snell H.M. (1994) Behavioral diversity as biological diversity: introduced cats and lava lizard wariness. *Conserv. Biol.* **8**, 569–573. [5.2.1]

Strauss R.E.(1982) Statistical significance of species clusters in association analysis. *Ecology* **63**, 634–639. [5.2.5]

Strayer D.L. & Likens, G.E. (1986) An energy budget for the zoobenthos of Mirror Lake, New Hampshire. *Ecology* **67**, 303–313 [5.4.1]

Terborgh J. & Robinson S. (1986) Guilds and their utility in ecology. In: *Community Ecology: pattern and process* (eds J. Kikkawa & J.J. Anderson), pp. 65–90. Blackwell, Palo Alto. [5.2.5]

Tilman D. & Downing J.A. (1994) Biodiversity and stability in grasslands. *Nature* **367**, 363–366. [5.2.1] [5.2.2] [5.2.3]

Toda M.J. (1984) Guild structure and its comparisons between two local drosophilid communities. *Physiol. Ecol. Jpn.* **21**, 131–172. [5.2.5]

Vitousek P.M. & Hooper D.U. (1993) Biological diversity and terrestrial ecosystem biogeochemistry. In: *Biodiversity and Ecosystem Function* (eds E-D. Schulze & H.A. Mooney), pp. 3–14. Springer-Verlag, Berlin. [5.2.2] [5.2.3] [5.2.4] [5.3.3] [5.5]

Walker B.H. (1992) Biodiversity and ecological redundancy. *Conserv. Biol.* **6**, 18–23. [5.2.1] [5.3.1] [5.5]

Warren P.H. (1990) Variation in food-web structure – the determinants of connectance. *Am. Nat.* **136**, 689–700. [5.4.3]

Warren P.H. (1994) Making connections in food webs. *Trends Ecol. Evol.* **9**, 136–141. [5.4.2] [5.4.3]

Westman W.E. (1977) How much are nature's services worth? *Science* **197**, 960–964. [5.2.1] [5.2.3]

Winemiller K.O. & Pianka E.R. (1990) Organization in natural assemblages of desert lizards and tropical fishes. *Ecol. Monogr.* **60**, 27–55. [5.2.5]

Woolf H.B. (ed.) (1979) *Webster's New Collegiate Dictionary.* Merriam-Webster, Springfield, Massachusetts. [5.2.1] [5.5]

Younès T. (1992) Ecosystem function of biodiversity. *Biol. Int.* **24**, 16–21. [5.1] [5.3]

Zak J.C., Willig M.R., Moorhead K.L. & Wildman H.G. (1994) Functional diversity of microbial communities: a quantitative approach. *Soil Biol. Biochem.* **26**, 1101–1108. [5.2.4] [5.2.5]

6: Diversity and higher levels of organization

JAMES A. DRAKE, CHAD L. HEWITT, GARY R. HUXEL
and JUREK KOLASA

6.1 Introduction

For diversity to be a meaningful concept, deserving of all its recent attention, it must somehow reflect the inner workings and dynamics of co-occurring species in space and time. It must be more than the simple measurement of this and that—variously accounting for the number of species and their relative commonness or rarity. Diversity is an ensemble construct, a reflection of past events and processes variously expressed in the presently observed state. It is dynamically variable in space and time—a synthesis of evolutionary change and real-time ecological constraints. So conceived, we seek to define diversity by considering the mechanics which occur during the assembly of biological systems.

6.2 Preliminaries

It is difficult to begin a discussion of the very notion of diversity at higher levels of organization without first appreciating the components of that diversity, the nature of higher levels of organization, and what is meant by diversity and how it is quantified (Huston, 1979, 1994; Brown, 1981; Ricklefs & Schluter, 1993). This is clearly a daunting task, one riddled with disputed though hopeful theories. Despite the analytically recalcitrant nature of biological systems, recent advances in the construction, evolution and behaviour of complex systems offer a promising avenue of inquiry (Bak *et al.*, 1988; Drake *et al.*, 1992, 1994; Kauffman, 1993; Luh & Pimm, 1993; Langton, 1994; Jorgensen, 1995; T. Keitt & P. Marquet, in prep.). The first part of this essay seeks to establish a framework and perspective from which we can detect significant levels of biological organization above the level of species. We offer the hypothesis that the dynamics which occur at those levels, as well as the processes responsible for the existence of those levels, are fundamental to the expression of biological diversity however conceived. While the *species* provides a natural frame of reference, our arguments apply to diversity regardless of the nature of the elements of that diversity (e.g. genetic diversity, species diversity, community diversity). All this brings us to consider how the production of biological structure, at whatever scale and

149

however expressed, impacts the number, relative abundance and distribution of species.

At the outset, we accept the premise that higher levels of organization occur in nature, and that the mechanics responsible for the production of those levels are variously capable of influencing, if not regulating, biodiversity. That higher levels of organization exist does not require a great leap of faith; an ecological community is clearly a higher level of organization. However, organization at higher levels involves more than an observational rescaling. Higher levels are pertinent only when the aggregate properties of the components gel into a functional unit. The information contained in a single gene or discrete set of genes is not the same as the information contained in those genes when behaviour is coordinated. Without recourse to group selection or Clementsian prose, the community is arguably a fundamental biological unit because membership therein can be limited (Clements, 1916; Wilbur & Alford, 1985; Gilpin *et al.*, 1986; Roughgarden, 1989; Drake, 1990; cf. Strong *et al.*, 1984; Underwood, 1986).

Despite the fact that ecological systems are complex, the predominant academic approach to ecology is one of reduction. We must be aware of the dangers inherent in reducing the system of interest to a point where significant controlling processes cannot be detected. Nevertheless, simplifying a system to a few manageable levels such as a single predator–prey interaction or a small spatial scale (e.g. a single patch within a landscape) is important, because the actions of mechanisms in real time are pertinent to current patterns and the development of subsequent community states. However, the action of such mechanisms must be viewed in the context under which they emerged; their function is historically contingent (Ricklefs, 1987; Drake, 1991; Brooks & McLennan, 1993; Drake *et al.*, 1993). Understanding cause is exceedingly difficult in historically contingent systems because the dynamical imprints of past mechanism and process are hidden within and adjusted by the contemporary state. The system had to arrive at its current state (state A) from some previous state (state B), a state which can play a significant role in defining the suite of plausible subsequent states and attendant patterns. The switches which initiate a transition from *state A* to *state B* represent an assembly step and include diverse processes and mechanisms which range in their dynamical character from the purely stochastic to the highly deterministic (Chesson & Case, 1986; DeAngelis & Waterhouse, 1987).

Consider, for example, the role that disturbance plays in initiating transitions from one state to another. Disturbance can function as a resource, variously acting to: (i) maintain the status quo (e.g. fire-maintained prairie; rocky-intertidal algal assemblages; see Pickett & White, 1985); (ii) radically alter the system trajectory; (iii) reset the system to an earlier state; or (iv) move the system to a state which cannot otherwise be reached. Assembly steps including those initiated by disturbance are a function of context. We view biological systems as a set of historically contingent assembly trajectories

150

(at ecological and evolutionary scales) variously influenced by the environment. Generally, ecological analyses focus on but a slice of those trajectories and only rarely can the slice provide insight into the nature of the system (e.g. a Poincaré section through a strange attractor). Contained within an assembly trajectory is a sequence of emergent properties which are a structural manifestation of an assembling hierarchy. Thus, emergent properties are a significant aspect of the context under which assembly proceeds. For example, the development of invasion resistance to colonization by specific organisms can be an emergent property of a given set of species (Robinson & Wellborn, 1988). Similarly, a population may derive some of its dynamics from complex age and spatial structure within its range, or it may be composed of several functional metapopulations (Harrison, 1994, in press; J. Kolasa & J.A. Drake, in prep.). If persistence of a population depends on its metapopulation geometry then indeed the metapopulation becomes a critical level of organization. Factors which exert control on assembly trajectories provide a framework for understanding organization, and ultimately the expression of diversity.

All biological structures, and perhaps all physical structures, are assembled. Pattern is derived through the interplay between the elements which comprise a structure, and the assembly processes which ultimately influence membership. Thus *the expression of mechanisms in space and time is under the control of assembly mechanics, operating against a variable environmental mosaic.* Presently expressed mechanisms and processes become the context for subsequent states. So conceived, we proceed by discussing assembly phenomena, focusing on the manner in which diversity is expressed as a function of assembly mechanics. The distinction between observed patterns with their attendant mechanisms (e.g. competition and predation) and assembly mechanics is critical to our arguments. Mechanics enable, regulate and tune the operation of the mechanism, variously influencing the manifestation of the mechanism. G. Huxel (in prep.), for example, found that it was relatively easy to manipulate assembly trajectories producing intransitive competitive dynamics (e.g. species A beats species B, species B beats species C, but species C beats species A). The mechanism is competition while the assembly mechanic is the specific sequence of invasions and the environmental context under which the invasion occurred. In such cases we note a duality in causal factors. Competition remains the mechanism but its expression is considerably richer than its action in real-time permits (e.g. the course of a typical field experiment).

We focus first on the changes in assembly states, their persistence, and resistance to invasion as a function of ecological, successional and evolutionary time. Next we extend our arguments spatially, considering the effects of patchiness and associated metastructures on assembly mechanics and thus diversity. Finally, we seek to integrate these mechanics with the fundamental processes of complex systems dynamics such as self-organization and approaches to critical states, the so-called 'edge of chaos'. We argue that commonality in the mechanics of system construction may be independent of the nature of the system and constituent elements.

As with most science, the terminology used in connection with issues of system organization is complex, sometimes unclear, and generally unfamiliar to many ecologists. Therefore we provide a brief explanation of the necessary jargon (Table 6.1).

6.3 Assembled systems

Assembly is a fundamental process which plays a powerful role in the construction, development and evolution of biological systems (Diamond, 1975; Drake, 1990, 1991; Nee, 1990; Luh & Pimm, 1993). Considering populations as basic elements of assembly, and restricting space for the moment, different sequences of species colonization can translate into different community assembly trajectories (cf. Drake *et al.*, 1994; Grover & Lawton, 1994). Much as the timing of genetic expression is critical to organism development, variation in colonization sequence can lead to alternative community states. All plausible assembly trajectories for a given set of species map onto a manifold which outlines the system's dynamical nature. Visually, this manifold is similar to the classic adaptive landscape. The manifold defines the plausible assembly space and forbidden combinations within that space, as well as a probabilistic density function. Some trajectories are rarely traversed and thus have a low probability of occurrence while others represent a common theme. The manifold functions as a kind of road map outlining where the system may go and where it has been, as well as the relationship among alternative assembly trajectories.

We can envision four types of assembly trajectories of direct interest to questions of diversity (Fig. 6.1). Either a given set of species within some environmental context has a single solution, multiple persistent solutions, a complex solution (e.g. community composition sits on a complex attractor), or no solution within the composite of possible assembly trajectories (Robinson & Dickerson, 1987; Tilman, 1988; Drake, 1991; Grover, 1994; Naeem *et al.*, 1994). Such dynamics lead us to conclude that the nature of assembly trajectories as expressed across the biosphere are intimately involved in the production, regulation, and expression of biodiversity.

As a starting point, we consider systems which possess biologically induced alternative states, whether these states are characterized by an approach to an invasion-resistant state or map to a more complex attractor. Several studies have noted that differences in the sequences of either species colonization or resource availability can lead to the production of alternative community states (Wilbur & Alford, 1985; Robinson & Dickerson, 1987; Drake, 1991; Drake *et al.*, 1993; see also Drake *et al.*, 1994; Grover & Lawton, 1994; G. Huxel *et al.*, in prep.). Priority effects (first-in wins type of dynamics)

Table 6.1 A terminology primer.

Organization	Dynamics persistence of structure: a combination of processes and forces responsible for glueing several components together. Examples of organized structures include a lichen, a termite nest, a pack of wolves, or a community in a pitcher plant. The main organizational forces include coordination, complementarity and information (see Kolasa & Pickett, 1989)
Structure	Identity and arrangement of elements showing organization: fungus and algal species association, termite caste, alpha, beta and other wolves and their offspring (see definition above)
$s^{-\omega}$ distribution	A typical distribution of magnitudes (f) of changes such as species extinctions; the power-law distribution being a hollow decreasing curve with few large and many small events
Self-similarity	An empirical phenomenon in which patterns observed at small scales resemble patterns observed at larger scales. Examples include the structure of snowflakes, fern leaves, or microhabitat mosaics
Topology	Figuratively speaking, an arrangement of objects in a three-dimensional or imaginary space
Manifold	An abstract surface (having dimension n) condensing many response surfaces and their interactions
Level of organization	A particular view of the systems structure determined by the resolution of criteria applied. If functional similarity is used as a criterion, guilds might emerge as a level of organization, or, if a taxonomical identity is used, species will become a resulting level of organization
Self-organization	An autonomous process of incorporating and rearranging/adjusting components in such a way that the organization of the system in the sense used earlier is preserved. Systems which self-organize develop pattern and structure in excess of that which can be understood based on a detailed knowledge of initial conditions

are a direct function of assembly and are capable of generating alternative community states (Wilbur & Alford, 1985; Barkai & McQuaid, 1988). Drake (1991), using a system of differential equations, detected a group of highly deterministic assembly trajectories which always lead to the production of repeatable community states. In other trajectories the outcome appeared indeterministic, multiple instances of a single trajectory diverging to distinct endpoints. When these dynamics are spatially extended, assembly can lead to either higher or lower diversity at higher levels of organization than might have otherwise been anticipated. The relative roles of determinism and indeterminism can make nature look *messy* and even *monotonous*.

Unlike systems with alternative endpoints, some sets of species and environments support assembly trajectories which appear to possess a single solution (Fig. 6.1A). When a set of species in a given environment exhibits a single or a few closely related alternative states, we view that community in classical succession terms. The observer sees a recurrent theme across space and time, albeit at the coarsest of levels—a handful of dominant species. Many of the paradigms of ecology as a science have been based on processes occurring in such equilibrial, deterministic, 'climax' systems. Of course without detailed information it is not easy to distinguish between cases with a single general solution (Fig. 6.1A), and cases in which alternative solutions exist but a single trajectory was dominant across space (Fig. 6.1B), or cases where multiple instances of a single trajectory are out of phase.

In other systems we fail to see directionality or even a multiplicity of recognizable states (Fig. 6.1C,D). In these cases, either rules of assembly do

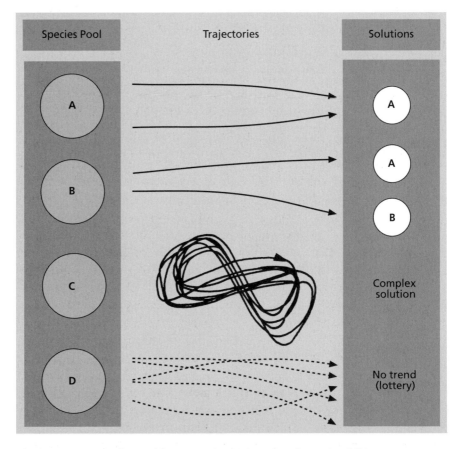

Fig. 6.1 Four types of assembly trajectories. Trajectories of type A exhibit convergence to the same solution, while trajectories of type B exhibit divergence. Type C trajectories exhibit a complex solution similar to a strange attractor, while type D trajectories are purely stochastic in nature. Hybrids of these four trajectory types are also likely to exist.

not exist or rules exist which create a lack of discernible pattern (Sale, 1977), some factor precludes the operation of rules, assembly is manifest at a level of scale which cannot be detected with the information at hand, or multiple scales map onto the spatial dimension producing patterns difficult to dissect (Roughgarden, 1989; Waltho & Kolasa, 1994). To further complicate the distinction between these scenarios, we can easily conceive of 'filler' species, species which are not engaged in or bounded by ongoing assembly processes—they are just there. Such species can be thought of as interchangeable occupants being added to the community in a lottery fashion.

Regardless of the specific nature of the assembly space, many assembled states have the emergent ability to resist invasion by one or more of the species available for colonization (Sugihara, 1985; Wilbur & Alford, 1985; Robinson & Dickerson, 1987; Drake, 1991; Drake *et al.*, 1992). Unless challenged by a truly exotic species (i.e. one not from the original species pool), this resistance may become complete in ecological time. However, invasion resistance is easily surmounted in evolutionary time as surely as new species arise (Ginzburg *et al.*, 1988). Once a new player enters the system, the aggregate of assembly trajectories or manifold can be fundamentally altered, making some trajectories or suites of trajectories inaccessible while creating others (see Vitousek, 1990). Thus evolutionary novelty can both be driven by and in turn drive the development of assembly trajectories. Evolution does not act without regard to assembly processes, in fact evolution is itself an assembly process (cf. Simberloff, 1974).

Alternatively, the assembly metaphor can be extended in space, resulting in a set of variously interconnected community patches within a metacommunity landscape. Clearly, the existence of alternative states, driven by whatever processes and dynamics, leads to enhanced regional diversity (cf. Hanski, 1991; Harrison, in press), sometimes at the expense of local (α) diversity—some invasion-resistant states contain fewer species than others. Is the frequency of occurrence of such states across a landscape a meaningful reflection of diversity at a higher level of organization? Are there specific interaction topologies (the geometric arrangement of locally available species pools) which serve as a recurrent theme across space and time?

Formally adding a spatial resource (e.g. discrete patches within a landscape) can lead to a complex geometry of assembling community patches. For example, it is easy to envision geometric arrangements of communities in space and time in which portions of the landscape contain vulnerable patches protected by invulnerable patches (Drake, 1990; Drake *et al.*, 1993). Further, the time to onset of invasion resistance varies among different trajectories, leading to the possibility of exploitation of this temporal resource by fugitive species (cf. Levins, 1969; Hanski, 1982; Gotelli & Simberloff, 1987; J. Kolasa & J.A. Drake, in prep.).

155

Spatial pattern in the distribution of assembling communities can significantly influence local and regional diversity, and consequently the occurrence of higher levels of organization. For example, the fractal nature of landscapes leads to increasing patchiness as the scaling exponent increases (Hastings & Sugihara, 1993). The occurrence of vulnerable and invulnerable states may be a recurrent theme derived largely from the fractal geometry of the landscape imposing a limit to the number of possible states. Within these limits, ecological factors provide further structure, causing vulnerability and invulnerability to break down above some spatial scales. Such geometric patterns can serve to further increase diversity by either accommodating species which can exploit such patchiness (e.g. the fractal nature itself) as a resource, or effectively increasing the total number of community states possible (cf. Levins, 1969). Thus, the existence of local alternative community states can lead to increased regional species richness (Whittaker, 1972; Waltho & Kolasa, 1994).

Alternative states are a manifestation of the combined influences of: (i) the rules of assembly contained within a given set of species; (ii) spatial and temporal dimensionality; (iii) environmental filtering; and (iv) evolutionary change. If, for example, disturbance (e.g. fire, predation, species invasion) occurs in a patchy fashion across a landscape, then removal of the disturbance or a change in its frequency and distribution of occurrence alters the rules of the game such that diversity changes as a function of alterations in assembly processes (e.g. Paine & Levin, 1981; Pickett & White, 1985). Disturbance can also function as a switch, sending a trajectory or set of trajectories onto different courses (Fig. 6.1B). Sometimes the trajectory is resilient in the face of disturbance (Fig. 6.1A), other times changes are irrecoverable. Trajectories can simply be reset to an earlier stage by disturbance, or disturbance can modify emergent properties expressed by the system, eliminating some and promoting the function of others (Clements, 1916). However, disturbance need not be physical in source—evolution can be a potent disturbance to the status quo.

Some higher levels of organization are detectable as a function of changing observational scales. A single sampled patch may represent a complete community and exhibit all of the attributes of that scale of observation. The composite (global) attributes of several sampled patches are not shared by the individual component patches: these are emergent properties. Yet these global attributes may not represent an *emergent structure*, this is a structure defined by the real processes and mechanisms not by observer-defined scale. For example, in a forest there are many local patches of differing successional states, each having the properties of a community. A sample of these patches will allow an estimate of some of the emergent *properties* of the next higher level of organization but will not necessarily be an emergent *structure* itself. The key to understanding emergent structures will not be found by the reductionist paradigm: a detailed understanding of the component structures. Rather the individual assembly trajectories and the synergistic interactions of components holds the key to understanding emergent structures. While it is relatively easy to rescale and observe a variety of 'levels of organization', detection of true emergent structures remains a difficult task.

6.4 Self-similarity in time and space: a signature of levels of organization?

Thus far we have focused purely on the biological manifestation of assembly (e.g. the production of alternative community states, invasion resistance), but what of the assembly process itself? Are there general mechanics at work here which provide insight into organization, particularly with respect to the emergence of higher levels of organization? Ordinarily, organized ecological systems involve several levels and a number of entities or compartments at each level. For example, a metapopulation consists of at least two levels in addition to the level of definition. It has an array of local populations with each local population consisting of individuals. This vertical nestedness and horizontal compartmentalization are only possible through varying levels of integration (see Kolasa & Pickett, 1989). Individuals would not be a part of a single population if they did not and could not interact, and the metapopulation would not exist if the component populations could not exchange individuals. The degree of those exchanges defines the level of integration and relative opportunity for the existence of emergent structures.

One interesting consequence of these relationships is that increasing integration is conducive to inclusion of a greater number of components. This is well illustrated at the kinship level where examples abound of highly integrated groups of individuals. Usually, such groups occur among invertebrates (social insects, spiders, colonial hydroids, bryozoans) and vertebrates (wrens, mole rats, elephants, lions), which are also significantly differentiated morphologically. While the evidence is not obvious, it is quite possible that highly organized and thus integrated multi-species assemblages will also support more different types of components (e.g. species; Connell, 1980). General observations comparing tropical forest and coastal communities against less integrated temperate or agricultural assemblages appear to support this idea (Connell, 1978). More interestingly, a build-up of nested structures provides the raw material for differential rates of extinction and invasion that, when analysed in a particular way, exhibit the general patterns discovered in other complex and organized systems, even non-biological systems (Fig. 6.2).

A recent wealth of studies in disparate fields of inquiry have reported strikingly similar spatial and temporal phenomena as a consequence of system construction and turnover (Bak *et al.*, 1987, 1988; Bak & Tang, 1989;

157

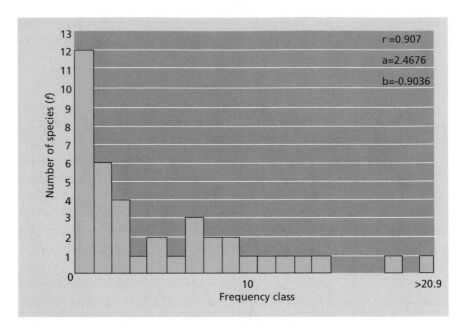

Fig. 6.2 Frequency distribution of extinction magnitude in invertebrate communities of 50 coastal rock pools, Discovery Bay, Jamaica. The extinctions were evaluated over three sample dates spanning 2 years.

Drake *et al.*, 1992; Lu & Hamilton, 1991; Olami *et al.*, 1992; Kauffman, 1993; Plotnick & McKinney, 1993; Hewitt *et al.*, in press; Jorgenson, 1995; T. Keitt & P. Marquet, in prep.). Recurrent themes found among many studies of these spatially extended dynamical systems are characteristic temporal signatures, the process of self-organization, and critical dynamics. Systems which self-organize develop complex pattern and structure in the absence of specific selective forces — that is without the detailed tuning of initial conditions. Can assembly in ecological and evolutionary time be viewed in the broader context of spatial and temporal self-similarity? Self-organization has been implicated in the development of emergent structures such as patterns of community membership and species extinction (Drake *et al.*, 1992; Hewitt *et al.*, in press; T. Keitt & P. Marquet, in prep.) offering new mechanisms capable of explaining phenomena such as punctuated equilibrium (Plotnick & McKinney, 1993). What implications do such dynamics have for biological diversity?

Consider first the elegant 'sand-pile' metaphor offered by Bak and colleagues (Bak *et al.*, 1987, 1988; Bak & Tang, 1989). They suggest that many spatially extended dissipative systems proceed until they reach a self-organized critical state with the system becoming self-similar in the time domain. Here, sand grains (species for comparative purposes) are added to a growing pile (a community, metacommunity, or landscape). Eventually, a critical point is reached where each addition of sand precipitates uncorrelated collapses in the pile (extinctions). The frequency histogram of collapse magnitude, duration of collapses, and time between collapses follows a power-law distribution (see Table 6.1) of the form:

$$D(s) \approx s^{-\omega}$$

Here, the frequency of occurrence of various collapse sizes $D(s)$ produces a familiar hollow curve. There are many collapses of a single grain, followed by fewer and fewer larger collapses. As the critical state is reached, here the angle of repose above which collapse events occur (Bak *et al.*, 1987, 1988), the structure of the sand pile is exposed to meaningful description and measurement.

To tie together these ideas with our earlier comments on organization and integration we could imagine adding water to our sand pile, increasing the cohesiveness between individual grains. The water serves the function of an integrative glue, increasing the integration and permitting a greater angle of repose before a new critical state is attained. A lesson from this exercise is that while the more highly integrated systems will hold more species (grains of sand), loss of the integrating glue will result in a catastrophic loss of species. Increasing and decreasing integration, and the mechanisms and mechanics responsible, are fundamental to the origination and persistence of biological systems.

Can this metaphor be extended to biological systems? If so, the implications of such dynamics may provide insight into many aspects of biological nature. At present there has been little direct experimental work dealing with the mechanics of biological self-organization and approaches to critical states. There is, however, information about the dynamical nature of various aspects of biological organization such as extinction trends in the fossil record and in community assembly models, which reveal compelling similarities (Raup, 1986; Drake *et al.*, 1992; Plotnick & McKinney, 1993). Do such signals point to the elements of biological diversity which are manifest as critical states of the system? Are there biological sandpiles which include emergent levels of organization, levels of organization detectable by a characteristic $s^{-\omega}$ signal? Preliminary data exist which are consistent with this general pattern (Fig. 6.2). Finally, can such signals become the basis for diversity measures which reflect the inner workings of biological nature, rather than simply account for the variety and abundance of species? We believe that the answer is yes. We envision a conceptual connection between the organization of ecological systems and the patterns of signals reflected by the $s^{-\omega}$ curve (Fig. 6.2) that could serve to link the functional organization of multiple species assemblages, signals of extinction and invasion and the number of species associated with each level of community organization.

It has been argued that information, whatever the nature of the elements of that information (e.g. genetic and species diversity, food web topology, expression of emergent properties), reaches a maximum at the critical state

159

between chaos and antichaos (Kauffman, 1991, 1993; Jorgensen, 1995). The critical state is the state furthest from thermodynamic equilibrium (Jorgenson, 1995). If higher biological diversity is akin to higher information content, diversity may very well be the result of dynamics which occur at the chaos–antichaos phase transition.

However, unlike the sandpile of Bak *et al.* (1987, 1988), biological systems effectively have a limit on the number of species (grains) available for colonization. This basic fact adds additional dynamics to biological systems not contained in the sandpile model. Frozen or invasion resistant states occur over relatively short time spans which are expressed as limited community membership, the $s^{-\omega}$ signal vanishes, and the system is effectively buffered from the chaos–antichaos boundary. Invasion, speciation and extinction are capable of eliminating this buffer, forcing the system back to the edge of chaos. The manner in which processes such as invasion and speciation influence the dynamics of community membership at the chaos–antichaos boundary should provide insight into the mechanisms behind the coexistence of species and the production of diversity.

6.5 A model of turnover and interacting signals

The historical record of the assembly process is obscured by the arrival and departure of species as well as the turnover of emergent structures. Hence, the driving forces and processes which produce community turnover may not be as significant as the dynamics of the turnover process itself. As discussed previously, patterns of biodiversity change have been linked to disturbance which induces a power-law or l/f signal of species turnover through time (Plotnick & McKinney, 1993; T. Keitt & P. Marquet, in prep.). These patterns have been discerned in studies at ecological scales such as gap dynamics in wave-stressed environments and wind blowdowns in forests, and palaeontological scale arguments such as patterns of species extinction and genesis. These patterns have typically been examined from the perspective of a single, dominant force which drives the system to exhibit power-law dynamics.

Biological systems, however, are subject to multiple sources of perturbation acting both in ecological and evolutionary time at a variety of levels of scale. Each source of perturbation may typically induce power-law dynamics alone. However, when these independent perturbations occur simultaneously across a landscape, the aggregate signal departs from the l/f distribution. Huxel *et al.* (in prep.) found that increasing the number of equivalent l/f perturbations resulted in a shift in the mode of the size–frequency histogram. While the underlying dynamics were l/f distributed, the emergent signal was best described by a gamma distribution. Thus at one level of scale the system ceases to be scale-independent, but instead can be described as self-similar over a portion of the range (self-affine). Discontinuities

160

in l/f signals, observed as a function of rescaling, point to important structural elements of the system which include any significant higher levels of organization.

6.6 Diversity and higher levels of organization

Other than those levels of organization which are glaringly obvious (e.g. organisms, populations), how do we detect where biological organization congeals into meaningful higher levels of organization? Above the species level, organization has generally been defined by either those sets of strongly interacting species which exhibit feedbacks (e.g. competition, indirect effects, top-down and bottom-up control scenarios), by phylogenetic relationships (e.g. kinship, conspecific, or congeneric relations), or by energy and material flow. These may very well represent significant levels of organization, but they must be observed not only by internal mechanism but also by the emergent properties and structures they generate. A higher level of organization is an emergent structure which exhibits processes and dynamics in excess of that produced by individual species (components). For example, species form variously interconnected food webs, which have properties the individual components do not. The food web topology, or perhaps a piece of it, can have the property of invasion resistance which emerges as a collective attribute (Robinson & Dickerson, 1987; Drake, 1991; Drake *et al.*, 1993, 1994; Grover & Lawton, 1994). The diversity of such elements, here invasion-resistant topologies and associated assembly trajectories across space and time, provide meaningful higher-level diversity information.

We suggest that levels of organization are detectable as a function of the signal which emerges with a systematic change in scale, that is, as we focus through the spectra of signals produced by the turnover of elements across space and time. If biological systems do indeed self-organize and approach a critical state our task is simplified because we look for recurrent signals such as power-law distributions (G. Huxel *et al.*, in prep.). Several available time series have revealed compelling l/f-like trends. Keitt and Marquet (in prep.), for example, have argued that extinction patterns among the Hawaiian avifauna exhibit a l/f trend, consistent with self-organizing internal dynamics as species invade, extinctions occur and the system changes. Is information, which is a function of the dynamics of approaching the critical state, expressed in patterns of biological diversity and system complexity?

Our model of disturbance-induced interacting l/f signals offers some insight into how diversity and pattern could be produced and change through time. As the exponent of the l/f curve changes the distribution of turnover changes and the system moves either away or toward the critical state (self-sustaining and self-similar at all scales). The higher the exponent (more negative) the more the curve flattens against both abscissa and ordinate. Multiple interacting l/f disturbances force the aggregate turnover signal toward a gamma distribution. Can we detect such appropriately signatured signals emerging from sets of species (and sets of sets of species) in space and time? Hastings and Sugihara (1993) have called for similar use of self-similarity in space. This may prove to be easier said than done in practice, as data sets are typically short and focus on a local subset of species.

We argue that an appropriate epistemological framework for understanding diversity must specify both the historical nature of the system, as well as the system-wide consequences of the operation of proximate mechanisms. Significant levels of organization above the population may be detected by l/f-like patterns in temporal flux. With reference to the trajectories mentioned above, we expect some systems to be forced by exogenous disturbance that self-organization never occurs. We also anticipate that some systems which look haphazard at one level of scale will exhibit self-similarity at other scales of observation (e.g. coral reefs, tropical forests). This points to a subtle yet critical aspect of dynamics exhibited by biological systems, there is not complete scale invariance. Rather, discontinuities point to important structural features of the system. For example, the production of fixed or frozen states adds complexity because the solution precludes further turnover at one level of scale – the configuration of the system (e.g. species composition; number of guild types).

6.7 Prospects

Because higher levels of organization are emergent structures, they can best be detected when they are either created (e.g. from a species invasion or disturbance) or extirpated from a system by strategic manipulations of species composition. Removal and addition of species are essential ingredients of many ecological studies, but such studies are generally limited in time. Studies which track the abundance of an entire ensemble of species are needed to examine the manner in which higher levels of organization develop. At the ensemble level we can observe changes in *diversity* in four ways: speciation; extinction; invasion; and changes in relative abundance. Variation in any of these processes clearly can lead to changes in the number of species coexisting in some areas, hence a change in local diversity by whatever measure. So conceived diversity is about the number of species coexisting in an area incorporating compelling geographical trends. Yet this is but one aspect of diversity; the meaning of diversity is enhanced at levels above the species. We may envision many levels of organization for which there are simple comparisons (e.g. the number of community types/unit area, the number of biomes in a province), yet there are also more subtle aspects of diversity. For example, the species which comprise a regional species pool contain a number of potential community states. Arrival at any one of those states can occur by chance, be forced to some state by the environment, or along some specific community assembly trajectory, or more likely a combination of all three. The number of assembly routes for a given set of species is variable (see Luh & Pimm, 1993). Different sets of species contain different assembly routes, ultimately regulating the number of species found on a given landscape. Certainly diversity is manifest as the number of trajectories possible after a disturbance or during assembly.

Evolutionary change if rapid enough could cause an assembling system to the virtual edge of chaos by altering or eliminating emergent properties such as resistance to invasion. In this case the frozen states disappear and the system exhibits l/f. We wonder whether a system's proximity to the edge of chaos can feed back on the rate of evolution, ensuring some degree of buffering from chaos. Such a process assumes that a collective attribute exists which dampens chaotic dynamics. Alternatively, one could argue that it is not in an individual population's best interest to push its system chaotic because of the increased risk of extinction. How then can a system near the edge of chaos feed back on the rate of evolution? What happens to selective pressures and processes of self-organization in systems near the edge of chaos?

During the course of this essay we have asked many more questions than we have answered. We offer no apologies for this uncertainty, rather we believe that such incertitude is a direct call for a combined theoretical and empirical programme which steps past the phenomenological boundaries of contemporary biology.

Acknowledgments

First we wish to thank Kevin Gaston for his willingness to entertain our ideas and put up with our endless delays and various levels of catastrophic collapse through which this essay has traversed. J.A.D. acknowledges the Department of Energy for support. C.L.H. acknowledges support by the Department of Energy Global Change Postdoctoral Fellowship and the Oak Ridge Institute for Science and Education. J.K. was supported by the Natural Sciences Research and Engineering Council of Canada. We also thank Professors N. Young and R. Kramel for encouragement.

References

Bak P. & Tang C. (1989) Earthquakes as self-organized critical phenomenon. *J. Geophys. Res.* **94(B)**, 15635–15637. [6.4]

Bak P., Tang C. & Wiesenfeld K. (1987) Self-organized criticality: an explanation of l/f noise *Phys. Rev. Lett.* **59**, 381–384. [6.3] [6.4]

Bak P., Tang C. & Wiesenfeld K. (1988) Self-organized criticality. *Phys. Rev.* **A 38**, 364–373. [6.2] [6.4]

Barkai A. & McQuaid C. (1988) Predator–prey role reversal in a marine benthic ecosystem. *Science* **242**, 62–64. [6.3]

Brooks D.R. & McLennan D.A. (1993) Historical ecology: examining phylogenetic components of community evolution. In: *Species Diversity in Ecological Communities* (eds R.E. Ricklefs & D. Schluter), pp. 267–280. University of Chicago Press. [6.2]

Brown J.H. (1981) Two decades of homage to Santa Rosalia: toward a general theory of diversity. *Am. Zool.* **21**, 877–888. [6.2]

Chesson P.L. & Case T.J. (1986) Overview: nonequilibrium community theories: chance, variability, history, and coexistence. In: *Community Ecology* (eds J.M. Diamond & T.J. Case), pp. 229–239. Harper & Row, New York. [6.2]

Clements F.E. (1916) *Plant Succession: analysis of the development of vegetation.* Carnegie Institute of Washington Publication No. 242. Washington DC. [6.2] [6.3]

Connell J.H. (1978) Diversity in tropical rain forests and coral reefs. *Science* **199**, 1302–1310. [6.4]

Connell J.H. (1980) Diversity and the coevolution of competitors, or the ghost of competition past. *Oikos* **35**, 131–138. [6.4]

DeAngelis D.L. & Waterhouse J.C. (1987) Equilibrium and nonequilibrium concepts in ecological models. *Ecol. Monogr.* **57**, 1–21. [6.2]

Diamond J.M. (1975) Assembly of species communities. In: *Ecology and Evolution of Communities* (eds M.L. Cody & J. Diamond), pp. 342–444. Harvard University Press, Cambridge, Massachusetts. [6.3]

Drake J.A. (1990) The mechanics of community assembly and succession. *J. Theor. Biol.* **147**, 213–233. [6.2] [6.3]

Drake J.A. (1991) Community assembly mechanics and the structure of an experimental species ensemble. *Am. Nat.* **137**, 1–26. [6.2] [6.3] [6.6]

Drake J.A., Witteman G.J. & Huxel G.R. (1992) Development of biological structure: critical states, and approaches to alternative levels of organization. In: *Biomedical Modelling and Simulation* (eds J. Eisenfeld, D.S. Levine & M. Witten), pp. 457–463. Elsevier, North Holland. [6.2] [6.3] [6.4]

Drake J.A., Flum T.E., Witteman G.J., Voskuil T., Hoylman A.M., Creson C., Kenney D.A., Huxel G.R., LaRue C.S. & Duncan J.R. (1993) The construction and assembly of an ecological landscape. *J. Anim. Ecol.* **62**, 117–130. [6.2] [6.3] [6.6]

Drake J.A., Flum T.E., & Huxel G.R. (1994) On defining assembly space: a reply to Grover and Lawton. *J. Anim. Ecol.* **63**, 488–489. [6.2] [6.3] [6.6]

Gilpin M.E., Carpenter M.P. & Pomerantz M.J. (1986) The assembly of laboratory communities: multispecies competition in *Drosophilia.* In: *Community Ecology* (eds J.M. Diamond & T.J. Case), pp. 23–40. Harper & Row, New York. [6.2]

Ginzburg L.R., Akçakaya H.R. & Kim J. (1988) Evolution of community structure: competition. *J. Theor. Biol.* **133**, 513–523. [6.3]

Gotelli N.J. & Simberloff D. (1987) The distribution and abundance of tallgrass prairie plants: a test of the core-satellite hypothesis. *Am. Nat.* **130**, 18–35. [6.3]

Grover J.P. (1994) Assembly rules for communities of nutrient-limited plants and specialist herbivores. *Am. Nat.* **143**, 258–282. [6.3]

Grover J. & Lawton J.H. (1994) Experimental studies on community convergence and alternative states: comments on a paper by Drake *et al. J. Anim. Ecol.* **63**, 484–487. [6.3] [6.6]

Hanski I. (1982) Dynamics of regional distribution: the core and satellite hypothesis. *Oikos* **38**, 210–221. [6.3]

Hanski I. (1991) Single-species metapopulation dynamics: concepts, models, and observations. In: *Metapopulation Dynamics* (eds M.E. Gilpin & I. Hanski), pp. 17–38. Academic Press, London. [6.3]

Harrison S. (1994) Metapopulations and conservation. In: *Large-scale Ecology and Conservation Biology* (eds P.J. Edwards, R.M. May, & N.R. Webb), pp. 111–128. Blackwell Scientific, Oxford. [6.2]

Harrison S. Do taxa persist as metapopulations in evolutionary time? In: *Biodiversity Dynamics Across Temporal Scales* (eds M.L. McKinney, C.L. Hewitt & J.A. Drake), in press. [6.2] [6.3]

Hastings H.M. & Sugihara G. (1993) *Fractals: a user's guide for the natural sciences.* Oxford University Press, New York. [6.3] [6.6]

Hewitt C.L., Huxel G.R. & Drake J.A. Colonization and extinction in a metacommunity assembly model. In: *Biodiversity Dynamics Across Temporal Scales* (eds M.L. McKinney, C.L. Hewitt & J.A. Drake), in press. [6.4]

Huston M. (1979) A general hypothesis of species diversity. *Am. Nat.* **113**, 81–101. [6.2]

Huston M. (1994) *Biological Diversity: the coexistence of species on changing landscapes.* Cambridge University Press. [6.2]

Jorgensen S.E. (1995) The growth rate of zooplankton at the edge of chaos, examined by ecological models. *J. Theor. Biol.* in press. [6.2] [6.4]

Kauffman S.A. (1991) Antichaos and adaptation. *Sci. Amer.* **265**, 78. [6.4]

Kauffman S.A. (1993) *The Origins of Order.* Oxford University Press. [6.2] [6.4]

Kolasa J. & Pickett S.T.A. (1989) Ecological systems and the concept of biological organization. *Proc. Natl. Acad. Sci, USA* **86**, 8837–8841. [6.3] [6.4]

Langton C. (ed.) (1994) *Artificial Life IV.* Addison-Wesley, Reading, Massachusetts. [6.2]

Levins R. (1969) Some demographic and genetic consequences of environmental heterogeneity for biological control. *Bull. Ent. Soc. Am.* **15**, 237–240. [6.3]

Lu E.T. & Hamilton R.J. (1991) Avalanches and the distribution of solar flares. *J. Astrophys.* **380 (L)**, 89–92. [6.4]

Luh H-K. & Pimm S.L. (1993) The assembly of ecological communities: a minimalist approach. *J. Anim. Ecol.* **62**, 749–765. [6.2] [6.3] [6.6]

Naeem S., Thompson L.J., Lawler S.P., Lawton J.H. & Woodfin R.M. (1994) Declining biodiversity can alter performance of ecosystems. *Nature* **368**, 734–737. [6.3]

Nee S. (1990) Community construction. *Trends Ecol. Evol.* **5**, 337–340.

Olami Z., Feder H.J.S. & Christensen K. (1992) Self organized criticality in a continuous nonconserved cellular automaton modelling earthquakes. *Phys. Rev. Lett.* **68**, 1244–1247. [6.4]

Paine R.T. & Levin S.A. (1981) Intertidal landscapes: disturbance and the dynamics of pattern. *Ecol. Monogr.* **51**, 145–178. [6.3]

Pickett S.T.A. & White P.S. (eds) (1985) *The Ecology of Natural Disturbance and Patch Dynamics.* Academic Press, New York. [6.2] [6.3]

Plotnick R.E. & McKinney M.L. (1993) Ecosystem organization and extinction dynamics. *Palaios* **8**, 202–212. [6.4] [6.5]

Raup D.M. (1986) Biological extinction in Earth history. *Science* **231**, 528–533. [6.4]

Ricklefs R.E. (1987) Community diversity: relative roles of local and regional processes. *Science* **235**, 167–171. [6.2]

Ricklefs R.E. & Schluter D. (eds) (1993) *Species Diversity in Ecological Communities.* University of Chicago Press. [6.2]

Robinson J.V. & Dickerson J.E. (1987) Does invasion sequence affect community structure. *Ecology* **68**, 587–595. [6.3] [6.6]

Robinson J.V. & Wellborn G.A. (1988) Ecological resistance to the invasion of a freshwater clam, *Corbicula fluminea*: fish predation effects. *Oecologia* **77**, 445–452. [6.2]

Roughgarden J. (1989) Community structure and assembly. In: *Perspectives in Ecological Theory* (eds J. Roughgarden, R.M. May & S.A. Levin), pp. 203–226. Princeton University Press, New Jersey. [6.2] [6.3]

Sale P.S. (1977) Maintenance of high diversity in coral reef fish communities. *Am. Nat.* **111**, 337–359. [6.3]

Simberloff D.S. (1974) Equilibrium theory of island biogeography and ecology. *Annu. Rev. Ecol. Syst.* **5**, 161–182. [6.3]

Strong D.R., Simberloff D., Abele L.G. & Thistle A.B. (eds) (1984) *Ecological Communities: conceptual issues and the evidence.* Princeton University Press, New Jersey. [6.2]

Sugihara G. (1985) Graph theory, homology and food webs. *Proc. Symp. Appl. Math.* **30**, 83–101. [6.3]

Tilman D. (1988) *Plant Strategies and the Dynamics and Structure of Plant Communities.* Princeton University Press, New Jersey. [6.3]

Underwood A.J. (1986) What is a community? In: *The Patterns and Processes in the History of Life* (eds D.M. Raup & D. Jablonski), pp. 351–368. Springer-Verlag, Berlin. [6.2]

Vitousek P.M. (1990) Biological invasions and ecosystem processes: towards an integration of population biology and ecosystem studies. *Oikos* **57**, 7–13. [6.3]

Waltho N. & Kolasa J. (1994) Organization of instabilities in multispecies systems: a test of hierarchy theory. *Proc. Natl. Acad. Sci., USA* **91**, 1682–1685. [6.3]

Whittaker R.H. (1972) Evolution and measurement of species diversity. *Taxonomy* **21**, 213–251. [6.3]

Wilbur H. & Alford R.A. (1985) Priority effects in experimental pond communities: responses of *Hyla* to *Bufo* and *Rana*. *Ecology* **66**, 1106–1114. [6.2] [6.3]

PART 2: PATTERNS IN BIODIVERSITY

7: Spatial and temporal patterns of genetic diversity within species

BRUNO BAUR and BERNHARD SCHMID*

7.1 Introduction

The Linnean concept conceived species as relatively constant with most of the variation occurring among them. Darwin's theory of evolution by gradual change, however, required that the variations among species must be generated from variation within species. Over geological times such divergence would have reached higher and higher taxonomic levels. During the first six decades of this century, heritable variation within species was found for many morphological characters. With the availability of new methods that could analyse variation in proteins (i.e. allozymes), overwhelming genetic variation even within populations was later detected, first for fruit flies by Hubby and Lewontin in 1966.

In spite of such large genetic variation that often does not seem to have any immediate function, species express relative constancy in overall appearance as exemplified by some phylogenetically old and today still abundant taxa such as the opossum *Didelphys virginiana* (>50 million years) or the giant redwood *Sequoiadendron giganteum* (>100 million years; Stebbins, 1977). Indeed, according to evolutionary theory, adaptation should generally lead to some optimal genetic composition of individuals of a single species. The question therefore is whether genetic variation simply results from imperfect adaptation in the face of disruptive random forces (mutation and drift) or whether there might be advantages for individuals to be genetically different from their neighbours within a population.

To address these questions we may look at species with low genetic variation. Some of them are very abundant and even invasive, for example aquatic weeds (Cook, 1986), or the grasses *Spartina anglica* (Thompson *et al.*, 1991) and *Bromus tectorum* (Novak *et al.*, 1991). Other species with low genetic variation are rare, such as *Luina serpentina*, a plant occurring in 19 populations over an area of 35 km² in Oregon (Alverson *et al.*, 1990) or the cheetah *Acinonyx jubatus* (O'Brien *et al.*, 1985; Caro & Laurenson, 1994). Overall, however, widespread plant species exhibit more polymorphic loci and alleles per polymorphic locus than do their geographically restricted

---

*Authors in alphabetical order.

congeners (Karron, 1987). It is assumed that low genetic variation is often correlated with reduced fitness, as for example indicated by large odontometric asymmetry in the cheetah (Kieser & Groeneveld, 1991), anatomical defects in the European lynx *Lynx lynx* (Bernhart & Zimmermann, pers. comm., 1994), or lower recruitment in genetically more uniform than less uniform populations of the plant *Trifolium hirtum* (Martins & Jain, 1979). For a more detailed discussion of interactions between fitness and genetic variation see Chapter 2.

This chapter deals with genetic variation within species. We first describe patterns of variation in space and time. Next we ask which factors are responsible for the currently observed rapid loss of genetic variation within and among populations in many species. Finally, we discuss potential strategies that could be employed to slow or prevent such reduction of genetic variation. This chapter is not intended as a comprehensive review of the literature on genetic diversity within species. It is an attempt to illustrate typical patterns with specific examples that can be interpreted with regard to possible processes.

7.2 Levels of genetic diversity

The genetic variation within species is hierarchically structured among intraspecific groups such as subspecies, varieties and populations (see Chapter 2). In essence, genetic diversity depends on the number and distinctness of genes within a species. For organisms without germ line (all plants and many clonal animals) it is important to recognize that the absolute number of nuclear genes within a group of individuals is not only a function of the number of families in the group and the number of individuals per family, but also of the number of individuals per clone (Jackson *et al.*, 1985; Buss, 1987), the number of cells per individual, the number of loci per cell and the number of alleles per locus (Table 7.1). That is, genetic diversity as measured by numbers and differences of genes can already be found within what is sometimes called a genetic individual or genet. For example, a single clone of the clover *Trifolium repens* usually covers many square metres (R. Sackville Hamilton, pers. comm., 1983) and consists of thousands of individual ramets with billions of cells each containing the diploid set of two alleles at each locus. Although at every locus many more than two different alleles may occur within a population, almost (but probably not absolutely) never more than two different ones can occur in a single clone (see Capossela *et al.*, 1992).

If variation occurs among alleles within loci and cells, then they are called heterozygous (cf. concept of heterozygosity discussed in Chapter 2). Such allelic variation can be larger in polyploid than in diploid cells. The distinctness among loci within cells can be large or small, depending on whether genes are repeated within the genome. In plants, 60% of the genome in typical cases consists of repeated sequences and the copy numbers can be as

Table 7.1 Levels of organization and their relation to the number of nuclear genes and differences among nuclear genes within species.

Structural level of organization	Number of structures (usual range)	Differences among randomly drawn genes	Examples
Subspecies	$1–10^2$ per species	Large, often geographic	Large number in apomictic groups of flowering plants, e.g. *Hieracium*
Varieties, races, morphs	$1–10^2$ per subspecies	Large, often ecological	30 morphs in lady beetle *Adalia bipunctata*
Populations	$1–10^6$ per variety	Large, spatial or temporal	Up to 14000 wildebeest *Connochaetes taurinus*
Subpopulations, demes, neighbourhoods	$1–10^6$ per population	Large, spatial or temporal	Neighbourhood size in land snails: 2800 in *Cepaea nemoralis* (Lamotte, 1951) and 5600 in *Arianta arbustorum* (Baur, 1993)
Families	$1–10^6$ per sub-population	Large, genealogical	50–800000 'effective females' per population in aquatic animals (mainly fish; Avise, 1992)
Individuals	$1–10^6$ per family	Large–small (within clones), genealogical	1–80 beetles per deme in *Bolitotherus cornutus* (Whitlock, 1992a)
Cells (nuclei)	$1–10^{20}$ per individual	Small, cell lineage	10^7 g of living cells in 100 m tall giant redwood tree (Bonner, 1988)
Loci (organized within linkage groups)	$10^4–10^9$ per cell	(Very) large, structural	$10^9–10^{10}$ nucleotide pairs ($10^2–10^3$ per locus) per plant genome (Scowcroft *et al.*, 1987)
Alleles	$1–10^1$ per locus	Large, heterozygosity	Up to 6 allozyme genes per locus in hexaploid plant cells

Notes The levels from subspecies to cells might be regarded as random factors in an analysis-of-variance approach to partitioning genetic variation. The second-last line would be considered a fixed factor for which no variance component would be estimated in such an analysis. The last line would be used as the comparative basis to test the significance against zero of the second-last variance component fitted above this line in the table.

high as 10^5–10^6 (Scowcroft *et al.*, 1987; Schaal *et al.*, 1991). The genetic variation among cells within individuals is of course very low, because in fact all cells, according to the cell-theory of life, contain all information to build an organism. As a consequence we would expect more active genes per cell, that is larger genomes, in more complex organisms (cf. Bonner, 1988).

The genetic variation within a species can be partitioned according to the hierarchy of structural levels presented in Table 7.1 (for techniques see Chapter 2). For quantitative genetic variation the preferred method of partitioning variation is a nested analysis-of-variance (ANOVA) approach (Shaw, 1987; see note to Table 7.1). A similar method (AMOVA, analysis of molecular variance) has recently been developed in which genetic variants cannot be related to alleles (RAPD technique; see Excoffier *et al.*, 1992; Huff *et al.*, 1993). Typically a very large amount of allelic variation and a smaller but still large amount of quantitative variation is contained within populations. This is illustrated in Fig. 7.1 by the example of humans and the annual plant *Phlox drummondii*.

7.3 Overview of evolutionary factors affecting genetic diversity

Given the observed genetic variation within populations and species, how can

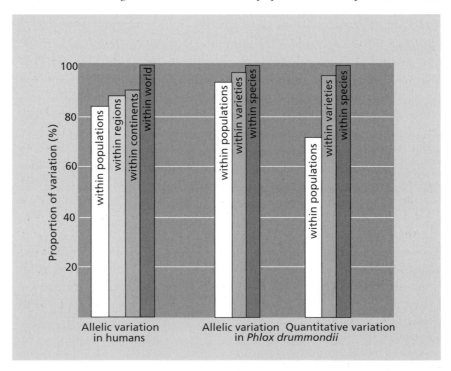

Fig. 7.1 Partitioning of genetic variation within humans *Homo sapiens* (Latter, 1980) and within the plant species *Phlox drummondii* (after Schwaegerle *et al.*, 1986).

we explain it? Ignoring the ultimate evolutionary causes, we may ask what the proximate mechanisms are that generate and maintain genetic variation. It is convenient to consider a population of haploid, asexually reproducing individuals, each of which accounts for one gene at a single locus in a gene pool.

7.3.1 Generation of genetic diversity

The primary mechanism for the generation of new genes at a single locus are point mutations occurring in cells giving rise to new individuals (Buss, 1987; Maynard Smith, 1989). Genetic variation among loci is further influenced by rearrangement of positions of genes within the genome (e.g. by chromosome mutations such as inversions or translocations; Stebbins, 1977), thereby creating new multi-locus gene combinations, and by recombination events if sexual reproduction occurs in the population. Basic mutation rates in the order of 10^{-9}–10^{-10} per base position in the DNA-molecules per cell cycle result in mutation rates of 10^{-6}–10^{-7} per locus (ca. 100–1000 bases) and in even higher rates for quantitative characters affected by genes at several loci (Maynard Smith, 1989). A typical recovery time from fixation to ordinary levels of genetic variation is 10^5–10^7 generations for characters affected by single loci and 10^2–10^3 for quantitative characters (Maynard Smith, 1989; Widén & Svensson, 1992). Because a single genome contains many genes (Table 7.1) the accumulation rate of mutant alleles can be as high as 0.5 per generation per individual in highly inbred plants (Charlesworth *et al.*, 1990). Blot and Arber (pers. comm., 1994) have recently demonstrated rapid generation of genetic variation by mutation in individual clones of *Escherichia coli* that had been cultivated by R. Lenski for 10000 generations.

There are several situations in which mutation rates are increased. The causes for these increases may come from the external environment (e.g. high temperature, chemicals, or high-energy radiation), or they may be internal (e.g. other genes (Bachmann & Low, 1980), genomic instability (particularly in hybridogeneous clones; Schmid, 1982; Kenton *et al.*, 1988), other presumably physiological instabilities (such as occur in cell cultures or in old seeds; Meins, 1983; Levin, 1990)). These latter examples, especially, raise the question of possible adaptive advantages of high mutation rates, although the mutations occur spontaneously and there is little indication that they can be directed (see Sarkar, 1992a; Meins & Seldran, 1994). It is important to note that much genetic variation may be generated by deleterious mutations of no use for the improvement of populations (see Houle, 1989 and references therein).

7.3.2 Maintenance of genetic diversity

If no further mutations were to occur in a population we might ask how a

173

certain level of genetic variation could be maintained. The three main factors affecting gene frequency in a breeding population are genetic drift, selection, and gene flow. Genetic drift can have large effects in small populations, where the probability of extinction for any mutant allele is particularly high (Spiess, 1989). Genetic diversity within and among local populations is primarily due to the dynamic balance between genetic drift (which causes the local population to lose genetic diversity but causes an increase in among-population differentiation) and gene flow (which brings new genetic diversity into the local population and reduces genetic differentiation among populations). Because qualitative characters such as isozymes often show genetic variation that can be interpreted in terms of population structures (Hamrick *et al.*, 1979; Loveless & Hamrick, 1984; Bazzaz & Sultan, 1986; Brown & Richardson, 1988; Brown & Schoen, 1992), these characters are presumed to be selectively neutral in many cases. Such allelic variation can be a bad predictor of genetic variation in characters of more direct relevance to fitness, in particular, quantitative morphological characters (Antonovics, 1984; Allendorf & Leary, 1986; Warwick *et al.*, 1987; Linhart *et al.*, 1989; Schaal *et al.*, 1991; Brown & Schoen, 1992; Schemske *et al.*, 1994).

For fitness-related characters, natural selection is an important factor influencing variation among and within groups (Endler, 1986). Selection can stabilize phenotypic characters such that the genetic variation within groups is reduced. Between groups, disruptive selection may lead to genetic divergence and therefore increase genetic variation among populations. Here the distinction between 'within-' and 'among-populations' becomes blurred because disruptive selection in itself can promote population division. Genetic differentiation is not correlated with population subdivision in all cases. Frequency-dependent selection can maintain genetic variation within populations in situations where minority genotypes have an adaptive advantage (e.g. if they can escape detection by pathogens or predators; Hamilton, 1980; Antonovics, 1984; Sarkar, 1992b). For example, individuals of the goldenrod *Solidago altissima* are less affected by the mildew fungus *Erisyphe chicoracearum* if they are surrounded by other individuals of different genotypes than of the same genotype (Schmid, 1994).

In summary, genetic variation within populations usually increases as a consequence of mutations and gene flow, or decreases because of random genetic drift and selection; whereas genetic variation among populations usually decreases as a consequence of mutations and gene flow or increases because of random genetic drift and selection. A qualitative example of this is shown in Fig. 7.2 (see Schmid, 1984, 1987).

There is generally large variation within populations, but much of this variation may be selectively neutral, especially the large allelic variation in molecular characters (DNA, isozymes). Fitness-related characters, to which

174

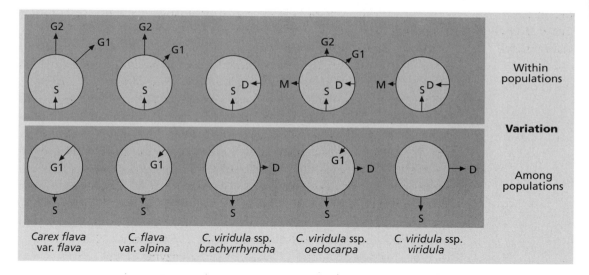

Fig. 7.2 Qualitative estimates of the relative magnitude of different factors increasing or decreasing genetic variation in five taxa of the sedge *Carex flava* s.l. (adapted from Schmid, 1980). M, mutation rate (i.e. generation length); G1, gene flow among populations; G2, gene flow among taxa; D, random genetic drift (i.e. population size); S, selection (i.e. mortality of clones).

probably the majority of the quantitative characters belong (see Chapter 2), can also vary within populations (Bradshaw, 1984a; Venable, 1984), but vary relatively more among populations than do characters less affected by selection (Schemske *et al.*, 1994). Even if genetic variation among populations appears to be small relative to variation within populations, its significance should not be underestimated. The null model without any structuring within species would always predict no variation among (arbitrary) populations and even a small sample may then contain most of the genetic variation (according to Brown (1989) 10% of an *ex situ* collection usually contains more than 70% of the alleles in the entire collection). A value of 20% among-population variation should be considered large, and may well be larger than genetic variation among species.

7.4 Spatial structure of genetic variation

Natural populations have structure in space as well as in time. The pattern of geographical subdivision and the extent of gene flow between local populations is important for the evolution and maintenance of genetic variability in any species. In this section, we first discuss the relationship between breeding systems and genetic diversity within species. Then we present models of genetic population structure and consider the processes that affect the spatial pattern of genetic variation within species. We present several examples of spatial genetic variation in natural populations.

7.4.1 Breeding system and genetic diversity

In the absence of other strong forces the genetic structures of animal and plant populations largely reflect the breeding system of species. Genetic variation is minimal among individuals within clones. In clonal organisms, genetic variation within a population is defined as the number of clones per individual. In invertebrates, clonal diversity often lies between 0.1 and 0.9 clones per individual (Hunter, 1993). In plants, clonal diversity is very variable and can range from <0.001 to 1 clone per individual (e.g. aquatic weeds, *Luina serpentina* and rhizomatous *Carex* species; Cook, 1986; Alverson *et al.*, 1990; McClintock & Waterway, 1993). The largest genetic component of variance is often due to differences among individuals (of different clones) within families; at higher levels progressively less additional variation may be found (see Fig. 7.1; Warwick & Briggs, 1979, 1980; Latter, 1980).

A lack of genetic variation (as assessed by enzyme electrophoresis) was found in several groups of invertebrates (e.g. Nevo *et al.*, 1984). For example, nine (15.8%) of 57 species of terrestrial gastropods surveyed completely lacked any variation in isozymes (Foltz *et al.*, 1984; Boato, 1988; Brown & Richardson, 1988; Baur & Klemm, 1989). The majority of these gastropods are frequent or obligate self-fertilizers.

In a comparative study, Apollonio and Hartl (1993) provide evidence that the isozyme variability in three families of mammals (Cervidae, Bovidae and Canidae) decreases as the degree of polygyny of the mating system increases (roving, territorial, harem-breeding and lek-breeding strategies). At the intra-specific level, however, Kurt *et al.* (1993) did not find any differences in the proportion of polymorphic loci and average heterozygosity between two groups of European roe deer *Capreolus capreolus* with different breeding strategies. Kurt *et al.* (1993) assumed that in forest-dwelling roe deer only territorial males are rutting, that is only about 20% of all males would contribute to the gene pool of the next generation and this might result in a lower genetic variability (and a certain degree of inbreeding) than in field and mountain-dwelling roe deer that were considered to be promiscuous. Similarly in plants, due to their large plasticity in size, a small proportion of individuals or clones may often produce the overwhelming majority of offspring (Silvertown & Lovett Doust, 1993).

Hamrick *et al.* (1979) and Loveless and Hamrick (1984) analysed correlations of life-history characteristics and seed dispersal with genetic population structure inferred from isozyme variation. In across-species comparisons, short generation length, animal pollination, low fecundity, limited seed dispersal and seed dormancy are correlated with restricted gene flow, low effective population size, small genetic variation within, and large genetic variation between (sub)populations. The breeding system, however, has the strongest correlation with genetic population structure in isozyme characters (Hamrick, 1982; Loveless & Hamrick, 1984). In a similar review, Nevo (1983) analysed correlations of demographic characters with allelic genetic variation across 33 animal species. He found that population size, fecundity, and niche width were positively related to genetic diversity.

Ledig (1986) reviewed the extensive literature on genetic variation in trees. Genetic diversity in conifers has been amply documented for seed isozymes. Tree species in general are highly outcrossing compared with herbaceous perennials and annuals (Schemske & Lande, 1985). As a result, heterozygosities in trees generally exceed those in herbaceous plants by twofold (Ledig, 1986). Nevertheless, exceptional tree species can be genetically depauperate. For example, no variation was detectable within populations of Torrey pine *Pinus torreyana* (Ledig, 1986).

In general, a lack of genetic variation is relatively uncommon in plants; only four (2.2%) out of 178 vascular plant species surveyed completely lacked genetic variation in isozymes (Hamrick *et al.*, 1979; Hamrick, 1983; Ellstrand & Roose, 1987; Lesica *et al.*, 1988). Outbreeding species generally have high variability within populations and low heterogeneity among populations, whereas self-fertilizing species often have low genetic variability within populations but significant differences among populations. *Howellia aquatilis* (Campanulaceae), a rare aquatic plant, is one of the species without isozyme variability (Lesica *et al.*, 1988). Anatomical studies of developing flowers indicated a restrictive breeding system approaching obligate self-fertilization. The small ecological niche (the plant occurs only in temporary ponds surrounded by trees) and the lack of genetic variability suggest that this species is prone to extinction (Lesica *et al.*, 1988).

Since other life-history characteristics and ecological factors can also be of significant importance, one should be cautious in only relating the genetic structure or levels of variability to the breeding system of species. Comparative studies often provide inconsistent evidence for most of the monocausal hypotheses attempting to explain allozyme variation by biological or ecological characteristics of the organisms (Baccus *et al.*, 1983; Apollonio & Hartl, 1993). For example, effects of the breeding system on the level of genetic variation might be confounded by differences in dispersal pattern. Experimental studies would have to be carried out to reveal the 'true' causal relationships.

7.4.2 Models of spatial structure

Wright (1943, 1946) developed different models to analyse the spatial structure in genetic variation of populations. In the island model the population is divided into groups (or demes) of randomly mating individuals that exchange a small fraction of migrants among them. In the 'stepping-stone' model the migrants always move to the next deme in line. In the continuous model there are no demes, but the dispersal distances are short in relation to the dimension of the population. Each model can be classified according to the number of spatial dimensions: there may be significant differentiation along linear structures (e.g. a roadside verge, the shore line of a lake, or a chain of islands), within a branching structure (e.g. a drainage system of an area (see Fig. 7.3)), or truly two-dimensional differentiation in all directions within an area (e.g. a large forest or lake). The two major parameters to be considered in the models are the number or density of breeding individuals and the dispersal distances of individuals and gametes. The distribution of these parameters within a species influences the potential patterns of population structure and therefore genetic variation.

Effective population size and neighbourhood size

In the island or stepping-stone models the genetically relevant size of a deme can be measured by the so-called effective population size (N_e). This is the size an ideal population would have if its genetic composition were influenced by random processes in the same way as the real population of size N (Wright, 1931). As a result of genetic drift, isolated populations lose variation at a rate

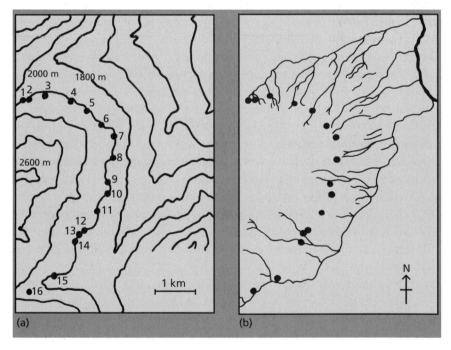

Fig. 7.3 Genetic differentiation in the land snail *Arianta arbustorum* on an alpine slope can be explained by a functional isolation-by-distance model that assumes gene flow over the drainage system: (a) sampling sites (1–16) situated along the contour line at an altitude of 2000 m on Mount Martegnas, Switzerland (snails occur from the valley bottom up to the peak of Mount Martegnas; contour interval 200 m); (b) drainage system. Genetic differentiation is correlated with the inter-population distance measured along streams (redrawn after Arter, 1990).

of $1/(2N_e)$ per generation. Nunney and Elam (1994) reviewed genetic and ecological methods that have been used to estimate N_e. Genetic estimates of N_e rely on frequency data from enzyme electrophoresis or molecular techniques. A general problem of genetic methods is the need to eliminate confounding effects of migration and population subdivision and the possibility that selection is acting at the marker loci or on loci linked to them. Ecological methods for estimations of N_e depend on theory linking demographic models and behavioural factors to changes in N_e (Nunney, 1993).

A number of factors may influence the relationship between population size and N_e: (i) fluctuations in population size, where the average N_e is determined by the harmonic mean; (ii) deviations from a 1:1 sex ratio; (iii) variation in reproductive success; (iv) mating system (e.g. polygyny; few males obtaining almost all matings); and (v) overlapping generations (Wright, 1938, 1969; Kimura & Crow, 1963; Nunney, 1993). N_e is almost always lower than the population number of adult individuals. For example, an N_e/N ratio of 0.59 was estimated for a grey squirrel *Sciurus carolinensis* population (Charlesworth, 1980), 0.59 for a geospizid finch *Geospiza scandens* population (Grant & Grant, 1992), 0.86 for a grizzly bear *Ursus arctos* population (Harris & Allendorf, 1989) and 0.78 for a moose *Alces alces* population (Ryman *et al.*, 1981). The N_e/N ratio could be increased above one, for example, in managed populations of endangered species if each family were only allowed to have a constant number of sons and daughters (see Simberloff, 1988).

In the continuum model the rate of dispersal is small compared with the area occupied by the entire population. Thus, isolation by distance will prevent panmixis and the effective population size will again be less than the actual number. Wright (1943, 1946) introduced the idea of a neighbourhood and defined it as an area from which the parents of central individuals may be treated as if drawn at random. Individuals are considered to be distributed at a uniform density either along a linear range, for example a roadside verge or a river bank, or throughout an area. The length or area of a neighbourhood depends upon σ^2, the variance of the parent–offspring dispersal distribution. If dispersal distances are normally distributed, a linear neighbourhood has the length $L = 2\sqrt{\pi}\sigma$ and is expected to contain about 92.4% of the parents of central individuals. If dispersal in a two-dimensional population is equal in all directions and the distance distribution follows a normal distribution, the neighbourhood is a circle of radius 2σ and has an area of $A = 4\pi\sigma^2$. The chances that the area contains a parent of a central individual are 86.5%. Neighbourhood size is the number of reproducing individuals that occur in the neighbourhood area ($N_b = 4\pi\sigma^2 d$, where d is the density of the breeding population).

Attempts to estimate the locally effective population size in plants have generally used Wright's model of genetic neighbourhood. Crawford (1984) extended this model by considering different components of dispersal in plants (pollen dispersal, seed dispersal). Crawford's model was further extended by Gliddon *et al.* (1987) who considered vegetative growth as a mechanism of gene dispersal (Handel, 1985) and examined populations of gynodioecious and dioecious species.

Crawford (1984) presented an extensive list of estimates of neighbourhood area and number for a variety of animals, trees, and herbaceous plants. In general, neighbourhood numbers in trees (range 4–253) and herbaceous plants (2–547) were smaller than in animals (151–25700). The neighbourhood size of plant species can be influenced by differences in dispersal capacity of different pollinators (e.g. bees versus butterflies).

Gene flow in demic and in continuous populations

Gene flow between randomly mating subpopulations or demes will tend to decrease genetic differences among and increase genetic variation within demes. The island and stepping-stone models (demic models) assume that there are distinct demes and that, in every generation, individuals or gametes move along demes according to the specification of the model. The rate of gene flow per generation from a source to a sink deme is quantified with the parameter *m*, which is the fraction of individuals or gametes that immigrate. *m* can be estimated directly from dispersal distances or indirectly from the gene frequencies in sink (q_1) and source demes (q_2) and the rate of gene frequency change in the source deme (Δq), i.e.:

$$m = \left(q_2 - q_1\right)\Big/\Delta q.$$

Another method was introduced by Slatkin (1981, 1985a,b; Slatkin & Barton, 1989), who showed by computer simulation that for rare alleles (p_i) that only occur in one deme, the following equation is little influenced by the spatial arrangement of demes (stepping-stone model) and by effects of selection:

$$\ln\left(\overline{p}\right) = -2.44 - 0.505 \cdot \ln\left(N_e m\right).$$

Often, it is more convenient to record instead of *m* the product $N_e m$, where N_e is the effective size of a deme in the island model. Population subdivision is frequently estimated by measuring F_{st} (for an introduction to F_{st} see Chapter 2). This model assumes, in addition to an island population, structure alleles that are selectively neutral and migrants that disperse randomly among all demes. It has been shown that estimates of $N_e m$ are robust to small deviations from complete neutrality and from the island model (Slatkin & Barton, 1989). The possible effects of other factors on estimates of the level of gene flow in indirect methods are discussed in Slatkin and Barton (1989). Demographic variation may invalidate the use of indirect methods for estimating gene flow (Whitlock, 1992b).

Slatkin (1985a) contrasted the direct methods to estimate gene flow which depend on observations of dispersing individuals or gametes with indirect methods that use isozyme or rare allele frequencies in samples from different populations. In direct methods the population structure (e.g. whether individuals occur in distinct demes or are more or less continuously distributed in space) and population sizes or densities must be known, while in indirect methods a detailed knowledge of population structure is not necessary. Allozymes have most frequently been used for this purpose, although there is some controversy whether natural selection on these characters can be strong and therefore distort the gene flow estimates (see below).

In some species both direct and indirect methods have been used to estimate gene flow. In *Gerris remigis*, a stream-dwelling waterstrider (Preziosi & Fairbairn, 1992), and in the terrestrial gastropod *Partula taeniata* (Murray & Clarke, 1984), the dispersal ability, as assessed by recording movements of marked individuals, agrees well with estimates of gene flow obtained by indirect methods. In other species, however, contradictory results from direct and indirect measures of gene flow were obtained (e.g. in the house mouse *Mus musculus* (Baker, 1981) and in carabid beetles (Liebherr, 1986)).

Gene flow in higher plants is accomplished by dispersal of seeds and pollen as well as by vegetative mobility (clonal growth, dispersal of plant fragments; Handel, 1985; Schmid, 1990; Parker & Hamrick, 1992). Gene flow by pollen dispersal is often less than 6 m in herbaceous plants (Widén & Svensson, 1992) and a distance of 2 km is usually sufficient to keep varieties isolated in plant breeding (Levin, 1984). Golenberg (1987) estimated rates of gene flow and neighbourhood area in wild emmer wheat *Triticum dicoccoides*, using a spatial hierarchical sampling design (seeds were collected 5–7 m, 15–17 m, and 100–120 m apart) and two types of indirect estimates (gene frequencies from allozymes and rare alleles). Both methods yielded similar $N_e m$ values. Using Slatkin's rare allele method the estimates of $N_e m$ were 4.68, 0.11, and 0.67 for the closest, intermediate, and most distant distance classes, respectively. The large gene flow value and the sharp decrease in gene flow beyond the first distance class (5–7 m) indicate that a genetic neighbourhood for this species may cover an area defined by a five metre radius. The results also suggest that gene flow between populations separated by ≥10 m may be quite limited. The bracken fern *Pteridium aquilinum* has a cosmopolitan distribution and is a noxious weed in many areas. Wolf *et al.* (1991) found that gene flow among British populations was extremely high ($N_e m = 36.51$), one of the highest estimates reported for plants. *F*-statistics based on allele frequencies of allozymes indicated that there is little genetic differentiation among seven British populations of bracken.

Unsuitable areas (lakes, rivers, desert, mountain ridges) can interrupt gene flow. Populations of the fence lizard *Sceloporus undulatus* separated by rivers were more differentiated than populations not separated by rivers

(Pounds & Jackson, 1981). Artificial habitats such as buildings, roads, railways, and agricultural areas can also act as partial or complete barriers to gene flow, e.g. for arthropods (Mader, 1984) or land snails (Selander & Kaufman, 1975; Baur & Baur, 1990). Furthermore, gene flow can be limited to particular types of habitats. On mountain slopes in the Alps, streams serve as gene flow paths for the land snail *Arianta arbustorum* (Arter, 1990). Adult snails moved on average 8 m per year along a stream (Baur, 1986). The spatial structure of allozyme variation could be best explained by a functional isolation-by-distance model (Arter, 1990), assuming gene flow along the drainage system (Fig. 7.3).

7.4.3 Spatial genetic structure in natural populations

The genetic structure of natural populations is mainly influenced by the combined effects of random genetic drift, restricted gene flow, and differential selection pressures. These effects lead to low within- and high among-population genetic variation in species consisting of small and isolated populations (Nevo, 1983; Waller *et al.*, 1987; Ellstrand & Elam, 1993; Holderegger & Schneller, 1994). In species consisting of large or less isolated populations, most of the genetic variation should occur within populations if drift and gene flow are the only important factors affecting gene frequencies. Especially in plants this often seems to be the case (Hamrick *et al.*, 1979; Loveless & Hamrick, 1984; Godt & Hamrick, 1993; Oyama, 1993).

To examine the spatial structure of natural populations, the variance in gene frequency (F_{st}) can be split into hierarchical components as a function of different distance scales (as determined by the sampling design): within localities, between localities within regions, between regions. For example, F statistics based on allozyme frequencies indicate that the extent of a local panmictic population of the tobacco budworm *Heliothis virescens* in the United States has an average diameter of 8 km or less (Korman *et al.*, 1993). More informative than F-statistics are spatial autocorrelation statistics (Epperson, 1990; Heywood, 1991; Renshaw, 1991). However, they have been relatively rarely used for the analysis of small-scale genetic variation. Spatial correlograms of isozyme data revealed moderate population structuring in *Drosophila buzzatii* and suggested that selection operated on scales ranging from local to continental (Sokal *et al.*, 1987). Spatial autocorrelations have also been criticized because the correlograms cannot be related to models of population structure and parameters like σ (Slatkin & Arter, 1991).

Small-scale genetic variation

Most studied plant species show microspatial genetic structure but also substantial gene flow among demes or neighbourhoods (Heywood, 1991).

182

Argyres and Schmitt (1991) described microgeographic variation in quantitative genetic characters within a 40m × 40m area in the plant *Impatiens capensis*. In herbaceous perennials the scale of genetic variation often depends on clonal architecture and whether different clones within a population intermingle or not. Spreading growth forms generally lead to polyclonal patches (e.g. in the bracken fern *Pteridium aquilinum*; Parks & Werth, 1993), whereas compact growth forms lead to monoclonal patches (e.g. in the grass *Spartina alterniflora*; Daehler & Strong, 1994).

Large and temporally constant heterogeneity in selection pressures can lead to genetic differences over very short distances even in the face of some gene flow (Levin, 1988; for examples see McNeilly, 1968; Antonovics *et al.*, 1971; Bradshaw, 1984a,b; Schmid, 1985). Local specialization of white clover *Trifolium repens* to specific grass neighbours (Turkington & Harper, 1979) and of pea aphids *Acyrthosiphon pisum* to specific 'home crops' (Via, 1991) demonstrate the importance of biotic interactions in maintaining genetic variation within species. Several studies provided evidence that the spatial patterns of colour morph frequencies in the land snail *Cepaea nemoralis* are a result of natural selection, and often change over distances of less than 100m (Cain & Sheppard, 1952; Cain & Currey, 1963; Wolda, 1969; Jones *et al.*, 1977; Clarke *et al.*, 1978).

Not all small-scale variation necessarily reflects natural selection processes. Genetic differentiation among populations in plants may often be due to divergent random genetic drift (Levin, 1988; see also Schmid, 1984). Especially in plants, species can often 'avoid' short-term adaptive genetic differentiation by being phenotypically plastic (Sultan, 1987; for examples see Antonovics & Primack, 1982; Schmid, 1985; Shaw, 1991).

Large-scale genetic variation

Ecotypic differentiation often occurs over longer distances (Clausen *et al.*, 1940). If the same selective forces operate in different parts of a species' range, 'polyphyletic' ecotypes may evolve, as was shown by the work of Turesson (1922). In many species, populations at the margin of the geographic distribution are genetically less diverse than central populations (see Brussard, 1984 for a review). For example, Eckert and Barrett (1993) found a reduced clonal diversity in marginal populations of the plant *Decodon verticillatus*.

If selection pressures change continuously with distance then clinal variation may evolve. Gregor (1930, 1938) described clines in coastal plants along decreasing salinity gradients (e.g. increasing leaf and seed size in *Plantago maritima*). Several lines of evidence suggest that natural selection and, to a minor extent, historical events shaped the geographical distribution of inversion frequencies in *Drosophila subobscura* (Menozzi & Krimbas, 1992). In Europe, the frequencies of several chromosomal arrangements in *D.*

183

subobscura are correlated with latitude (Menozzi & Krimbas, 1992). Recently, *D. subobscura* colonized South and North America. Three years after the species was detected in South America, and 4 years after its detection in North America, latitudinal clines were already established in chromosomal-arrangement frequencies (Prevosti *et al.*, 1985, 1988). The similarity of these clines in both American colonized areas and the Old World provides strong evidence for the adaptive nature of this polymorphism with respect to environmental factors which change with latitude (Prevosti *et al.*, 1990). A similar situation occurs in two goldenrod species introduced to Europe from North America about 300 years ago and now occurring over a latitudinal gradient from Italy to Scandinavia, which express continuous genetic variation in life-history characters such as age and size at maturity (Fig. 7.4; Weber, 1994).

Genetic variation without spatial structure

If genetic variation within a population cannot be related to patterns of spatial differentiation, we may ask why such polymorphism is not being removed from the population by stabilizing or directional selection, since all individuals have to deal with the same general environment. The classic view held that there was a mutation–selection balance, but after the detection of large allelic variation this was replaced by the view that much of this variation is non-adaptive in the short term (see Section 7.1). Neutralists argue that the variability that characterizes most organisms is merely a reflection of stochastic processes, i.e. mutation and drift in gene frequencies (Kimura & Ohta, 1971). However, there are several situations in which one can also envisage an adaptive advantage of genetic variation *per se*.

First, if there are limits or costs of plasticity, a single genotype may not be able to occupy the entire niche of the species. Genetic variation within populations would then be expected to be large in habitats with spatial and temporal variation in environmental conditions. This is more likely to occur with regard to coevolving biotic factors such as pathogens than with regard to abiotic factors such as soil conditions (Antonovics, 1984). The coevolution of genetic variation within populations of interacting organisms may be the most important factor in maintaining adaptive genetic variation within populations (Sarkar, 1992b; see e.g. Turkington & Harper, 1979; Dirzo & Harper, 1982; Via, 1991; Schmid, 1994).

Second, the sib-competition hypothesis (Waller, 1980; Willson *et al.*, 1987) suggests that genetically variable offspring differ slightly in their niches and therefore show reduced resource competition (genetically identical offspring with identical resource requirements should compete more strongly). The evidence for this hypothesis seems weak and contrasting cases, where the most abundant genotypes also have very similar resource require-

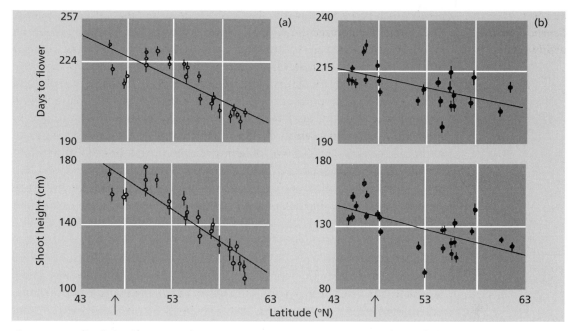

Fig. 7.4 Latitudinal clinal variation in two introduced species of goldenrod in Europe when grown in the common environment of an experimental garden (the plants were raised from cuttings for 2 consecutive years and then measured): (a) *Solidago altissima*, (b) *S. gigantea*. Population means ± standard error are represented together with fitted regression lines ($P < 0.05$), the arrows below the X-axis indicate the latitude of the experimental garden (from Weber & Schmid, unpublished, 1994).

ments, have been described (e.g. in the earthworm *Octolasion tyrtaeum*; Jaenike *et al.*, 1980).

These two situations correspond to the case of frequency-dependent selection mentioned in Section 7.3. A third explanation is provided by the balance theory of the 'selectionists' school. According to this view, genetic diversity is maintained by natural selection that favours variation within individuals, i.e. heterozygosity (see Table 7.1). If one allele at a locus is negatively associated the other one may still be positively associated with fitness in some instances, and vice versa at others (Spiess, 1989). Unfortunately, it is difficult to show balance without experimentally perturbing it (Antonovics, 1984) and a potentially large segregation load must be explained (the recombination of lethal recessive alleles produces offspring that die before they reach the reproductive stage). Segregation load may, for example, be detected if plants are cultivated from seeds or cuttings in a non-selective, controlled environment. If there is a segregation load due to characters exposed to selection during or after seed germination in the wild, then in culture the plants propagated from seeds should show more genetic variation than the vegetatively propagated plants, because these would already have suffered natural selection. We know of only one experiment that has found such an effect (Warwick & Briggs, 1978; see also Schmid & Bazzaz, 1990). In a similar way to heterozygosity, negative genetic correlations among quantitative characters (e.g. Sultan & Bazzaz, 1993) may also maintain genetic variation in fitness components, without affecting the combined fitness value of the characters.

In spite of these problems there are good examples of the advantages of heterozygosity at loci affecting phenotypic fitness components. In animals, individuals heterozygous at a particular locus can be larger, grow faster, live longer, be metabolically more efficient, have more offspring, or be more resistant to diseases than individuals homozygous at that locus. Individuals with more heterozygous loci are often superior to individuals with fewer heterozygous loci (for reviews see Allendorf & Leary, 1986; Zouros & Foltz, 1987). Heterozygosity can also be positively associated with developmental homeostasis; individuals with more heterozygous loci may actually show less non-adaptive phenotypic variability than individuals with fewer heterozygous loci (see Van Valen, 1962; Kieser & Groeneveld, 1991). There are actually only few cases in which no positive (e.g. Koehn *et al.*, 1988; Booth *et al.*, 1990) or even negative relationships (Allendorf & Leary, 1986) were found between heterozygosity and fitness in animals. For example, Booth *et al.* (1990) found no significant relationships between the number of heterozygous loci in individuals and measures of the same individual's shell morphology and phenotypic variability in a West Indian land snail *Cerion bendalli*. High segregation load due to heterozygosity in the palaeoendemic shrub *Dedeckera eurekensis* may actually lead to the extinction of this rare species (Wiens *et al.*, 1989). The evidence for adaptive advantages of heterozygosity in populations of wild plants is very scant (Schemske *et al.*, 1994).

Although 'neutralists' and 'selectionists' have different views concerning the short-term significance of most genetic variation, both schools consider that the evolutionary potential of a species depends on the presence of genetic variation, and that the long-term value of genetic diversity is most likely proportional to its amount (Lewontin, 1974).

7.5 Temporal variation in genetic diversity

Studies of genetic change over time in natural populations can give many insights into the processes of evolution (e.g. industrial melanism in the moth *Biston betularia*; Kettlewell, 1973). However, in many cases the results of observational studies are of limited explanatory power and are difficult to interpret.

7.5.1 Short-term variation in genetic diversity

Seasonal variations in population structures and selection pressures can induce cyclic changes in the genetic composition of populations. In a classical study of *Drosophila pseudoobscura*, Dobzhansky (1943, 1947) found that a cycle of the two gene arrangements ST versus CH regularly took place at two warm localities, leading him to the assumption that these changes must be of adaptive significance. Carvalho and Crisp (1987) found that a population of *Daphnia magna* in a Welsh lake had a much higher clonal diversity in summer, when food variety presumably was greatest, than in winter. The cyclic temporal changes may continually shift competitive advantages of individual clones, thereby maintaining genetic diversity in the long run (Carvalho & Crisp, 1987).

Studies on changes in morph frequencies in the land snails *Cepaea nemoralis* and *C. hortensis* range over periods from a few to 60 years (see review by Cameron, 1992). Most of the studies found little temporal variation in morph frequencies (Wolda, 1969; Williamson *et al.*, 1977; Cain & Cook, 1989; Cain *et al.*, 1990). A gradual change in morph frequencies resulting from gene flow was observed in one study (Goodhart, 1973), while in two studies systematic changes in morph frequencies were observed in several populations (Clarke & Murray, 1962; Murray & Clarke, 1978; Wall *et al.*, 1980). Cameron (1992) examined changes in morph frequencies in 71 *C. hortensis* populations over 25 years, corresponding to approximately 10 generations. Some populations were stable in shell colour and banding patterns. However, populations in valley bottoms showed a consistent and significant decline in the frequency of yellow shells. Cameron (1992) interpreted this change in morph frequency as a possible case of climatic selection.

In forked fungus beetles *Bolitotherus cornutus* the effects of extinction and recolonization increased genetic variation among young demes compared with that among old demes (Whitlock, 1992a). In plant populations, short-term changes in genetic variation are often correlated with successional phases of colonization, establishment, and stand maturity. Many herbaceous perennials, but also annuals or trees, may start off with a high genetic variation in the initial seed population, that is then continually being depleted by thinning processes (Turkington, 1985; Schmid, 1990; Silvertown & Lovett Doust, 1993; Watkinson & Powell, 1993). It is not clear, however, to what extent these thinning processes are random or reflect differential survival of genotypes. Thinning may continue for a long time, even after a closed vegetation has developed in plants with extensive clonal growth (Schmid & Harper, 1985).

If genet mortality risk after initial establishment declines rapidly, then a comparatively high level of genetic variation may be retained in a plant population (Pleasants & Wendel, 1989). Genetic diversity in a now asexual

187

population of the dwarf birch *Betula glandulosa* may be residual from a once sexually reproducing population (Hermanutz *et al.*, 1989) and the same explanation has been suggested for high levels of genetic diversity within and among populations of the moss *Sphagnum pulchrum* (Daniels, 1982). Genetic variation (e.g. between colonization events) can be 'stored' in seed, bulb, and tuber banks that may buffer plant populations against dramatic changes in genetic composition (Ellstrand & Elam, 1993). These banks can contain dormant plant propagules of different age and reflect genetic variation over time. For example, genetic variation in the seed bank of *Luzula parviflora* may result from random drift from year to year (Bennington *et al.*, 1991).

7.5.2 Long-term variation in genetic diversity

Land snails again play an important role in long-term studies on genetic variation (Cain, 1983). Using subfossil samples, Currey and Cain (1968) and Cain (1971) showed a major change in morph frequencies in *Cepaea nemoralis* over the last 6000 years in southern England. The change in morph frequency was associated with known changes in climate, from unbanded to five-banded shells in valleys and lowlands, and, to a minor extent, to banded shells on uplands. During the same period populations of *C. hortensis* changed the proportions of unbanded and banded morphs, but the species also expanded its area of occurrence (Cain, 1971).

7.6 Loss of genetic diversity

Habitat destruction, fragmentation, and environmental stresses such as air pollution and increasing temperatures limit or reduce the size of animal and plant populations (see, e.g. Parsons, 1992). For over a decade, much attention has focused on the potential genetic risks associated with small population size, particularly from inbreeding and genetic drift, but also from gene flow (Allendorf, 1983; Schonewald-Cox *et al.*, 1983; Templeton, 1986; Simberloff, 1988; Thornhill, 1993). The theoretical risks are often straightforward extensions of population genetics theory, but relevant data are scarce and sometimes conflicting. Furthermore, the relative importance of genetics in conservation efforts is questioned by some, on the grounds that ecological factors may be more important (Lande, 1988; Schemske *et al.*, 1994).

In small populations (<100 individuals), allele frequencies may undergo large and unpredictable fluctuations due to random genetic drift. In large populations, changes in allele frequency due to drift are generally small. As pointed out above, drift decreases the genetic variation within populations (loss of heterozygosity and eventual fixation of alleles) and increases the differentiation among populations. The effects of drift are more pronounced in decreasing populations than in increasing populations of the same size.

Wright (1931) predicted that drift will substantially alter the organization of genetic variation of populations when $1/(4N_e)$ is much greater than the mutation rate and the selection coefficients. The loss of genetic variation may significantly limit the adaptability of populations in changing environments in future generations.

Bijlsma *et al.* (1991) examined the amount of genetic variation (allozymes) in relation to the population size of two plant species, *Salvia pratensis* and *Scabiosa columbaria*, in the Netherlands. For both species, Bijlsma *et al.* (1991) found significant correlations between population size and both the proportion of polymorphic loci and the mean observed number of alleles. Large populations were genetically more variable than small populations. In addition, substantial genetic differentiation was observed between populations. The differentiation was more pronounced among small populations than among large populations. The results indicated that small populations were depauperate in genetic variation. Similarly, Billington (1991) found associations between levels of genetic variation (expected heterozygosity, percentage of polymorphic loci, mean number of alleles) and population size in the dioecious conifer *Halocarpus bidwillii*. The relationship was non-linear and the variation may reach an asymptote at population sizes above approximately 8000 individuals. Below this population size all three measures of genetic variation sharply decreased.

Habitat fragmentation is generally considered to be harmful to animals and plants. For example, specialized plant-pollinator systems are sensitive to disturbances of any kind, and the pollinator as well as the plant decrease if either of the two changes in abundance (Janzen, 1974). Jennersten (1988) examined the effects of habitat fragmentation on the pollination success of a butterfly-pollinated herb, the maiden pink *Dianthus deltoides*. *Dianthus* flowers received fewer visits in the fragmented area than in the mainland area, and the seed set was much lower in the fragmented area. In a fragmented area with low pollinator visitation *D. deltoides* is more likely to produce selfed seeds which may further decrease the genetic variation of small isolated populations (Jennersten, 1988).

A metapopulation is an extreme form of spatial structure in which loosely coupled local populations suffer extinction followed by recolonization from elsewhere within the metapopulation (Gilpin, 1991). Habitat fragmentation, which can create a metapopulation from a formerly continuously distributed species, may have unappreciated large genetic consequences for species impacted by human development. Gilpin (1991) showed with a simulation study that a metapopulation could lose a considerable amount of genetic variation without going through bottlenecks.

In small populations, individuals are more likely to be related to one another and consequently inbreeding may occur. Due to the fixation of deleterious alleles, inbreeding often results in a significant loss in fitness (Charlesworth & Charlesworth, 1987; Thornhill, 1993; Templeton, 1994).

A decrease in fitness as a result of inbreeding in captive animals is a commonplace observation (Falconer, 1989; Lacy *et al.*, 1993). A long-term study of a song sparrow *Melospiza melodia* population demonstrated deleterious effects of inbreeding on survivorship in natural environments (Keller *et al.*, 1994). The population on Mandarte Island, British Columbia, fluctuated from 10 to 140 breeding individuals. The wide variation in population size was mainly due to two population crashes associated with severe winter conditions. During the second crash, the more distantly related the parents of the birds were, the higher was their probability of survival, resulting in an immediately post-crash population of less inbred individuals. Levin (1988) estimated inbreeding coefficients within neighbourhoods at equilibrium (with gene flow $m = 0.01$) for several plant species. Whereas very little inbreeding (0.037) was expected in *Liatris cylindracea, Primula veris* had a value of 0.772.

Genetic bottlenecks lead to a loss of rare alleles and thus to a decrease in the proportion of polymorphic loci. This is especially problematic if population recovery after the bottleneck is slow (Burgman *et al.*, 1993). A remarkable genetic homogeneity in allozymes was found in the northern elephant seal *Mirounga angustirostrus* (Bonnell & Selander, 1974). A recent survey confirmed the lack of allozyme diversity and showed that even the mtDNA sequence variation was very low in this species (Hoelzel *et al.*, 1993). The northern elephant seal was heavily exploited during the 19th century and it experienced an extreme population bottleneck. Using a simulation model, Hoelzel *et al.* (1993) showed that the population was most likely less than 30 individuals during 20 years or, if hunting was the primary pressure on the population, a single-year bottleneck of less than 20 individuals.

7.7 Conservation and management of genetic diversity

Although the spatial structure of genetic variation is complex and interesting, our understanding is too preliminary to use it as a basis for manipulating populations for conservation (Boyce, 1992; Schemske *et al.*, 1994). Attempts to manage species by transplanting individuals, seedlings or pollen between local populations might be an effective tool to maintain or increase genetic variation within populations, but may decrease variation among populations (Dobson *et al.*, 1991).

Ex situ conservation of wild plant species through seed banks has been suggested as a conservation strategy to help preserve the biological and genetic diversity of wild plants (Brown & Briggs, 1991; Hamrick *et al.*, 1991). However, this method halts adaptive evolution and does not prevent genetic erosion (Marshall, 1990). Further, the exclusive use of allelic models and measures of genetic diversity ignores several important issues in evolutionary genetics, including quantitative genetic variation, genetic correlations and genotype–environment interactions (Hamilton, 1994). Hamilton argued that *ex situ* collections may be ineffective at preserving genetic diversity and the evolutionary potential of populations for adaptive or neutral evolution. Different populations will respond differently to the action of natural selection and are therefore unique evolutionary entities. Careful consideration of these factors may show that the successful preservation of genetic diversity is extremely difficult.

The optimal breeding strategies for *ex situ* conservation of rare species will depend on the specific biology of the taxa involved (Simberloff, 1988; Burgman *et al.*, 1993), especially on their mating systems and breeding history (e.g. inbreeding). Whereas in some cases a maximum-inbreeding avoidance strategy may be optimal (Thornhill, 1993), temporary population subdivision may be so in other cases because lethal alleles may be purged, bottlenecks may release additive genetic variation 'caught' in dominance variation, different alleles may be preserved in different groups, and most mutants can be accumulated in small and growing (sub)populations (Simberloff, 1988; Spiess, 1989; Barrett & Charlesworth, 1991; Burgman *et al.*, 1993).

Although Caro and Laurenson (1994) have recently questioned the importance of genetic variation as a factor decreasing the risk of short-term population extinction, the cases of increased susceptibility to pathogens in natural populations that are low in genetic variation indicate the importance of genetic variation (O'Brien *et al.*, 1985). There is growing evidence that individual genetic variation (i.e. heterozygosity) is positively associated with components of fitness in outbreeding species (Allendorf & Leary, 1986; Burgman *et al.*, 1993). Further, the evolutionary potential of any species might depend upon the amount of genetic variation it contains (Allendorf & Leary, 1986). Once genetic variation is lost in all populations of a species it can only be replaced by the process of mutation (see above), which is particularly slow in declining populations (Spiess, 1989). Genetic diversity must be preserved in order to increase both the probability of the short- and long-term survival of existing species and the potential for continued evolution of new species.

Acknowledgments

We thank K.J. Gaston and J. Mallet for comments on the manuscript. Our research is supported by the Priority Programme Environment (Module Biodiversity) of the Swiss National Science Foundation (grants 5001-035229 to B.S. and 5001-035241 to B.B.).

References

Allendorf F.W. (1983) Isolation, gene flow, and genetic differentiation among populations. In: *Genetics and Conservation* (eds C.M. Schonewald-Cox, S.M. Chambers, B. MacBryde & L. Thomas), pp. 51–65. Benjamin/Cummings, Menlo Park, California. [7.6]

Allendorf F.W. & Leary R.F. (1986) Heterozygosity and fitness in natural populations of animals. In: *Conservation Biology: the science of scarcity and diversity* (ed. M.E. Soulé), pp. 57–76. Sinauer Associates, Sunderland, Massachusetts. [7.3.2] [7.4.3] [7.6] [7.7]

Alverson E.R., Meinke R.J. & Ranker T.A. (1990) Reproductive biology and allozyme variation in *Luina serpentina*, a clonal endemic plant. *Bull. Ecol. Soc. Am.* **71**, 76. [7.1] [7.4.1]

Antonovics J. (1984) Genetic variation within populations. In: *Perspectives on Plant Population Ecology* (eds R. Dirzo & J. Sarukhán), pp. 229–241. Sinauer Associates, Sunderland, Massachusetts. [7.3.2] [7.4.3]

Antonovics J. & Primack R.B. (1982) Experimental ecological genetics in *Plantago*. IV. The demography of seedling transplants of *P. lanceolata*. *J. Ecol.* **70**, 55–75. [7.4.3]

Antonovics J., Bradshaw A.D. & Turner R.G. (1971) Heavy metal tolerance in plants. *Adv. Ecol. Res.* **7**, 1–85. [7.4.3]

Apollonio M. & Hartl G.B. (1993) Are biochemical-genetic variation and mating systems related in large mammals? *Acta Theriol.* **38** Suppl. 2, 175–185. [7.4.1]

Argyres A.Z. & Schmitt J. (1991) Microgeographic genetic structure of morphological and life history traits in a natural population of *Impatiens capensis*. *Evolution* **45**, 178–189. [7.4.3]

Arter H.E. (1990) Spatial relationship and gene flow paths between populations of the alpine snail *Arianta arbustorum* (Pulmonata: Helicidae). *Evolution* **44**, 966–980. [7.4.2]

Avise J.C. (1992) Molecular population structure and the biogeographic history of a regional fauna: a case history with lessons for conservation biology. *Oikos* **63**, 62–76. [7.2]

Baccus R., Ryman N., Smith M.H., Reuterwall C. & Cameron D. (1983) Genetic variability and differentiation of large grazing mammals. *J. Mammalogy* **64**, 109–120. [7.4.1]

Bachmann B.J. & Low K.B. (1980) Linkage map of *Escherichia coli* K-12, edition 6. *Microbiol. Rev.* **44**, 1–56. [7.3.1]

Baker A.E.M. (1981) Gene flow in house mice: introduction of a new allele into free-living populations. *Evolution* **35**, 243–258. [7.4.2]

Barrett S.C.H. & Charlesworth D. (1991) Effects of a change in the level of inbreeding on the genetic load. *Nature* **352**, 522–524. [7.7]

Baur A. & Baur B. (1990) Are roads barriers to dispersal in the land snail *Arianta arbustorum*? *Can. J. Zool.* **68**, 613–617. [7.4.2]

Baur B. (1986) Patterns of dispersion, density and dispersal in alpine populations of the land snail *Arianta arbustorum* (L.) (Helicidae). *Holarct. Ecol.* **9**, 117–125. [7.4.2]

Baur B. (1993) Population structure, density, dispersal and neighbourhood size in *Arianta arbustorum* (Linnaeus, 1758) (Pulmonata: Helicidae). *Ann. Naturhist. Mus. Wien* **94/95B**, 307–321. [7.2]

Baur B. & Klemm M. (1989) Absence of isozyme variation in geographically isolated populations of the land snail *Chondrina clienta*. *Heredity* **63**, 239–244. [7.4.1]

Bazzaz F.A. & Sultan S.E. (1986) Ecological variation and the maintenance of plant diversity. In: *Differentiation Patterns in Higher Plants* (ed. K. Urbanska), pp. 69–93. Academic Press, London. [7.3.2]

Bennington C.C., McGraw J.B. & Vavrek C. (1991) Ecological genetic variation in seed banks. II. Phenotypic and genetic differences between young and old subpopulations of *Luzula parviflora*. *J. Ecol.* **79**, 627–643. [7.5.1]

Bijlsma R., Ouborg N.J. & van Treuren R. (1991) Genetic and phenotypic variation in relation to population size in two plant species: *Salvia pratensis* and *Scabiosa columbaria*. In:

Species Conservation: a population-biological approach (eds A. Seitz & V. Loeschcke), pp. 89–101. Birkhäuser Verlag, Basel. [7.6]

Billington H.L. (1991) Effect of population size on genetic variation in a dioecious conifer. *Conserv. Biol.* **5**, 115–119. [7.6]

Boato A. (1988) Microevolution in *Solatopupa* land snails (Pulmonata, Chondrinidae): genetic diversity and founder effects. *Biol. J. Linn. Soc.* **34**, 327–348. [7.4.1]

Bonnell M.L. & Selander R.K. (1974) Elephant seals: genetic variation and near extinction. *Science* **184**, 908–909. [7.6]

Bonner J.T. (1988) *The Evolution of Complexity.* Princeton University Press, New Jersey. [7.2]

Booth C.L., Woodruff D.S. & Gould S.J. (1990) Lack of significant associations between allozyme heterozygosity and phenotypic traits in the land snail *Cerion. Evolution* **44**, 210–213. [7.4.3]

Boyce M.S. (1992) Population viability analysis. *Annu. Rev. Ecol. Syst.* **23**, 481–506. [7.7]

Bradshaw A.D. (1984a) Ecological significance of genetic variation between populations. In: *Perspectives on Plant Population Ecology* (eds R. Dirzo & J. Sarukhán), pp. 213–228. Sinauer Associates, Sunderland, Massachusetts. [7.3.2] [7.4.3]

Bradshaw A.D. (1984b) The importance of evolutionary ideas in ecology and vice versa. In: *Evolutionary Ecology* (ed. B. Shorrocks), pp. 1–25. Blackwell Scientific Publications, Oxford. [7.4.3]

Brown A.D.H. (1989) The case for core collections. In: *The Use of Plant Genetic Resources* (eds A.H.D. Brown, O.H. Frankel, D.R. Marshall & J.T. Williams), pp. 136–156. Cambridge University Press. [7.3.2]

Brown A.D.H. & Briggs J.D. (1991) Sampling strategies for genetic variation in *ex situ* collections of endangered plant species. In: *Genetics and Conservation of Rare Plants* (eds D.A. Falk & K.E. Holsinger), pp. 99–119. Oxford University Press, New York. [7.7]

Brown A.D.H. & Schoen D.J. (1992) Plant populations genetic structure and biological conservation. In: *Conservation of Biodiversity for Sustainable Development* (eds O.T. Sandlund, K. Hindar & A.D.H. Brown), pp. 88–104. Scandinavian University Press, Oslo. [7.3.2]

Brown K.M. & Richardson T.M. (1988) Genetic polymorphism in gastropods: a comparison of methods and habitat scales. *Am. Malacol. Bull.* **6**, 9–17. [7.3.2] [7.4.1]

Brussard P.F. (1984) Geographic patterns and environmental gradients: the central-marginal model in *Drosophila* revisited. *Annu. Rev. Ecol. Syst.* **15**, 25–64. [7.4.3]

Burgman M.A., Ferson S. & Akçakaya H.R. (1993) *Risk Assessment in Conservation Biology.* Chapman & Hall, London. [7.6] [7.7]

Buss L.W. (1987) *The Evolution of Individuality.* Princeton University Press, New Jersey. [7.2] [7.3.1]

Cain A.J. (1971) Colour and banding morphs in subfossil samples of the snail *Cepaea.* In: *Ecological Genetics and Evolution* (ed. E.R. Creed), pp. 65–92. Blackwell Scientific, Oxford. [7.5.2]

Cain A.J. (1983) Ecology and ecogenetics of terrestrial molluscan populations. In: *The Mollusca*, Vol. 6 (ed. W.D. Russell-Hunter), pp. 597–647. Academic Press, London. [7.5.2]

Cain A.J. & Cook L.M. (1989) Persistence and extinction in some *Cepaea* populations. *Biol. J. Linn. Soc.* **38**, 183–190. [7.5.1]

Cain A.J. & Currey J.D. (1963) Area effects in *Cepaea. Phil. Trans. R. Soc. Lond.* B **246**, 1–81. [7.4.3]

Cain A.J. & Sheppard P.M. (1952) Natural selection in *Cepaea. Genetics* **39**, 89–116. [7.4.3]

Cain A.J., Cook L.M. & Currey J.D. (1990) Population size and morph frequency in a long-term study of *Cepaea nemoralis. Proc. R. Soc. Lond.* B **240**, 231–250. [7.5.1]

Cameron R.A.D. (1992) Change and stability in *Cepaea* populations over 25 years: a case of climatic selection. *Proc. Roy. Soc. Lond.* B **248**, 181–187. [7.5.1]

Capossela A., Silander J.A., Janson R.K., Bergen B. & Talbot D.R. (1992) Nuclear ribosomal DNA variation among ramets and genets of white clover. *Evolution* **46**, 1240–1247.

[7.2]

Caro T.M. & Laurenson M.K. (1994) Ecological and genetic factors in conservation: a cautionary tale. *Science* **263**, 485–486. [7.1] [7.7]

Carvalho G.R. & Crisp D.J. (1987) The clonal ecology of *Daphnia magna* (Crustaceae: Cladocera). I. Temporal changes in the clonal structure of a natural population. *J. Anim. Ecol.* **56**, 453–468. [7.5.1]

Charlesworth B. (1980) *Evolution in Age-structured Populations.* Cambridge University Press. [7.4.2]

Charlesworth B., Charlesworth D. & Morgan M.T. (1990) Genetic loads and estimates of mutation rates in highly inbred plant populations. *Nature* **347**, 380–382. [7.3.1]

Charlesworth D. & Charlesworth B. (1987) Inbreeding depression and its evolutionary consequences. *Annu. Rev. Ecol. Syst.* **18**, 237–268. [7.6]

Clarke B. & Murray J. (1962) Changes of gene frequency in *Cepaea nemoralis* (L.). *Heredity* **17**, 445–465. [7.5.1]

Clarke B., Arthur W., Horsley D.T. & Parkin D.T. (1978) Genetic variation and natural selection in pulmonate molluscs. In: *Pulmonates,* Vol. 2A (eds V. Fretter & J.F. Peake), pp. 219–270. Academic Press, London. [7.4.3]

Clausen J., Keck D.D. & Hiesey W.M. (1940) Experimental studies on the nature of species. I. The effect of varied environments on western North American plants. *Carnegie Inst. Wash. Publ.* **520**, Washington DC. [7.4.3]

Cook C.D.K. (1986) Vegetative growth and genetic mobility in some aquatic weeds. In: *Differentiation Patterns in Higher Plants* (ed. K.M. Urbanska), pp. 217–225. Academic Press, London. [7.1] [7.4.1]

Crawford T.J. (1984) What is a population? In: *Evolutionary Ecology* (ed. B. Shorrocks), pp. 135–173. Blackwell Scientific, Oxford. [7.4.2]

Currey J.D. & Cain, A.J. (1968) Studies on *Cepaea*. IV. Climate and selection of banding morphs in *Cepaea* from the climatic optimum to the present day. *Phil. Trans. Roy. Soc., Lond. B* **253**, 483–498. [7.5.2]

Daehler C.C. & Strong D.R. (1994) Variable reproductive output among clones of *Spartina alterniflora* (Poaceae) invading San Francisco Bay, California: the influence of herbivory, pollination, and establishment site. *Am. J. Bot.* **81**, 307–313. [7.4.3]

Daniels R.E. (1982) Isozyme variation in British populations of *Sphagnum pulchrum* (Braithw.) Warnst. *J. Bryology* **12**, 65–76. [7.5.1]

Dirzo R. & Harper J.L. (1982) Experimental studies on slug–plant interactions. *J. Ecol.* **70**, 101–117. [7.4.3]

Dobson A.P., Mace G.M., Poole J. & Brett R.A. (1991) Conservation biology: the ecology and genetics of endangered species. In: *Genes in Ecology* (eds T.J. Crawford & G.M. Hewitt), pp. 405–430. Blackwell Scientific, Oxford. [7.7]

Dobzhansky T. (1943) Genetics of natural populations: IX. Temporal changes in the composition of populations of *Drosophila pseudoobscura*. *Genetics* **28**, 162–186. [7.5.1]

Dobzhansky T. (1947) Adaptive changes induced by natural selection in wild populations of *Drosophila*. *Evolution* **1**, 1–16. [7.5.1]

Eckert C.G. & Barrett S.C.H. (1993) Clonal reproduction and patterns of genotypic diversity in *Decodon verticillatus* (Lythraceae). *Am. J. Bot.* **80**, 1175–1182. [7.4.3]

Ellstrand N.C. & Elam D.R. (1993) Population genetic consequences of small population size: implications for plant conservation. *Annu. Rev. Ecol. Syst.* **24**, 217–242. [7.4.3] [7.5.1]

Ellstrand N.C. & Roose M.L. (1987) Patterns of genotypic diversity in clonal plant species. *Am. J. Bot.* **74**, 123–131. [7.4.1]

Endler J.A. (1986) *Natural Selection in the Wild.* Princeton University Press, New Jersey. [7.3.2]

Epperson B.K. (1990) Spatial patterns of genetic variation within plant populations. In: *Plant Population Genetics, Breeding, and Genetic Resources* (eds A.H.D. Brown, M.T. Clegg, A.L. Kahler & B.S. Weir), pp. 229–253. Sinauer Associates, Sunderland, Massachusetts.

[7.4.3]

Excoffier L., Smouse P.E. & Quattro J.M. (1992) Analysis of molecular variance inferred from metric distances among DNA haplotypes: application to human mitochondrial DNA restriction sites. *Genetics* **131**, 479–491. [7.2]

Falconer D.S. (1989) *Introduction to Quantitative Genetics* (3rd edn). Longman, New York. [7.6]

Foltz D.W., Ochman H. & Selander R.K. (1984) Genetic diversity and breeding systems in terrestrial slugs of the families Limacidae and Arionidae. *Malacologia* **25**, 593–605. [7.4.1]

Gilpin M. (1991) The genetic effective size of a metapopulation. *Biol. J. Linn. Soc.* **42**, 165–175. [7.6]

Gliddon C., Belhassen E. & Gouyon P-H. (1987) Genetic neighbourhoods in plants with diverse systems of mating and different patterns of growth. *Heredity* **59**, 29–32. [7.4.2]

Godt M.J. & Hamrick J.L. (1993) Genetic diversity and population structure in *Tradescantia hirsuticaulias* (*commelinaceae*). *Am. J. Bot.* **80**, 959–966. [7.4.3]

Golenberg E.M. (1987) Estimation of gene flow and genetic neighborhood size by indirect methods in a selfing annual, *Triticum dicoccoides*. *Evolution* **41**, 1326–1334. [7.4.2]

Goodhart C.B. (1973) A 16 year survey of *Cepaea* on the Hundred Foot Bank. *Malacologia* **14**, 327–331. [7.5.1]

Grant P.R. & Grant R. (1992) Demography and genetically effective sizes of two populations of Darwin's Finches. *Ecology* **73**, 766–784. [7.4.2]

Gregor J.W. (1930) Experiments on the genetics of wild populations, I. *Plantago maritima*. *J. Genetics* **22**, 15–25. [7.4.3]

Gregor J.W. (1938) Experimental taxonomy. 2. Initial population differentiation in *Plantago maritima* in Britain. *New Phytol.* **37**, 15–49. [7.4.3]

Hamilton M.B. (1994) *Ex situ* conservation of wild plant species: time to reassess the genetic assumptions and implications of seed banks. *Conserv. Biol.* **8**, 39–49. [7.7]

Hamilton W.D. (1980) Sex versus non-sex versus parasite. *Oikos* **35**, 282–290. [7.3.2]

Hamrick J.L. (1982) Plant population genetics and evolution. *Am. J. Bot.* **69**, 1685–1693. [7.4.1]

Hamrick J.L. (1983) The distributions of genetic variation within and among natural plant populations. In: *Genetics and Conservation* (eds C.M. Schönewald-Cox, S.M. Chambers, B. MacBryde & L. Thomas), pp. 335–348. Benjamin Cummings, Menlo Park, California. [7.4.1]

Hamrick J.L., Linhart Y.B. & Mitton J.B. (1979) Relationship between life-history characteristics and electrophoretically detectable genetic variation in plants. *Annu. Rev. Ecol. Syst.* **10**, 173–200. [7.3.2] [7.4.1] [7.4.3]

Hamrick J.L., Godt M.J.W., Murawski D.A. & Loveless M.D. (1991) Correlations between species traits and allozyme diversity: implications for conservation biology. In: *Genetics and Conservation of Rare Plants* (eds D.A. Falk & K.E. Holsinger), pp. 75–86. Oxford University Press, New York. [7.7]

Handel S.N. (1985) The intrusion of clonal growth patterns on plant breeding systems. *Am. Nat.* **125**, 367–384. [7.4.2]

Harris R.B. & Allendorf F.W. (1989) Genetically effective population size of large mammals: an assessment of estimators. *Conserv. Biol.* **3**, 181–191. [7.4.2]

Hermanutz L.A., Innes D.J. & Weis I.M. (1989) Clonal structure of arctic dwarf birch (*Betula glandulosa*) at its northern limit. *Am. J. Bot.* **76**, 755–761. [7.5.1]

Heywood J.S. (1991) Spatial analysis of genetic variation in plant populations. *Annu. Rev. Ecol. Syst.* **22**, 335–355. [7.4.3]

Hoelzel A.R., Halley J., O'Brien S.J., Campagna C., Arnbom T., Le Boeuf B., Ralls K. & Dover G.A. (1993) Elephant seal genetic variation and the use of simulation models to investigate historical population bottenecks. *J. Heredity* **84**, 443–449. [7.6]

Holderegger R. & Schneller J.J. (1994) Are small isolated populations of *Asplenium septentrionale* variable? *Biol. J. Linn. Soc.* **51**, 377–385. [7.4.3]

Houle D. (1989) The maintenance of polygenic variation in finite populations. *Evolution* **43**,

1767–1780. [7.3.1]

Hubby J.L. & Lewontin R.C. (1966) A molecular approach to the study of genic heterozygosity in natural populations. I. The number of alleles at different loci in *Drosophila pseudoobscura*. *Genetics* **54**, 577–594. [7.1]

Huff D.R., Peakall R. & Smouse P.E. (1993) RAPD variation within and among natural populations of outcrossing buffalograss (*Buchloë dactyloides* (Nutt.) Engelm.). *Theor. Appl. Genetics* **86**, 927–934. [7.2]

Hunter C.L. (1993) Genotypic variation and clonal structure in coral populations with different disturbance histories. *Evolution* **47**, 1213–1228. [7.4.1]

Jackson J.B.C., Buss L.W. & Cook, R.E. (1985) *Population Biology and Evolution of Clonal Organisms*. Yale University Press, New Haven. [7.2]

Jaenike J., Parker E.D. & Selander R.K. (1980) Clonal niche structure in the parthenogenetic earthworm *Octolasion tyrtaeum*. *Am. Nat.* **116**, 196–205. [7.4.3]

Janzen D.H. (1974) The deflowering of Central America. *Nat. Hist.* **83**, 48–53.

Jennersten O. (1988) Pollination in *Dianthus deltoides* (Caryophyllaceae): effects of habitat fragmentation on visitation and seed set. *Conserv. Biol.* **2**, 359–366. [7.6]

Jones J.S., Leith B.H. & Rawlings P. (1977) Polymorphism in *Cepaea*: a problem with too many solutions? *Annu. Rev. Ecol. Syst.* **8**, 109–143. [7.4.3]

Karron J.D. (1987) A comparison of levels of genetic polymorphism and self-compatibility in geographically restricted and widespread plant congeners. *Evol. Ecol.* **1**, 47–58. [7.1]

Keller L.F., Arcese P., Smith J.N.M., Hochachka W.M. & Stearns S.C. (1994) Selection against inbred song sparrows during a natural population bottleneck. *Nature* **372**, 356–357. [7.6]

Kenton A., Langton D. & Coleman J. (1988) Genomic instability in a clonal species, *Tradescantia commelinoides* (Commelinaceae). *Genome* **30**, 734–744. [7.3.1]

Kettlewell H.B.D. (1973) *The Evolution of Melanism*. Oxford University Press, London. [7.5]

Kieser J.A. & Groeneveld H.T. (1991) Fluctuating odontometric asymmetry, morphological variability, and genetic monomorphism in the cheetah *Acinonyx jubatus*. *Evolution* **45**, 1175–1183. [7.1] [7.4.3]

Kimura M. & Crow J.F. (1963) The measurement of effective population number. *Evolution* **17**, 279–288. [7.4.2]

Kimura M. & Ohta T. (1971) *Theoretical Aspects of Population Genetics*. Princeton University Press, New Jersey. [7.4.3]

Koehn R.K., Diehl W.J. & Scott T.M. (1988) The differential contribution by individual enzymes of glycolysis and protein catabolism to the relationship between heterozygosity and growth rate in the coot clam, *Mulinia lateralis*. *Genetics* **118**, 121–130. [7.4.3]

Korman A.K., Mallet J., Goodenough J.L., Graves J.B., Hayes J.L., Hendricks D.E., Luttrell, R., Pair S.D. & Wall M. (1993) Population structure in *Heliothis virescens* (Lepidoptera: Noctuidae): an estimate of gene flow. *Ann. Entomol. Soc. Am.* **86**, 182–188. [7.4.3]

Kurt F., Hartl G.B. & Völk G. (1993) Breeding strategies and genetic variation in European roe deer *Capreolus capreolus* populations. *Acta Theriol.* **38**, Suppl. 2, 187–194. [7.4.1]

Lacy R.C., Petric A. & Warneke M. (1993) Inbreeding and outbreeding in captive populations of wild animals. In: *The Natural History of Inbreeding and Outbreeding* (ed. N.W. Thornhill), pp. 352–374. University of Chicago Press. [7.6]

Lamotte M. (1951) Recherches sur la structure génétique des populations naturelles de *Cepaea nemoralis* (L.). *Bull. biol. France Belg. Suppl.* **35**, 1–239. [7.2]

Lande R. (1988) Genetics and demography in biological conservation. *Science* **241**, 1455–1460. [7.6]

Latter B.D.H. (1980) Genetic differences within and between populations of the major human subgroups. *Am. Nat.* **116**, 220–237. [7.2] [7.4.1]

Ledig F.T. (1986) Heterozygosity, heterosis, and fitness in outbreeding plants. In: *Conservation Biology: the science of scarcity and diversity* (ed. M.E. Soulé), pp. 77–104. Sinauer

Associates, Sunderland, Massachusetts. [7.4.1]

Lesica P., Leary R.F., Allendorf F.W. & Bilderback D.E. (1988) Lack of genic diversity within and among populations of an endangered plant, *Howellia aquatilis. Conserv. Biol.* **2**, 275–282. [7.4.1]

Levin D.A. (1984) Immigration in plants: an exercise in the subjunctive. In: *Perspectives on Plant Population Ecology* (eds R. Dirzo & J. Sarukhán), pp. 242–260. Sinauer Associates, Sunderland, Massachusetts. [7.4.2]

Levin D.A. (1988) Local differentiation and the breeding structure in plant populations. In: *Plant Evolutionary Biology* (eds L.D. Gottlieb & S.K. Jain), pp. 305–329. Chapman & Hall, London. [7.4.3] [7.6]

Levin D.A. (1990) The seed bank as a source of genetic novelty in plants. *Am. Nat.* **135**, 563–572. [7.3.1]

Lewontin R.C. (1974) *The Genetic Basis of Evolutionary Change.* Columbia University Press, New York. [7.4.3]

Liebherr J.K. (1986) Comparison of genetic variation in two Carabid beetles (Coleoptera) of differing vagility. *Ann. Entomol. Soc. Am.* **79**, 424–433. [7.4.2]

Linhart Y.B., Grant M.C. & Montazer P. (1989) Experimental studies in Ponderosa pine. I. Relationship between variation in proteins and morphology. *Am. J. Bot.* **76**, 1024–1032. [7.3.2]

Loveless M.D. & Hamrick J.L. (1984) Ecological determinants of genetic structure in plant populations. *Annu. Rev. Ecol. Syst.* **15**, 65–95. [7.3.2] [7.4.1] [7.4.3]

Mader H.-J. (1984) Animal habitat isolation by roads and agricultural fields. *Biol. Conserv.* **29**, 81–96. [7.4.2]

Marshall D.R. (1990) Crop genetic resources: current and emerging issues. In: *Plant Population Genetics, Breeding, and Genetic Resources* (eds A.H.D. Brown, M.T. Clegg, A.L. Kahler & B.S. Weir), pp. 367–389. Sinauer Associates, Sunderland, Massachusetts. [7.7]

Martins P.S. & Jain S.K. (1979) Role of genetic variation in the colonizing ability of rose clover (*Trifolium hirtum* All.). *Am. Nat.* **114**, 591–595. [7.1]

Maynard Smith J. (1989) *Evolutionary Genetics.* Oxford University Press. [7.3.1]

McClintock K.A. & Waterway M.J. (1993) Patterns of allozyme variation and clonal diversity in *Carex lasiocarpa* and *C. pellita* (Cyperaceae). *Am. J. Bot.* **80**, 1251–1263. [7.4.1]

McNeilly T. (1968) Evolution in closely adjacent plant populations. III. *Agrostis tenuis* on a small copper mine. *Heredity* **23**, 99–108 [7.4.3]

Meins F. (1983) Heritable variation in plant cell culture. *Annu. Rev. Plant Physiol.* **34**, 327–346. [7.3.1]

Meins F. & Seldran M. (1994) Pseudodirected variation in the requirement of cultured plant cells for cell-division factors. *Development* **120**, 1163–1168. [7.3.1]

Menozzi P. & Krimbas C.B. (1992) The inversion polymorphism of *Drosophila subobscura* revisited: synthetic maps of gene arrangement frequencies and their interpretation. *J. Evol. Biol.* **5**, 625–641. [7.4.3]

Murray J. & Clarke B. (1978) Changes of gene-frequency in *Cepaea nemoralis* over 50 years. *Malacologia* **17**, 317–330. [7.5.1]

Murray J. & Clarke B. (1984) Movement and gene flow in *Partula taeniata. Malacologia* **25**, 343–348. [7.4.2]

Nevo E. (1983) Population genetics and ecology: the interface. In: *Evolution from Molecules to Men* (ed. D.S. Bendall), pp. 287–321. Cambridge University Press. [7.4.2]

Nevo E., Beiles A. & Ben-Shlomo R. (1984) The evolutionary significance of genetic diversity: ecological demographic and life-history correlates. *Lecture Notes Biomath.* **53**, 13–213. [7.4.1] [7.4.3]

Novak S.J., Mack R.N. & Soltis D.E. (1991) Genetic variation in *Bromus tectorum* (Poaceae): population differentiation in its North American range. *Am. J. Bot.* **78**, 1150–1161. [7.1]

Nunney L. (1993) The influence of mating system and overlapping generations on effective population size. *Evolution* **47**, 1329–1341. [7.4.2]

Nunney L. & Elam D.R. (1994) Estimating the effective population size of conserved populations. *Conserv. Biol.* **8**, 175–184. [7.4.2]

O'Brien S.J., Roelke M.E., Marker L., Newman A., Winkler D., Meltzer D., Colly L., Evermann J.F., Bush M. & Wildt D.E. (1985) Genetic basis for species vulnerability in the cheetah. *Science* **227**, 1428–1434. [7.1] [7.7]

Oyama K. (1993) Conservation biology of tropical trees: demographic and genetic considerations. *Environ. Update* **1**, 17–32. [7.4.3]

Parker K.C. & Hamrick J.L. (1992) Genetic diversity and clonal structure in a columnar cactus, *Lophocereus schottii. Am. J. Bot.* **79**, 86–96. [7.4.2]

Parks J.C. & Werth C.R. (1993) A study of spatial features of clones in a population of bracken fern, *Pteridium aquilinum* (Dennstaedtiaceae). *Am. J. Bot.* **80**, 537–544. [7.4.3]

Parsons P.A. (1992) Biodiversity and climatic change. In: *Conservation of Biodiversity for Sustainable Development* (eds O.T. Sandlund, K. Hindar & A.D.H. Brown), pp. 155–167. Scandinavian University Press, Oslo. [7.6]

Pleasants J.M. & Wendel J.F. (1989) Genetic diversity in a clonal narrow endemic, *Erythronium propullans*, and its widespread progenitor, *Erythronium albidum. Am. J. Bot.* **76**, 1136–1151. [7.5.1]

Pounds J.A. & Jackson J.F. (1981) Riverine barriers to gene flow and the differentiation of fence lizard populations. *Evolution* **35**, 516–528. [7.4.2]

Prevosti A., Serra L., Ribo G., Aguade M., Segarra C., Monclus M. & Garcia M.P. (1985) The colonization of *Drosophila subobscura* in Chile. II. Clines in the chromosomal arrangements. *Evolution* **39**, 838–844. [7.4.3]

Prevosti A., Ribo G., Serra L., Aguade M., Balana J., Monclus M. & Mestres F. (1988) Colonization of America by *Drosophila subobscura*: experiment in natural populations that supports the adaptive role of chromosomal-inversion polymorphism. *Proc. Natl. Acad. Sci., USA* **85**, 5597 5600. [7.4.3]

Prevosti A., Serra L., Segarra C., Aguade M., Ribo G. & Monclus M. (1990) Clines of chromosomal arrangements of *Drosophila subobscura* in South America evolve closer to old world patterns. *Evolution* **44**, 218–221. [7.4.3]

Preziosi R.F. & Fairbairn D.J. (1992) Genetic population structure and levels of gene flow in the stream dwelling waterstrider, *Aquarius* (= *Gerris*) *remigis* (Hemiptera: Gerridae). *Evolution* **46**, 430–444. [7.4.2]

Renshaw E. (1991) *Modelling Biological Populations in Space and Time.* Cambridge University Press. [7.4.3]

Ryman N.R., Baccus C., Reuterwall C. & Smith M.H. (1981) Effective population size, generation interval, and potential loss of genetic variability in game species under different hunting regimes. *Oikos* **36**, 257–266. [7.4.2]

Sarkar S. (1992a) Neo-Darwinism and the problem of directed mutations. *Evol. Trends Plants* **6**, 73–79. [7.3.1]

Sarkar S. (1992b) Sex, disease, and evolution—variations on a theme from J.B.S. Haldane. *BioScience* **42**, 448–454. [7.3.2] [7.4.3]

Schal B.A., O'Kane S.L. & Rogstadt S.H. (1991) DNA variation in plant populations. *Trends Ecol. Evol.* **6**, 329–333. [7.2] [7.3.2]

Schemske D.W. & Lande R. (1985) The evolution of self-fertilization and inbreeding depression in plants. II. Empirical observations. *Evolution* **39**, 41–52. [7.4.1] [7.4.3]

Schemske D.W., Husband B.C., Ruckelshaus M.H., Goodwillie C., Parker I.M. & Bishop J.G. (1994) Evaluating approaches to the conservation of rare and endangered plants. *Ecology* **75**, 584–606. [7.3.2] [7.6] [7.7]

Schmid B. (1980) *Carex flava* L. s.l. im Lichte der r-Selektion. PhD dissertation, University of Zurich. [7.3.2]

Schmid B. (1982) Karyology and hybridization in the *Carex flava* complex in Switzerland. *Feddes Repertorium* **93**, 23–59. [7.3.1]

Schmid B. (1984) Niche width and variation within and between populations in colonizing species (*Carex flava* group). *Oecologia* **63**, 1–5. [7.4.3]

Schmid B. (1985) Clonal growth in grassland perennials. III. Genetic variation and plasticity between and within populations of *Bellis perennis* and *Prunella vulgaris. J. Ecol.* **73**, 819–

830. [7.4.3]

Schmid B. (1987) Patterns of variation and population structure in the *Carex flava* group. *Symbolae Botanicae Upsalienses* **27**, 113–126. [7.3.2]

Schmid B. (1990) Some ecological and evolutionary consequences of modular organization and clonal growth in plants. *Evol. Trends Plants* **4**, 25–34. [7.4.2] [7.5.1]

Schmid B. (1994) Effects of genetic diversity in experimental stands of *Solidago altissima*: evidence for the potential role of pathogens as selective agents in plant populations. *J. Ecol.* **82**, 165–175. [7.3.2] [7.4.3]

Schmid B. & Bazzaz F.A. (1990) Plasticity in plant size and architecture in rhizome-derived vs. seed-derived *Solidago* and *Aster*. *Ecology* **71**, 523–535. [7.4.3]

Schmid B. & Harper J.L. (1985) Clonal growth in grassland perennials. I. Density and pattern dependent competition between plants with different growth form. *J. Ecol.* **73**, 793–808. [7.4.3] [7.5.1]

Schonewald-Cox C.M., Chambers S.M., MacBryde B. & Thomas L. (eds) (1983) *Genetics and Conservation*. Benjamin/Cummings, Menlo Park, California. [7.6]

Schwaegerle K.E., Garbutt K. & Bazzaz F.A. (1986) Differentiation among nine populations of *Phlox*. I. Electrophoretic and quantitative variation. *Evolution* **40**, 506–517. [7.2]

Scowcroft W.R., Brettell R.I.S., Ryan S.A., Davies P.A. & Pallotta M.A. (1987) Somaclonal variation and genomic flux. In: *Plant Tissue and Cell Culture* (eds C.E. Green, D.A. Somers, W.P. Hackett & D.D. Biesboer), pp. 275–286. Alan R. Liss, New York. [7.2]

Selander R.K. & Kaufman D.W. (1975) Genetic structure of populations of the brown snail (*Helix aspersa*). I. Microgeographic variation. *Evolution* **29**, 385–401. [7.4.2]

Shaw A.J. (1991) Ecological genetics of serpentine tolerance in the moss, *Funaria flavicans*: variation within and among haploid sib families. *Am. J. Bot.* **78**, 487–1493. [7.4.3]

Shaw R.G. (1987) Maximum-likelihood approaches applied to quantitative genetics of natural populations. *Evolution* **41**, 812–826. [7.2]

Silvertown J. & Lovett Doust J. (1993) *Introduction to Plant Population Biology* (3rd edn). Blackwell Scientific, Oxford. [7.4.1] [7.5.1]

Simberloff, D. (1988) The contribution of population and community biology to conservation science. *Annu. Rev. Ecol. Syst.* **19**, 473–511. [7.4.2] [7.6] [7.7]

Slatkin M. (1981) Estimating levels of gene flow in natural populations. *Genetics* **99**, 323–335. [7.4.2]

Slatkin M. (1985a) Gene flow in natural populations. *Annu. Rev. Ecol. Syst.* **16**, 393–430. [7.4.2]

Slatkin M. (1985b) Rare alleles as indicators of gene flow. *Evolution* **39**, 53–65. [7.4.2]

Slatkin M. & Arter H.E. (1991) Spatial autocorrelation methods in population genetics. *Am. Nat.* **138**, 499–517. [7.4.3]

Slatkin M. & Barton N.H. (1989) A comparison of three indirect methods for estimating average levels of gene flow. *Evolution* **43**, 1349–1368. [7.4.2]

Sokal R.R., Oden N.L. & Barker J.S.F. (1987) Spatial structure in *Drosophila buzzatii* populations: simple and directional spatial autocorrelation. *Am. Nat.* **129**, 122–144. [7.4.3]

Spiess E.B. (1989) *Genes in Populations* (2nd edn). Wiley & Sons, New York. [7.3.2] [7.4.3] [7.7]

Stebbins G.L. (1977) *Processes of Organic Evolution* (3rd edn). Prentice-Hall, Englewood Cliffs, New Jersey. [7.1] [7.3.1]

Sultan S.E. (1987) Evolutionary implications of phenotypic plasticity in plants. *Evol. Ecol.* **21**, 127–178. [7.4.3]

Sultan S.E. & Bazzaz F.A. (1993) Phenotypic plasticity in *Polgonum persicaria* I. Diversity and uniformity in genotypic norms of reaction to light. *Evolution* **47**, 1009–1031. [7.4.3]

Templeton A.R. (1986) Coadaptation and outbreeding depression. In: *Conservation Biology: the science of scarcity and diversity* (ed. M.E. Soulé), pp. 105–116. Sinauer Associates, Sunderland, Massachusetts. [7.6]

Templeton A.R. (1994) Inbreeding: one word, several meanings, much confusion. In: *Genetics and Conservation* (eds V. Loeschcke, J. Tomiuk & S.K. Jain), pp. 91–105.

Birkhäuser Verlag, Basel. [7.5.1]

Thompson J.D., McNeilly T. & Gray A.J. (1991) Population variation in *Spartina anglica* C.E. Hubbard I. Evidence from a common garden experiment. *New Phytologist* **117**, 115–128. [7.1]

Thornhill N.W. (ed.) (1993) *The Natural History of Inbreeding and Outbreeding: theoretical and empirical perspectives*. University of Chicago Press. [7.6] [7.7]

Turesson G. (1922) The genotypical response of the plant species to habitat and climate. *Hereditas* **3**, 211–350. [7.4.3]

Turkington R. (1985) Variation and differentiation in populations of *Trifolium repens* in permanent pastures. In: *Studies on Plant Demography* (ed. J. White), pp. 69–82. Academic Press, London. [7.5.1]

Turkington R. & Harper J.L. (1979) The growth, distribution and neighbour relationships of *Trifolium repens* in a permanent pasture. III. Fine-scale biotic differentiation. *J. Ecol.* **67**, 245–254. [7.4.3]

Van Valen L. (1962) A study of fluctuating asymmetry. *Evolution* **16**, 125–142. [7.4.3]

Venable D.L. (1984) Using intraspecific variation to study the ecological significance and evolution of plant life-histories. In: *Perspectives on Plant Population Ecology* (eds R. Dirzo & J. Sarukhán), pp. 167–187. Sinauer Associates, Sunderland, Massachusetts. [7.3.2]

Via S. (1991) The genetic structure of host plant adaptation in a spatial patchwork: demographic variability among reciprocally transplanted pea aphid clones. *Evolution* **45**, 827–852. [7.4.3]

Wall S., Carter M.A. & Clarke B. (1980) Temporal changes of gene-frequencies in *Cepaea hortensis*. *Biol. J. Linn. Soc.* **14**, 303–318. [7.5.1]

Waller D.M. (1980) Environmental determinants of outcrossing in *Impatiens capensis* (Balsaminaceae). *Evolution* **34**, 747–761. [7.4.3]

Waller D.M., O'Malley D.M. & Gawler S.C. (1987) Genetic variation in the extreme endemic *Pedicularis furbishiae* (Scrophulariaceae). *Conserv. Biol.* **1**, 335–340. [7.4.3]

Warwick S.I. & Briggs D. (1978) The genecology of lawn weeds. II. Evidence for disruptive selection in *Poa annua* L. in a mosaic environment of bowling green lawns and flower beds. *New Phytol.* **81**, 725–737. [7.4.3]

Warwick S.I. & Briggs D. (1979) The genecology of lawn weeds. III. Cultivation experiments with *Achillea millefolium* L., *Bellis perennis* L., *Plantago lanceolata* L., *Plantago major* L. and *Prunella vulgaris* L. collected from lawns and contrasting grassland habitats. *New Phytol.* **83**, 509–536. [7.4.1]

Warwick S.I. & Briggs D. (1980) The genecology of lawn weeds. IV. Adaptive significance of variation in *Bellis perennis* L. as revealed in a transplant experiment. *New Phytol.* **85**, 275–288. [7.4.1]

Warwick S.J., Thompson B.K. & Black L.D. (1987) Genetic variation in Canadian and European populations of the colonizing weed species *Apera spica-venti*. *New Phytol.* **106**, 301–317. [7.3.2]

Watkinson A.R. & Powell J.C. (1993) Seedling recruitment and the maintenance of clonal diversity in plant populations—a computer simulation of *Ranunculus repens*. *J. Ecol.* **81**, 707–717. [7.5.1]

Weber E. (1994) *Evolutionary trends in European neophytes: a case study of two Solidago species*. PhD dissertation, University of Basel. [7.4.3]

Whitlock M.C. (1992a) Nonequilibrium population structure in forked fungus beetles: extinction, colonization, and the genetic variance among populations. *Am. Nat.* **139**, 952–970. [7.5.1] [7.2]

Whitlock M.C. (1992b) Temporal fluctuations in demographic parameters and the genetic variance among populations. *Evolution* **46**, 608–615. [7.4.2]

Widén B. & Svensson L. (1992) Conservation of genetic variation in plants—the importance of population size and gene flow. In: *Ecological Principles of Nature Conservation* (ed. L. Hansson), pp. 113–161. Elsevier Applied Science, London. [7.3.1] [7.4.2]

Wiens D., Nickrent D.L., Davern C.I., Calvin C.L. & Vivrette N.J. (1989) Developmental failure and loss of reproductive capacity in the rare palaeoendemic shrub *Dedeckera eurekensis. Nature* **338**, 65–67. [7.4.3]

Williamson P., Cameron R.A.D. & Carter M.A. (1977) Population dynamics of the landsnail *Cepaea nemoralis* (L.): a six year study. *J. Anim. Ecol.* **46**, 181–194. [7.5.1]

Willson M.F., Thomas P.A., Hoppes W.G., Katusic-Malmborg P.L., Goldman D.A. & Bothwell J.L. (1987) Sibling competition in plants: an experimental study. *Am. Nat.* **129**, 304–311. [7.4.3]

Wolda H. (1969) Stability of a steep cline in morph-frequencies in the snail *Cepaea nemoralis* near Groningen. *J. Anim. Ecol.* **38**, 623–635. [7.4.3] [7.5.1]

Wolf P.G., Sheffield E. & Haufler C.H. (1991) Estimates of gene flow, genetic substructure and population heterogeneity in bracken (*Pteridium aquilinum*). *Biol. J. Linn. Soc.* **42**, 407–423. [7.4.2]

Wright S. (1931) Evolution in Mendelian populations. *Genetics* **16**, 97–159. [7.4.2] [7.6]

Wright S. (1938) Size of population and breeding structure in relation to evolution. *Science* **87**, 430–431. [7.4.2]

Wright S. (1943) Isolation by distance. *Genetics* **28**, 114–138. [7.4.2]

Wright S. (1946) Isolation by distance under diverse systems of mating. *Genetics* **31**, 39–59. [7.4.2]

Wright S. (1969) *The Theory of Gene Frequencies,* Vol. 2. University of Chicago Press. [7.4.2]

Zouros E. & Foltz D.W. (1987) The use of allelic isozyme variation for the study of heterosis. In: *Isozymes: current topics in biological and medical research,* Vol. 13 (eds M.C. Rattazzi, J.G. Scandalios & G.S. Whitt), pp. 1–59. Alan R. Liss, New York. [7.4.3]

8: Spatial patterns in taxonomic diversity

KEVIN J. GASTON and PAUL H. WILLIAMS

8.1 Introduction

Biodiversity is spatially heterogeneous. Examples of this simple observation tend to be derived from measures of taxonomic diversity, which we define here broadly to encompass numbers of species, numbers of supra-specific taxa, and taxonomically based measures of character diversity (Chapter 3). Indeed, the vast majority of examples of this heterogeneity simply concern species richness. In this chapter we review some of the basic spatial patterns in taxonomic diversity, and provide pictorial examples of many of them (Figs 8.1–8.6, Plates 8.1, 8.2 (between pp. 214 & 215)). We are concerned principally with patterns between meso- and macro-scale (regional) units, and in the main ignore those at more local scales. We essentially regard regional and local scales as those at which biogeographic and ecological processes, respectively, predominate (Cornell & Lawton, 1992). An understanding of large scale patterns in diversity is increasingly seen as a key to understanding patterns and processes at more local scales (Ricklefs, 1987; Ricklefs & Schluter, 1993).

Although some indication of the relevant literature is provided, we pay little detailed attention to the mechanisms generating observed regional patterns in taxonomic diversity, taking a view of biodiversity that is predominantly concerned with pattern rather than process. In general, mechanisms remain contentious. There are several reasons. First, they continue frequently to be framed in terms of single factors when often several may plausibly contribute. Second, the importance of different factors is difficult to disentangle, because they tend often to co-vary in space, because correlation and causation remain confused, and because of the problem of relating regional-scale processes to the results of experiments which have of necessity to be carried out at far smaller scales; exploration of the subsidiary predictions associated with general mechanisms may enable escape from some of these limitations (e.g. Turner et al., 1987, 1988, in press). Third, it is difficult to integrate mechanisms operating on ecological and evolutionary timescales. Fourth, it is difficult to quantify the role of history (but see Ricklefs & Schluter, 1993).

Constraining considerations to large spatial scales has some interesting consequences. It serves to emphasize: (i) how comparatively few quantitative, as opposed to qualitative, studies actually have approached the topic at this level when compared to the numbers addressing patterns in local diversity (although of course it is a matter of debate, and of perception, when a local study becomes a regional one); (ii) that amongst terrestrial systems, most attention has been paid to North America and Australia, some to Africa, and comparatively little to Europe, Asia or South America; (iii) the unevenness of sampling effort in different areas (Chapter 4; e.g. Nelson *et al.*, 1990; Rich & Woodruff, 1992; Prendergast *et al.*, 1993a); (iv) the enormous differences in the extent and intensity of sampling of different taxa; (v) the sampling difficulties encountered in studying many marine, and especially deep sea, assemblages, where small scale work is essentially all that is possible; our knowledge of the sediment dwelling infauna of the deep sea, for example, comes from the study of less than 2000 quantitative cores, an estimated area of $500\,m^2$ (Paterson, 1993); and (vi) that most data for terrestrial systems concern geopolitical units (e.g. states, countries; see for example the collations of Slud, 1976; Tsuda, 1984; Davis *et al.*, 1986; Stuart & Adams, 1991; Groombridge, 1992).

Two broad generalizations, in particular, emerge from this survey of geographic patterns. First, many patterns in taxonomic diversity remain essentially similar when that diversity is quantified in different ways. Species richness and numbers of supra-specific taxa tend, at least approximately, to be positively correlated across space (e.g. Chapters 3 & 4; Stehli *et al.*, 1967; Wilson, 1974; Rabinovich & Rapoport, 1975; Taylor & Taylor, 1977; Eggleton *et al.*, 1994; but see below), as do species richness and taxonomically based measures of character diversity when species richness is moderately high (Chapter 3). Thus, whilst emphasis on regional patterns in taxonomic diversity has been placed on species richness, the conclusions may often be generalized more widely.

Second, differences in the sizes of areas has, with some important exceptions, a pervasive influence on many of those spatial patterns in taxonomic diversity which have been documented at regional scales. Area effects may contribute, amongst others, to latitudinal, altitudinal, and productivity-related gradients in diversity, to differences in the diversity associated with different habitat types, biomes and biogeographic areas, and to patterns of endemism.

We divide spatial patterns in taxonomic diversity into a consideration of areas with extreme values of high and low taxonomic diversity (Section 8.2), gradients of taxonomic diversity (Section 8.3), co-variance in diversity among different taxa (Section 8.4), and relationships between patterns of regional taxonomic diversity, environmental factors and local diversity (Section 8.5).

8.2 Extremes of high and low diversity

The identification of areas of extreme taxonomic diversity (especially those of high diversity) has garnered much interest. Such areas may result from the spatial coincidence of particular states of many different factors, some of these conditions may otherwise have only small effects on general trends in diversity.

8.2.1 Biological realms

Comparison of terrestrial and marine realms provides perhaps the best example where relative richness does not appear to be replicated at different taxonomic levels. Whilst many more phyla are known from marine systems than from terrestrial (32 marine, 12 land), fewer than 15% of species currently named are found in the ocean, despite the vastly greater area covered by the sea (May, 1994). The probable change in the proportion of global species that are marine, were all species to have been described, remains controversial. However, it seems unlikely that anything like parity with the terrestrial system would be achieved (for discussion see Grassle & Maciolek, 1992; May, 1992, 1994; Angel, 1994; Briggs, 1994).

May (1994) lists five sets of factors which might help to explain the contrast in the diversities of land and sea: (i) life began in the sea; (ii) continental environments are more heterogeneous than marine ones; (iii) the ocean-bed environment is less 'architecturally elaborate' than the terrestrial environment; (iv) patterns of herbivory differ between land and sea; and (v) there are differences in the body size distributions of marine and terrestrial species assemblages. Whilst difficult to test, the effect of the first of these factors seems likely to have been profound, and the difference in diversity of terrestrial and marine realms may, in part at least, rest on history and species selection (in the sense of, e.g. Vrba, 1980). Early metazoan diversification was predominantly marine, with only some lineages succeeding in colonizing the land, so that high-ranking taxa are more numerous in the oceans. Subsequently, the configuration of large oceans has been relatively stable in the long term in comparison with the continental blocks, where there have been complex re-arrangements of many small fragments. This fragmentation may be favourable to much higher rates of allopatric speciation on land, where self-powered dispersal may be limited, at least in comparison with long-range dispersal of planktonic larvae by ocean currents.

8.2.2 Biogeographic regions

Comparison of the relative diversity of different biogeographic regions has been obscured by the variety of schemes that exist for their delimitation, and which may profoundly influence observed patterns. Nonetheless, some generalities have emerged. Of the six or more biogeographic regions commonly delimited in the terrestrial realm, the Neotropics is generally considered to comprise the greatest overall levels of taxonomic diversity. A figure of around 50% of species is often cited. Though this is probably something of an over-estimate, the three 'tropical' regions tend to decline in richness from the Neotropics to the Indotropics to the Afrotropics (Gaston & Hudson, 1994). The mammals represent an important example at variance with this pattern of diversity, with the Afrotropical region being unusually diverse. On the basis of the most recent compilation, there are 52 families, 298 genera and 1045 species of mammals in the Afrotropical region (sub-Saharan Africa + Madagascar), compared with 50 families, 309 genera and 1096 species in the Neotropics, and 50 families, 312 genera and 995 species in the Oriental region (Cole *et al.*, 1994). For mammalian genera, the faunal similarity among the continents is a decreasing function of the overland distance between regions, the degree of longitudinal separation, and the difference in area (Flessa, 1981). There are many examples of higher taxa for which the Indotropics is more diverse, or not markedly less diverse, than the Neotropics. These include swallowtail butterflies (Papilionidae) and dragonflies (Sutton & Collins, 1991); swallowtails are far more speciose in the Indotropics (Indotropics 225 species, Neotropics 166 species, Afrotropics 53 species), dragonflies have similar levels of diversity in the Indotropics and Neotropics (Indotropics 1435 species, Neotropics 1473 species, Afrotropics 881 species).

Marine diversity is highest around coral reefs (Groombridge, 1992). The Indo-West Pacific is the richest region both for species of reef fish (Ehrlich, 1975; McAllister *et al.*, 1994) and for genera of corals (Fig. 8.1).

8.2.3 Biogeographic provinces

The identification of areas of high taxonomic diversity at more moderate scales than that of biogeographic regions has been a topic of some concern to conservationists. Given that appropriate data tend to be organized on the basis of geopolitical units, it was perhaps inevitable therefore that a set of 'mega-diversity' countries would come to be recognized. Some six to 12 countries are believed to harbour 50–80% of the world's biodiversity, expressed in terms of species richness (Mittermeier, 1988; McNeely *et al.*, 1990; Mittermeier & Werner, 1990). In a similar vein, several other projects have sought to identify areas of high diversity amongst or within countries (e.g. Schueler & McAllister, 1991; Pomeroy, 1993; Väisänen & Heliövaara, 1994). Thus, the IUCN Plant Conservation Programme is identifying several hundred major centres of plant diversity world-wide, places particularly rich in plant life (Groombridge, 1992). Increasingly, cognizance is being taken of the turnover of species identities between these areas.

The need for explicit quantitative analyses of patterns of taxonomic

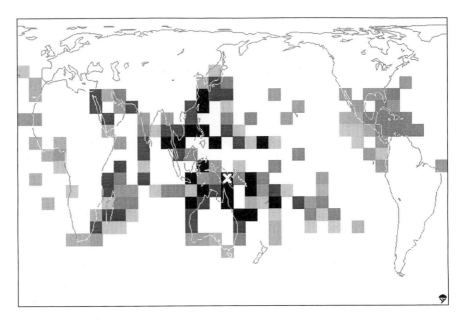

Fig. 8.1 Richness in genera of hermatypic corals. Equal-area map projection, divided into grid cells of equal area (c. 611 000 km²) based on intervals of 10° longitude (Williams, 1993). Values are represented by logarithmic grey-scale intensities, in classes of approximately equal size by the frequency of values between minimum, light grey, and maximum, black with a white cross, with white for no data. Data from Veron (1993).

diversity has been highlighted by the persistence of various misconceptions about its geographic distribution at these spatial scales. Mares (1992) has demonstrated, for example, that contrary to many statements, the taxonomic diversity (number of species and higher taxa) of mammals in the lowland Amazon forest of the Neotropics is less than that of the dryland areas. (See also Redford *et al.* (1990) and Henderson *et al.* (1991) who likewise suggest that the flora of the northern Andes is as rich or richer than that of the Amazon basin.) The Amazon lowlands support fewer species, endemic species, genera, endemic genera, families and endemic families than the drylands. In large part, such differences between habitats, and likewise differences in the relative diversity of the same habitat in different regions, can be explained on the basis of the areas of those habitats (e.g. Flessa, 1975; Haffer, 1990; Mares, 1992; Rosenzweig, 1992). Thus, much of the difference in the mammal diversity of the drylands of the Neotropics compared to other habitats can be explained in terms of their greater extent (Mares, 1992). Haffer (1990) finds that the most extensive biomes of tropical South America are richest in total number of bird species.

Comparison of the biodiversity of different areas will be greatly facilitated by the growing number of reviews of the taxonomic diversity associated with particular habitat types and biomes (e.g. aeolian systems — Swan, 1992; arctic and alpine systems — Chapin & Körner, 1994; rangelands — West, 1993; deep terrestrial subsurfaces—Fliermans & Balkwill, 1989; soil—Lee, 1991; ground waters—Marmonier *et al.*, 1993; running waters—Allan & Flecker, 1993; coastal waters—Ray, 1991; pelagic ocean—Angel, 1993; deep sea benthos—Grassle, 1989, 1991; islands—Groombridge, 1992).

8.2.4 Endemism

The term 'hotspots' has been used recently to refer to areas of extreme taxonomic richness (Prendergast *et al.*, 1993a,b). In previous studies of diversity the term had, however, been used to refer to areas where high levels of richness, threat and endemism coincide (Myers, 1988, 1990).

A taxon is endemic to an area if it occurs there and nowhere else. The area of endemism can be very large or very small, and the proportion of taxa in an area that are endemic to it tends to be an increasing function of the size of the area (Major, 1988; Anderson, 1994). However, for areas of similar size, the proportion of taxa which are endemic may be very variable. Regions can be identified which are particularly rich in endemics, or endemics of given kinds (e.g. Davis *et al.*, 1986; Gentry, 1986, 1992; ICBP, 1992). Endemism can also be viewed as a form of range-size rarity (Rabinowitz, 1981). Rather than being restricted to a categorical assignment of taxa as endemic or non-endemic at a particular scale, measures of range size can be applied as continuous weightings to measures of taxon richness in order to describe patterns of endemism (Usher, 1986; Lahti *et al.*, 1988; Daniels *et al.*, 1991; Howard, 1991; Williams, 1993).

If taxon distributions of different extent showed a high degree of concentric 'nestedness' (Patterson & Atmar, 1986; Cutler, 1991) or 'orderedness' (Ryti & Gilpin, 1987), then areas of extreme endemism would indicate areas of extreme richness. This relationship is expected to be strongest at larger spatial scales (Prendergast *et al.*, 1993b; Curnutt *et al.*, 1994), although empirically, this is not always so, as in the case of seed-plant family richness (Plate 8.1b) and endemism (Fig. 8.2) (Williams *et al.*, 1994) and in the case of bumble bee species richness (Plate 8.2a) and endemism (Plate 8.2b).

The statistics of endemism in the tropics, particularly, are of importance to an understanding of many questions, pure and applied, such as how many species there are, and where to expend conservation resources. Unfortunately, for most taxa these statistics are poorly understood. The extent of species occurrences is on the one hand under-estimated as a consequence of low levels of sampling and recording, and parochial taxonomy (leading to widespread species being described under multiple names), and on the other hand is over-estimated through failure to identify cryptic species complexes. The net effect remains, in general, one of substantial under-estimation.

A comprehensive understanding of endemism necessitates understanding

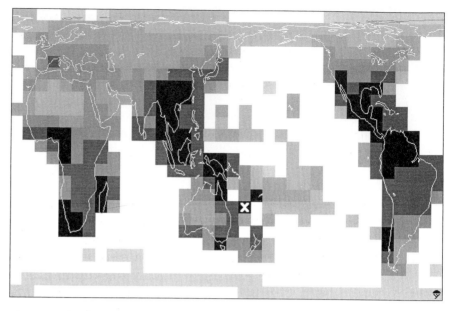

Fig. 8.2 Endemism in families of seed-plants. Endemism is measured as the sum for each grid cell of the inverse of the range sizes for the families represented within it. Map and grey scale as for Fig. 8.1. Data compiled by Williams *et al.* (1994).

broad differences in the typical extent of distribution of species in different higher taxa. For example, whilst most work on endemism has been carried out on plants, various authors have suggested that these may on average be more narrowly distributed than species in many other groups (Gentry, 1986; Gaston, 1994). Estimates of the mean geographic range size of species are lacking for virtually all groups (but see Solem (1984) for a prediction for land snails), although several recent studies have compared range sizes within particular continents (e.g. Anderson & Marcus, 1992; Smith *et al.*, 1994).

8.3 Gradients in diversity

8.3.1 Latitudinal gradients in levels of diversity

By far the most widely cited example of a direct gradient in overall taxonomic diversity is that associated with latitude. The taxonomic diversity of all groups taken together is low towards the poles and high towards the tropics. It has been most extensively documented at the species level (e.g. Simpson, 1964; Tramer, 1974; Wilson, 1974; Rabinovich & Rapoport, 1975; Stevens, 1989; Currie, 1991; Pagel *et al.*, 1991; Willig & Sandlin, 1991; Dennis, 1992), but has also been reported at higher taxonomic levels (e.g. Stehli *et al.*, 1967; Stehli, 1968; Cook, 1969; Wilson, 1974; Rabinovich & Rapoport, 1975; Taylor & Taylor, 1977; Eggleton *et al.*, 1994; Williams *et al.*, 1994). Comparison and interpretation of results needs to account for factors such as the shape of continents, major habitat discontinuities, and variation in methods of analysis (e.g. McCoy & Connor, 1980; Willig & Sandlin, 1991).

A large number of possible mechanisms for latitudinal gradients in taxonomic diversity have been proffered and tested (Pianka, 1966; Brown, 1988; Stevens, 1989, 1992a; Price, 1991; Rohde, 1992; Rosenzweig, 1992; Jablonski, 1993; Colwell & Hurtt, 1994). These include the effects of competition, mutualism, predation, patchiness, environmental stability, environmental predictability, productivity, area, number of habitats, ecological time, evolutionary time, and solar energy (Rohde, 1992). Empirical tests of mechanisms remain inadequate, both in number and in their lack of ambiguity. Substantial variation remains in the support which they receive, with, for example, Rohde (1992) concluding that 'greater species diversity is due to greater "effective" evolutionary time (evolutionary speed) in the tropics, probably as the result of shorter generation times, faster mutation rates, and faster selection at greater temperatures', and Rosenzweig (1992) stating that 'Latitudinal gradients arise because the tropics cover more area than any other zone. Their greater area stimulates speciation and inhibits extinction.'

The vast majority of examples of latitudinal gradients in taxonomic diversity are for terrestrial, freshwater, and coastal marine assemblages. Evidence for gradients in the deep-sea benthos, for example, has only emerged recently, and for logistical reasons is based on small scale (although widely distributed) samples (Rex *et al.*, 1993). Debate persists as to how general such gradients are in the sea (Clarke, 1992; Angel, 1993, 1994; Lambshead, 1993; Vincent & Clarke, 1995).

Perhaps as a consequence of the fact that most studies have been performed on data-sets from the northern hemisphere (notably for North America, and seldom extending as far as the Equator) it has tended tacitly to be assumed that latitudinal gradients in taxonomic diversity are symmetrical in the two hemispheres. This assumption has been challenged by Platnick (1991, 1992). Using data for New World spiders, he provides some evidence that taxonomic diversity is much higher in the southern hemisphere than the northern, likening the pattern of diversity across the earth more to a pear (increasing rapidly from northern regions to the Equator and declining slowly from the Equator to southern regions) than to an egg. Further support for this viewpoint exists at a global scale for termite genera (Fig. 8.3) (Eggleton, 1994), and at a local scale for deep-sea benthos (Poore & Wilson, 1993; Rex *et al.*, 1993; Brey *et al.*, 1994), and possibly for plants (see Fig. 2 in Gentry, 1988). However, there are also analyses which seem to provide little support for marked asymmetry of this kind (e.g. Woodward, 1987; Willig & Sandlin, 1991; Macpherson & Duarte, 1994).

The work of Willig and colleagues (Willig & Selcer, 1989; Willig & Sandlin, 1991) on spatial patterns in the numbers of New World bat species serves to

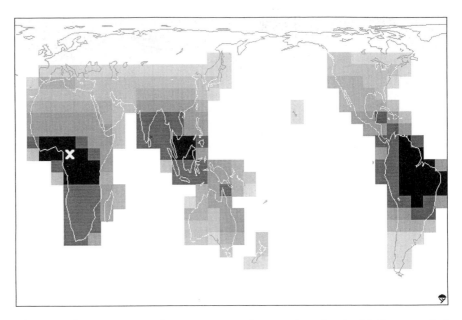

Fig. 8.3 Richness in genera of termites. Map and grey scale as for Fig. 8.1. Data compiled by Eggleton *et al.* (1994).

highlight additional considerations with regard to the symmetry of latitudinal gradients. Whilst they found broad similarity in the numbers of species at different latitudes both north and south of the Equator, they found variation in the model which best describes the relation between bat species numbers and latitude. The simple polynomial models of choice for South America, North America, and the continental New World as a whole, were first, second, and second order equations, respectively; with each model accounting for more than 80% of the variation in species numbers (Willig & Selcer, 1989). Severely non-linear gradients in species richness from low to high latitudes may themselves not be unusual (e.g. Wood & Olmstead, 1984).

Whilst overall taxonomic diversity tends to increase towards low latitudes, within individual higher taxa it is well known that this pattern is not always replicated. For most taxa this is a trivial observation. It says little more than that some higher taxa are adapted to conditions at higher latitudes (or altitudes) (e.g. bumble bees in Plate 8.2a, or cod in Fig. 8.4), or have not had time or opportunity to spread appropriately. However, for major taxa it is less trivial, and much discussion has surrounded possible explanations. As yet, it is unclear what proportion of higher taxa at a given level fail to show increases in diversity with declining latitude. Aphids (Aphidoidea, Hemiptera), sawflies (Symphyta, Hymenoptera), ichneumonid wasps (Ichneumonidae, Hymenoptera), and bees (Apidae, Hymenoptera) are all examples of major terrestrial insect taxa which either are most speciose

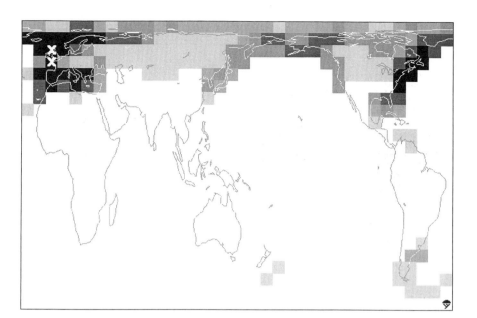

Fig. 8.4 Richness in species of gadid codfish. Map and grey scale as for Fig. 8.1. Data from Cohen *et al.* (1990).

outside of the tropics, or do not markedly increase in richness from mid to low latitudes (Owen & Owen, 1974; Michener, 1979; Janzen, 1981; Gauld, 1986; Dixon *et al.*, 1987; Roubik, 1992; Kouki *et al.*, 1994). Resource fragmentation, the decline in the average density of potential hosts, appears to be a particularly significant force in generating such patterns, by making host specialization more difficult at low latitudes (Janzen, 1981; Dixon *et al.*, 1987; Gauld & Gaston, 1994). Atypical latitudinal changes in diversity may often be associated with substantial shifts in the relative proportions of taxa in a group which exhibit particular life histories or ecologies (e.g. Gauld, 1986; Roubik, 1992).

Exceptions to the general latitudinal gradient in taxonomic diversity may be associated with particular areas. Thus, primarily as a consequence of the patterns of environmental gradients, the species richness of a number of groups does not increase monotonically with declining latitude in Australia, indeed, some groups are most speciose towards higher latitudes (e.g. Fig. 8.5; Cogger & Heatwole, 1981; Kitching, 1981; Pianka & Schall, 1981; Shiel & Williams, 1990; Anderson & Marcus, 1992).

8.3.2 Latitudinal gradients in turnover

A corollary of much discussion surrounding the general increase in the taxonomic diversity of an area towards low latitudes is that there is an associated increase in turnover between areas in the identities of taxa (beta

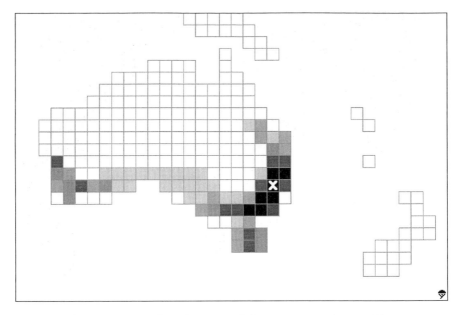

Fig. 8.5 Richness in species of eucalypt trees of the genus *Monocalyptus*. Grid map representing 2° × 2° grid covering Australia. Grey scale as for Fig. 8.1. Data from Ladiges *et al.* (1983a, 1983b, 1989, 1992), Ladiges and Humphries (1986), and references therein.

diversity). Given the importance of the topic generally, the number of studies which have explored any patterns in beta diversity is pitifully small. Direct attempts to explore the interaction between beta diversity and latitude are sufficiently few and heterogeneous (beta diversity can be measured in a variety of ways) as to prevent any generalizations from being drawn (see, e.g. Lahti *et al.*, 1988; Willig & Sandlin, 1991; Harrison *et al.*, 1992; Lawton *et al.*, 1994a).

In contrast to beta diversity *per se*, growing attention is being paid to latitudinal gradients in a related statistic, the mean (or median) geographic range size of species in an area. This has been found to decline towards low latitudes in studies of a sufficient variety of higher taxa for the pattern to have been termed a rule (Rapoport's rule; Plate 8.2c; Rapoport, 1982; Stevens, 1989, 1992a; Pagel *et al.*, 1991; France, 1992; Lawton *et al.*, 1994b; Letcher & Harvey, 1994; Macpherson & Duarte, 1994; Ruggiero, 1994). Numbers of examples are, however, being documented in which the pattern does not hold (e.g. mammal families in Fig. 8.6), raising questions about its generality, but also providing potentially useful tests of mechanisms for those higher taxa for which the interaction is observed (Ricklefs & Latham, 1992; Rohde *et al.*, 1993; Macpherson & Duarte, 1994; Ruggiero, 1994; Smith *et al.*, 1994; Roy *et al.*, 1994). For the bumble bees (Plate 8.2c), mean range-size picks out certain island faunas in Southeast Asia and montane faunas of southern Eurasia and meso-America.

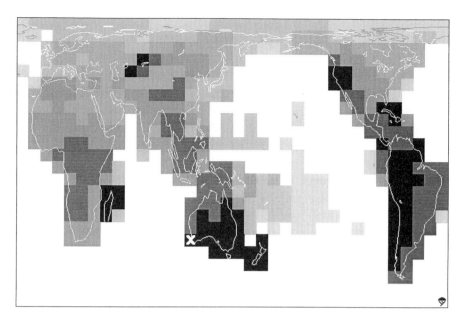

Fig. 8.6 Mean range size for families of land mammals. Mean range size is scored as the sum for each grid cell of the inverse of the range sizes for the families represented, divided by the number of families represented within it. Map and grey scale as for Fig. 8.1. Data compiled from Macdonald (1985) and Anderson and Jones (1984).

8.3.3 Longitudinal gradients

In stark contrast to latitudinal gradients in taxonomic diversity, longitudinal patterns have been largely ignored. This is doubtless because typically they are substantially more complicated, although it has been suggested that ultimately they may prove more illuminating because they are less confounded by systematic changes in multiple environmental parameters than are latitudinal clines (Pagel *et al.*, 1991).

Broad gradients in taxonomic diversity with changing longitude have been reported. These include increases in insect, avian and mammalian species density towards the west of North America (Tramer, 1974; Pagel *et al.*, 1991; Danks, 1993, 1994), an increase in water bird species richness towards the east of southern Africa (Guillet & Crowe, 1986), and an increase in butterfly species richness towards the east of the Iberian peninsula (Martin & Gurrea, 1990). Other examples have been explored by Kiester (1971), McAllister *et al.* (1986, 1994), Willig and Sandlin (1991), O'Brien (1993), Brener and Ruggiero (1994) and Cotgreave and Harvey (1994). Longitude often enters into multiple regression models in which species richness is the dependent variable, though seldom ahead of variables such as latitude (e.g. Willig & Sandlin, 1991; Myklestad & Birks, 1993; Cotgreave & Harvey, 1994).

Common longitudinal patterns across continents for some taxa of temperate northern latitudes might be expected as the result of the predominant west-to-east flow of air masses. These include the tendency of areas near the north-western coasts of both the North American and European continents to have less extreme seasonal climatic variation and generally higher rainfall in comparison with areas near the north-eastern coasts (*Times Atlas of The World*, 1987). Similarly, repeating patterns of marine diversity with longitude might be expected to result from large-scale patterns of ocean water circulation. These factors may change seasonally. For example, in Europe, January mean isotherms generally run north–south, whereas July mean isotherms run more west–east (Wallen, 1970).

Associated with longitude and latitude, an effect of insularity on terrestrial taxonomic diversity can often be observed, with diversity being relatively low in the vicinity of, or increasing away from, coasts (Simpson, 1964; Kiester, 1971; Currie & Paquin, 1987; Currie, 1991). As Simpson (1964) observes, for analyses based on land in grid cells, this may be (at least in part) an artefact of the fact that those cells nearer the coast tend to have less land area on average.

8.3.4 Altitudinal and depth gradients

Spatial measures of taxonomic diversity refer not strictly to areas, but to volumes. In the terrestrial realm the third dimension is commonly construed as the altitude of land. Altitude could arguably be ignored when considering large areas, because its magnitude is small compared with those of longitude or latitude, although it must be remembered that a moderate increase in altitude has, for example, an associated temperature change corresponding to a latitudinal change of several hundred kilometres.

Indeed, most interest in the relationship between taxonomic diversity and altitude for terrestrial systems has concerned small spatial scales and has received particular attention with regard to insects (e.g. Lawton *et al.*, 1987; Wolda, 1987; Fernandes & Price, 1988; McCoy, 1990; Williams, 1991). Here, two broad patterns have been documented, a decline in species richness from low to high altitudes, and a peak of richness at mid-altitudes. Ecological interactions, latitude, disturbance and sampling regime have all been postulated as possible determinants of the relationship actually observed in a given study (Lawton *et al.*, 1987; Wolda, 1987; McCoy, 1990). In the main, simple declines in taxonomic diversity with altitude have been documented for other groups of organisms (Brown, 1988; Haffer, 1990), though there is some evidence for a more complex picture, more akin to that for insects (e.g. Stevens, 1992b).

As at local scales, considerable variation in relationships between elevation and taxonomic diversity has been found at regional scales. Positive or negative relationships have been documented in some cases, with the absence of a relationship in others (e.g. Rabinovich & Rapoport, 1975; Schall &

Plate 2.1 Sex-limited Batesian mimicry of three Malaysian female morphs of *Papilio memnon agenor* (right) and their models (left). Top: *Atrophaneura nox erebus*, female (Perak); *P. memnon*, tailless female f. 'esperi' (Perak). Centre: *A. coon*, female (Malacca); *P. memnon*, tailed female f. 'distantius' (Penang). Bottom: *A. varuna varuna*, female (King Is., Margui); *P. memnon*, tailless female f. 'butlerianus' (Malacca). Female colour pattern and tails are inherited as a single 'supergene'. Male *P. memnon* (not shown) are tailless and black with iridescent blue-green rays.

Plate 2.2 *Heliconius* from Ecuador and N. Peru, showing sympatric Müllerian mimicry and geographic divergence. Top, from near Puyo, E. Ecuador: left, *H. erato notabilis*, male; right, *H. melpomene plesseni*, male. Second row, from W. Ecuador: left, *H. e. cyrbia*, male; right, *H. m. cythera*, female. Third row, from near Zamora, S.E. Ecuador: left, *H. e. etylus*, male; right, *H. m. ecuadorensis*, male. Fourth row, from Andean valleys in S. Ecuador and N. Peru: *H. himera*. Fifth row, from the lower Huallaga, N.E. Peru: left, *H. e. emma*, male; right, *H. m. aglaope*, male. Bottom, from the upper Huallaga and Mayo valleys, N.E. Peru: left, *H. e. favorinus*, male; right, *H. m. amaryllis*, male. *Heliconius himera* is closely related to, and hybridizes with *H. erato* wherever the two species meet, but is rarely sympatric with *H. melpomene*. Although *H. erato* mimics *H. melpomene* wherever they co-occur, there is no form of *H. melpomene* that mimics *H. himera* well.

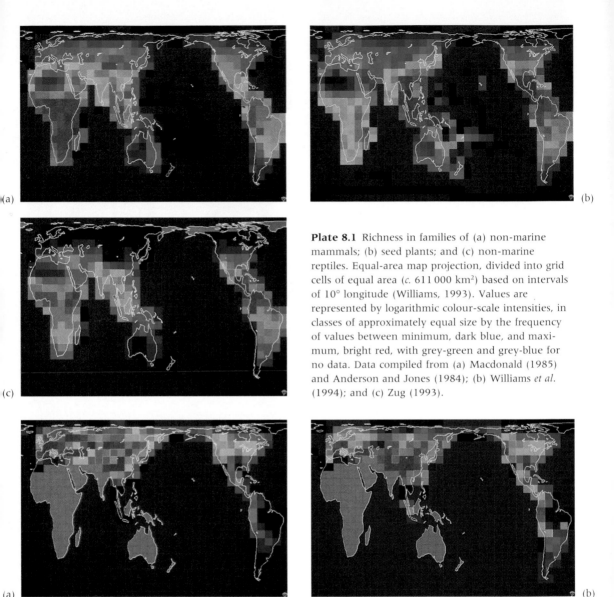

(a)

(b)

(c)

Plate 8.1 Richness in families of (a) non-marine mammals; (b) seed plants; and (c) non-marine reptiles. Equal-area map projection, divided into grid cells of equal area (*c.* 611 000 km²) based on intervals of 10° longitude (Williams, 1993). Values are represented by logarithmic colour-scale intensities, in classes of approximately equal size by the frequency of values between minimum, dark blue, and maximum, bright red, with grey-green and grey-blue for no data. Data compiled from (a) Macdonald (1985) and Anderson and Jones (1984); (b) Williams *et al.* (1994); and (c) Zug (1993).

(a)

(b)

(c)

Plate 8.2 Richness and endemism in species of bumble bees: (a) species richness; (b) species endemism, measured as the sum of the inverse range sizes per grid cell; and (c) mean species endemism, measured as endemism divided by richness for each grid cell. Map and colour scale as for Plate 8.1.

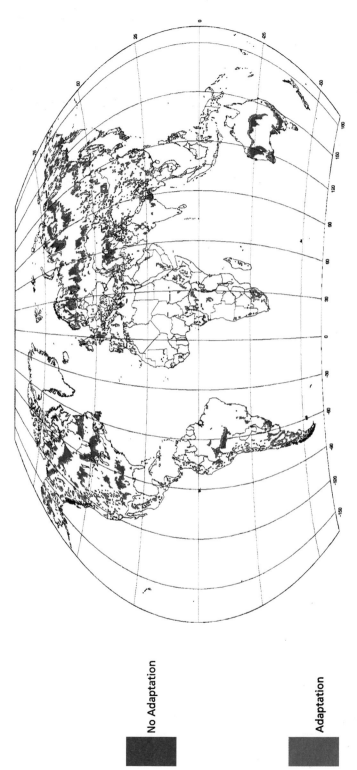

No Adaptation

Adaptation

Plate 14.1 World map with the changing biomes for a climate based on the GFDL model scenario. Red areas are those that do not have a similar vegetation type within 100 km distance and could have problems in adapting to rapid climate change.

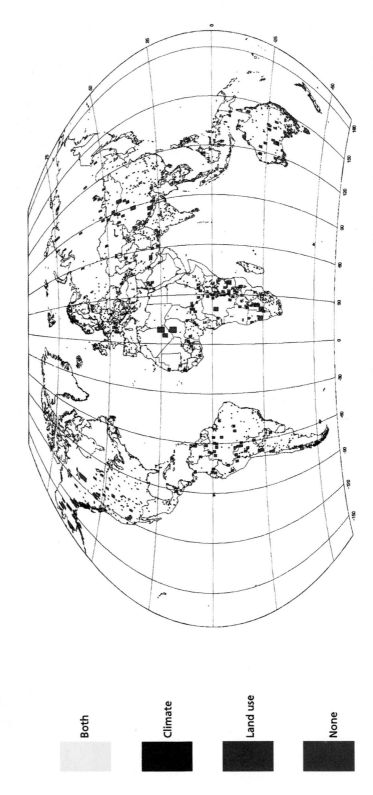

Plate 14.2 Worldwide threats to biodiversity by looking at land-cover change, due to changes in climate and land-use using a 'business-as-usual' future scenario with the integrated global change model, IMAGE (Alcamo, 1994).

Pianka, 1978; White *et al.*, 1984; Currie & Paquin, 1987; White & Miller, 1988; Martin & Gurrea, 1990; Currie, 1991; Linder, 1991; O'Brien, 1993). The extent of more complex interactions is difficult to ascertain, because only correlation and linear regression coefficients are usually presented. The elevation of a region has been characterized in a variety of ways, including lowest, highest, median and mean altitude. Differences between areas in some of these measures probably reflect differences in their topographic diversity, perhaps explaining positive relationships with taxonomic diversity. The topographic and taxonomic diversity of areas have frequently been found to be correlated, with the former sometimes explaining a high proportion of variation in the latter (Holland, 1978; Richerson & Lum, 1980; Adams & Woodward, 1989; Owen, 1989).

When records of species occurrences at different altitudes are collated across large areas, either declines in numbers of species or the hump-shaped relationship with increasing altitude are recovered (Samways, 1989; Olmstead & Wood, 1990; Stevens, 1992b; Fernandes & Lara, 1993; Hunter & Yonzon, 1993).

In some sense, depth can be regarded as the marine equivalent of altitude. However, plainly there are limitations to the parallel because few species are able to achieve a purely aerial existence; distinction must be drawn between the effects of depth on benthic and pelagic assemblages, and in some circumstances a closer analogy may lie between vegetation height in terrestrial systems and depth in marine ones. Again, insights into the interaction of depth and marine taxonomic diversity have largely to be derived from analyses at local scales. Here, in the pelagic and benthic realms, diversity tends to peak at intermediate depths; species richness peaks at depths of 1000–1500 m for pelagic assemblages, and in many taxa increases downslope to a maximum at 1000–2000 m for megabenthos and 2000–3000 m for macrobenthic infauna (Rex, 1981; Etter & Grassle, 1992; Angel, 1993, 1994; Lambshead, 1993; see also Macpherson & Duarte, 1994); maximum diversity tends to occur at shallower depths for higher trophic levels (Huston, 1994). The peak in species richness at intermediate depths may be associated with a peak in the vertical depth range of individual species; the vertical depth range of deep-sea benthic species tends to peak at intermediate depths and decreases towards either shallow or abyssal boundaries (Pineda, 1993).

Whether terrestrial or marine, analyses of the interaction between altitude and taxonomic diversity can be seriously confounded by area effects; the available space at different elevations can vary drastically. Detailed studies of these effects would be valuable.

8.3.5 Peninsulas and bays

The shapes of land masses and water bodies can have profound effects on the taxonomic diversity associated with them, leading to gradients in that diversity. The so-called 'peninsula effect', the reduction in diversity towards the end of a peninsula (i.e. from the mainland base towards the distal tip), provides a classic, if much debated example. Although initially claimed as a quite general phenomenon, it has proven to be more occasional, demonstrated by some higher taxa on some peninsulas (Simpson, 1964; Due & Polis, 1986; Brown, 1987; Means & Simberloff, 1987; Schwartz, 1988; Brown & Opler, 1990; Martin & Gurrea, 1990; Rapoport, 1994). Likewise, some taxa show a marine equivalent to the peninsular effect, the 'bay effect', a decline in diversity with distance from the open sea (Rapoport, 1994).

8.4 Co-variance in diversity among different taxa

8.4.1 Coincidence of extreme values among different taxa

The extent to which hotspots of species richness for different higher taxa coincide has only recently begun to be explored quantitatively. Prendergast *et al.* (1993b) showed that within Britain species-rich areas frequently do not coincide for different taxa. Williams and Gaston (1994) reviewed evidence showing that the same is broadly true within continents (Africa – MacKinnon & MacKinnon, 1986; Australia – Schall & Pianka, 1978; Pianka & Schall, 1981; North America – Schall & Pianka, 1978; Currie, 1991; South America – Gentry, 1992). Here, this outcome can be illustrated at the global scale, with Africa as the centre of richness for families of non-marine mammals (Plate 8.1a), Asia as the centre for families of seed plants (Plate 8.1b), and the Americas as the centre for families of non-marine reptiles (Plate 8.1c).

Largely missing from analyses of such patterns has been any consideration of what is or is not a significant level of coincidence of hotspots. Two issues need to be disentangled. First, is whether the level of coincidence is greater or less than expected by chance. Second, is whether the level of coincidence is of use for the purposes of conservation. The former is testable, with reference to an appropriate null model. The latter is based on a more arbitrary decision. There is some evidence that coincidence of hotspots is greater than expected under a random draw model (Gaston & David, 1994). The practical value of such coincidence is, however, probably less than has often been assumed.

8.4.2 Co-variance in gradients among different taxa

A moderate sized literature exists on the subject of broader relationships between the taxonomic diversity of pairs of groups of animals or plants in space. Until recently, it has remained scattered and lacking in synthesis. Gaston (in press) provides an extensive list of studies documenting or analysing patterns of spatial co-variance in species richness (additions include Pomeroy, 1993). Much store seems to have been placed by the observation that most higher taxa tend to have low diversity in arctic regions, at high altitudes, and in arid environments, and to have high diversity at low latitudes, at low to middle elevations, and in forest environments. In consequence, positive co-variance in the diversity of pairs of higher taxa has been widely assumed, tacitly or explicitly, particularly in the context of conservation and the identification of areas of high biodiversity. Perusal of species richness maps for different groups at the continental level serves, nonetheless, to demonstrate that broad generalizations about spatial patterns of diversity do not necessarily translate into such. Analysis of relationships between the species richness of different higher taxa in different areas reveals a full range of interactions, from the strongly positive, through the weak and non-existent, to the strongly negative or complementary (Fig. 8.7; e.g. Kiester, 1971; Rogers, 1976; Schall & Pianka, 1977, 1978; Pianka & Schall, 1981; Currie, 1991; Gaston, 1992).

Four sets of mechanisms can potentially explain relationships between the diversity of pairs of higher taxa, where these are observed (Gaston, in press). These are: (i) random draw—patterns of co-variance result because assemblages are a random draw of taxa from a larger pool of available colonists (see below); (ii) interactions—patterns of co-variance result from trophic, competitive or mutualistic interactions between taxa; (iii) common determinants of diversity—patterns of co-variance result from independent but similar interactions of both groups of taxa with a third variable; and (iv) spatial covariance in different factors—factors determining the diversity of both groups of taxa are not shared, but themselves show spatial co-variance.

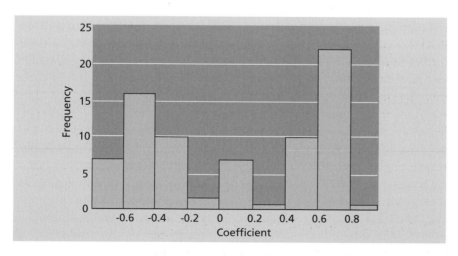

Fig. 8.7 Frequency distribution of correlation coefficients between the species richness of pairs of groups of amphibians, reptiles, birds and mammals in grid squares (240 × 240 km) across Australia. Comparisons are not independent. From data in Pianka and Schall (1981).

Emphasis has been given to trophic and competitive interactions as determinants of patterns of spatial co-variance in the diversity of groups of taxa. However, it seems probable that common determinants of diversity may explain a greater number of such relationships.

8.5 Relationships to environmental variables and local richness

8.5.1 Environmental variables

Relationships between taxonomic richness and environmental variables (e.g. evapotranspiration, hours of sunshine, precipitation, primary productivity, soil fertility, temperature) have been explored both for heuristic reasons and for the practical purpose of attempting to predict the levels of species richness in different areas (Chapter 4). In broad terms, the taxonomic diversity of an area will usually be found to alter predictably with changes in several environmental variables. For example, O'Brien (1993) finds significant linear relationships (Pearson's product correlations ranging between -0.36 and 0.78) between the species richness of woody plants in 65 grid cells of $20\,000$ km^2 in southern Africa and precipitation (annual, maximum monthly, minimum monthly, and seasonal variability), actual evapotranspiration (annual, maximum monthly, and minimum monthly), Thornwaite's moisture index, number of months with precipitation greater than or equal to 2.5 mm, annual water surplus, annual water deficit, months with water deficit, and potential evapotranspiration (maximum monthly, minimum monthly, and seasonal variability). In general, those environmental factors which are energy-related (food or limiting nutrient availability, productivity) have more explanatory power than other factors (Wright *et al.*, 1993).

The study of relationships between taxonomic diversity and environmental variables is complicated by several analytical difficulties. First, there is frequently spatial co-variance between pairs, and more complex combinations, of environmental variables. For example, Schall and Pianka (1978) report correlation coefficients between five climatic variables, for 895 quadrats of $1°$ latitude by $1°$ longitude across the continental United States, which range from 0.15 to 0.87 (e.g. annual temperature $\times$ frost free days $= 0.87$, hours of sunshine $\times$ coefficient of variation of precipitation $= 0.77$). Second, environmental variables tend to be strongly spatially autocorrelated. Third, interactions between taxonomic diversity and individual environmental variables, especially those which are energy-related, may be seriously non-linear (even under transformation). These problems may be especially acute when attempting to derive a model which explains the most variation in taxonomic diversity in terms of combinations of environmental variables.

218

Here, spatial autocorrelation is almost invariably ignored, and pair-wise relationships treated as if they were linear.

Two additional issues have served further to complicate and sometimes confuse interpretation of general patterns between taxonomic diversity and environmental variables. The first of these has been differences in the size of the areas that are sampled. The second has been differences in the size of the arena within which the sampled areas are distributed (the size of the larger area within which study areas are embedded). For example, when study areas are large, species richness tends to rise monotonically with indices of available energy, and when study areas are small the interaction tends to be hump-shaped or peaked, with relatively low richness when energy is low or very high (Wright *et al.*, 1993). Because the range of values of energy availability tends to increase with the spatial extent of a study, this extent may influence whether the increasing or decreasing phase of a hump-shaped relationship, or the full pattern is observed. Such hump-shaped interactions may themselves result, in part or otherwise, from how widespread the respective conditions are, intermediate ones perhaps being generally more widespread (Tilman & Pacala, 1993).

8.5.2 Regional and local diversity patterns

Although this chapter is concerned primarily with spatial patterns in taxonomic diversity at the regional scale, local and regional patterns in diversity are not independent. Indeed, it is becoming increasingly evident that the structures of regional species pools can exert a powerful interaction with the structure of more local assemblages (see Ricklefs, 1987; Tonn *et al.*, 1990; Cornell & Lawton, 1992; Eriksson, 1993; and references therein). For example, simple random draw models (drawing species at random from regional species pools) have been found to describe, or contribute to the description of, various patterns in the structure of species assemblages at smaller spatial scales. These patterns include body size ratios (e.g. Greene, 1987; Williams, 1988; Zwölfer & Brandl, 1989), predator–prey ratios (e.g. Cole, 1980; Warren & Gaston, 1992), levels of floristic or faunistic similarity (e.g. Rice & Bellard, 1982), numbers of higher taxa (e.g. genera, families; see Järvinen, 1982, for a historical review; Rogers, 1976; Gotelli & Abele, 1982; Tokeshi, 1991; Williams & Gaston, 1994), and co-variance in the species richness of pairs of higher taxa (Gaston, in press).

Local taxonomic diversity tends to be an increasing function of regional diversity, such that local areas in very diverse regions tend to have greater levels of diversity than do local areas in regions of low diversity (e.g. Cornell, 1985a, 1985b; Ricklefs, 1987, 1989; Lawton, 1990; Tonn *et al.*, 1990; Cornell & Lawton, 1992; Hawkins & Compton, 1992; Gaston & Gauld, 1993). Both theory and data suggest that unsaturated local assemblages are likely to be

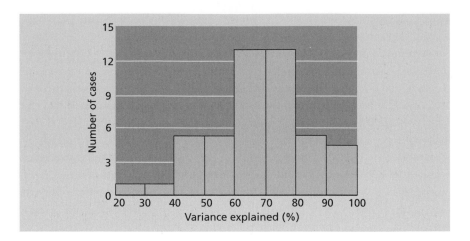

Fig. 8.8 The percentage of geographic variation in regional species richness of higher taxa explained by independent variables in a selection of published studies.

ubiquitous, and thus that local richness does not asymptote with increasing regional richness (i.e. proportional sampling is the norm). If local richness is not often saturated, then regional richness is freed from local constraint, and other limits on regional richness, such as phylogenetic diversification over evolutionary time scales (Chapter 9), become important (Cornell & Lawton, 1992).

8.6 Conclusion

A large proportion of geographic variation in the taxonomic diversity of animal and plant assemblages can be 'explained' in terms of a few variables. For a selection of studies, the proportion ranges from *c.* 30 to *c.* 90%, with a mode of around 60–80% (Fig. 8.8). In general, the set of variables which will explain most of the geographic variation in the taxonomic diversity of most groups of organisms is known. However, whilst it may often also be possible to predict accurately something about their rank importance, it is seldom, if ever, possible to predict accurately their absolute contribution. This remains an important test of our understanding of spatial patterns in taxonomic diversity, and one which has still to be met.

Acknowledgments

K.J.G. is a Royal Society University Research Fellow. This work was funded in part by a NERC grant to K.J.G. We are grateful to Paul Eggleton, Chris Humphries, Natasha Loder, and Dick Vane-Wright for discussion, and for allowing us to make use of data they have compiled.

References

Adams J.M. & Woodward F.I. (1989) Patterns in tree species richness as a test of the glacial extinction hypothesis. *Nature* **339**, 699–701. [8.3.4]

Allan J.D. & Flecker A.S. (1993) Biodiversity conservation in running waters. *BioScience* **43**, 32–43. [8.2.3]

Anderson S. (1994) Area and endemism. *Quart. Rev. Biol.* **69**, 451–471. [8.2.4] [8.3.1]

Anderson S. & Jones J.K. (eds) (1984) *Orders and Families of Recent Mammals of the World.* Wiley, New York. [Plate 8.1] [8.3.2]

Anderson S. & Marcus L.F. (1992) Areography of Australian tetrapods. *Aust. J. Zool.* **40**, 627–651. [8.2.4] [8.3.1]

Angel M.V. (1993) Biodiversity of the pelagic ocean. *Conserv. Biol.* **7**, 760–772. [8.2.3] [8.3.1] [8.3.4]

Angel M.V. (1994) Spatial distribution of marine organisms: patterns and processes. In: *Large-scale Ecology and Conservation Biology* (eds P.J. Edwards, R.M. May & N.R. Webb), pp. 59–109. Blackwell Scientific, Oxford. [8.2.1] [8.3.1] [8.3.4]

Brener A.G.F. & Ruggiero A. (1994) Leaf-cutting ants (*Atta* and *Acromyrmex*) inhabiting Argentina: patterns in species richness and geographical range sizes. *J. Biogeog.* **21**, 391–399. [8.3.3]

Brey T., Klages M., Dahm C., Gorny M., Gutt J., Hain S., Stiller M., Arntz W.E., Wägele J-W. & Zimmermann A. (1994) Antarctic benthic diversity. *Nature* **368**, 297. [8.3.1]

Briggs J.C. (1994) Species diversity: land and sea compared. *Syst. Biol.* **43**, 130–135. [8.2.1]

Brown J.H. (1988) Species diversity. In: *Analytical Biogeography: an integrated approach to the study of animal and plant distributions* (eds A.A. Myers & P.S. Giller), pp. 57–89. Chapman & Hall, London. [8.3.1] [8.3.4]

Brown J.W. (1987) The peninsular effect in Baja California: an entomological assessment. *J. Biogeog.* **14**, 359–365. [8.3.4]

Brown J.W. & Opler P.A. (1990) Patterns of butterfly species density in peninsular Florida. *J. Biogeog.* **17**, 615–622. [8.3.4]

Chapin III F.S. & Körner C. (1994) Arctic and alpine biodiversity: patterns, causes and ecosystem consequences. *Trends Ecol. Evol.* **9**, 45–47. [8.2.3]

Clarke A. (1992) Is there a latitudinal diversity cline in the sea? *Trends Ecol. Evol.* **7**, 286–287. [8.3.1]

Cogger H.G. & Heatwole H. (1981) The Australian reptiles: origins, biogeography, distribution patterns and island evolution. In: *Ecological Biogeography of Australia* (ed. A. Keast), pp. 1331–1373. Junk, The Hague. [8.3.1]

Cohen D.M., Ianada T., Iwamoto T. & Scialabba N. (1990) *FAO Species Catalogue: gadiform fishes of the world (order Gadiformes). An annotated and illustrated catalogue of cods, hakes, grenadiers and other gadiform fishes known to date.* FAO Fisheries Synopsis no. 125, vol. 10. Food & Agriculture Organization of the United Nations, Rome. [8.3.1]

Cole B.J. (1980) Trophic structure of a grassland insect community. *Nature* **288**, 76–77. [8.5.2]

Cole F.R., Reeder D.M. & Wilson D.E. (1994) A synopsis of distribution patterns and the conservation of mammal species. *J. Mammal.* **75**, 266–276. [8.2.2]

Colwell R.K. & Hurtt G.C. (1994) Nonbiological gradients in species richness and a spurious Rapoport effect. *Am. Nat.* **144**, 570–595. [8.3.1]

Cook R.E. (1969) Variation in species density of North American birds. *Syst. Zool.* **18**, 63–84. [8.3.1]

Cornell H.V. (1985a) Local and regional richness of cynipine gall wasps on California oaks. *Ecology* **66**, 1247–1260. [8.5.2]

Cornell H.V. (1985b) Species assemblages of cynipine gall wasps are not saturated. *Am. Nat.* **126**, 565–569. [8.5.2]

Cornell H.V. & Lawton J.H. (1992) Species interactions, local and regional processes, and limits to the richness of ecological communities: a theoretical perspective. *J. Anim. Ecol.* **61**, 1–12. [8.1] [8.5.2]

Cotgreave P. & Harvey P.H. (1994) Associations among biogeography, phylogeny and bird species diversity. *Biodiv. Lett.* **2**, 46–55. [8.3.3]

Curnutt J., Lockwood J., Lu H.-K., Nott P. & Russell G. (1994) Hotspots and species diversity. *Nature* **367**, 326–327. [8.2.4]

Currie D.J. (1991) Energy and large-scale patterns of animal- and plant-species richness. *Am. Nat.* **137**, 27–49. [8.3.1] [8.3.4] [8.4.1] [8.4.2]

Currie D.J. & Paquin V. (1987) Large-scale biogeographical patterns of species richness of trees. *Nature* **329**, 326–327. [8.3.3] [8.3.4]

Cutler A. (1991) Nested faunas and extinction in fragmented habitats. *Conserv. Biol.* **5**, 496–505. [8.2.4]

Daniels R.J.R., Hegde M., Joshi N.V. & Gadgil M. (1991) Assigning conservation value: a case study from India. *Conserv. Biol.* **5**, 464–475. [8.2.4]

Danks H.V. (1993) Patterns of diversity in the Canadian insect fauna. *Mem. Ent. Soc. Can.* **165**, 51–74. [8.3.3]

Danks H.V. (1994) Regional diversity of insects in North America. *Am. Entomol.* **40**, 50–55. [8.3.3]

Davis S.D., Droop S.J.M., Gregerson P., Henson L., Leon C.J., Villa-Lobos J.L., Synge H. & Zantovska J. (1986) *Plants in Danger: what do we know.* IUCN, Cambridge. [8.1] [8.2.4]

Dennis R.L.H. (1992) Islands, regions, ranges and gradients. In: *The Ecology of Butterflies in Britain* (ed. R.L.H. Dennis), pp. 1–21. Oxford University Press. [8.3.1]

Dixon A.F.G., Kindlmann P., Leps J. & Holman J. (1987) Why there are so few species of aphids, especially in the tropics. *Am. Nat.* **129**, 580–592. [8.3.1]

Due A.D. & Polis G.A. (1986) Trends in scorpion diversity along the Baja California peninsula. *Am. Nat.* **128**, 460–468. [8.3.4]

Eggleton P. (1994) Termites live in a pear-shaped world: a response to Platnick. *J. Nat. Hist.* **28**, 1209–1212. [8.3.1]

Eggleton P., Williams P.H. & Gaston K.J. (1994) Explaining global termite diversity: productivity or history? *Biodiv. Conserv.* **3**, 318–330. [8.1] [8.3.1]

Ehrlich P.R. (1975) The population ecology of coral reef fishes. *Annu. Rev. Ecol. Syst.* **6**, 211–247. [8.2.2]

Eriksson O. (1993) The species-pool hypothesis and plant community diversity. *Oikos* **68**, 371–374. [8.5.2]

Etter R.J. & Grassle J.F. (1992) Patterns of species diversity in the deep sea as a function of sediment particle size diversity. *Nature* **360**, 576–578. [8.3.4]

Fernandes G.W. & Lara A.C.F. (1993) Diversity of Indonesian gall-forming herbivores along altitudinal gradients. *Biodiv. Lett.* **1**, 186–192. [8.3.4]

Fernandes G.W. & Price P.W. (1988) Biogeographical gradients in galling species richness: tests of hypotheses. *Oecologia* **76**, 161–167. [8.3.4]

Flessa K.W. (1975) Area, continental drift and mammalian diversity. *Paleobiology* **1**, 189–194. [8.2.3]

Flessa K.W. (1981) The regulation of mammalian faunal similarity among the continents. *J. Biogeog.* **8**, 427–437. [8.2.2]

Fliermans C.B. & Balkwill D.L. (1989) Microbial life in deep terrestrial subsurfaces. *BioScience* **39**, 370–377. [8.2.3]

France R. (1992) The North American latitudinal gradient in species richness and geographical range of freshwater crayfish and amphipods. *Am. Nat.* **139**, 342–354. [8.3.2]

Gaston K.J. (1992) Regional numbers of insect and plant species. *Functional Ecol.* **6**, 243–247. [8.4.2]

Gaston K.J. (1994) *Rarity.* Chapman & Hall, London. [8.2.4]

Gaston K.J. Spatial covariance in the species richness of higher taxa. In: *The Genesis and Maintenance of Biological Diversity* (eds M. Hochberg, M.E. Clobert & R. Barbault). Oxford University Press, in press. [8.4.2] [8.5.2]

Gaston K.J. & David R. (1994) Hotspots across Europe. *Biodiv. Lett.* **2**, 108–116. [8.4.1]

Gaston K.J. & Gauld I.D. (1993) How many species of pimplines are there in Costa Rica? *J. Trop. Ecol.* **9**, 491–499. [8.5.2]

Gaston K.J. & Hudson E. (1994) Regional patterns of diversity and estimates of global insect species richness. *Biodiv. Conserv.* **3**, 493–500. [8.2.2]

Gauld I.D. (1986) Latitudinal gradients in ichneumonid species-richness in Australia. *Ecol. Entomol.* **11**, 155–161. [8.3.1]

Gauld I.D. & Gaston K.J. (1994) The taste of enemy-free space: parasitoids and nasty hosts. In: *Parasitoid Community Ecology* (eds B.A. Hawkins & W. Sheehan), pp. 279–299. Oxford University Press. [8.3.1]

Gentry A.H. (1986) Endemism in tropical versus temperate plant communities. In: *Conservation Biology: the science of scarcity and diversity* (ed. M.E. Soulé), pp. 153–181. Sinauer Associates, Sunderland, Massachusetts [8.2.4]

Gentry A.H. (1988) Changes in plant community diversity and floristic composition on environmental and geographical gradients. *Ann. Missouri Bot. Gard.* **75**, 1–34. [8.3.1]

Gentry A.H. (1992) Tropical forest biodiversity: distribution patterns and their conservational significance. *Oikos* **63**, 19–28. [8.2.4] [8.4.1]

Gotelli N.J. & Abele L.G. (1982) Statistical distributions of West Indian land bird families. *J. Biogeog.* **9**, 421–435. [8.5.2]

Grassle J.F. (1989) Species diversity in deep-sea communities. *Trends Ecol. Evol.* **4**, 12–15. [8.2.3]

Grassle J.F. (1991) Deep-sea benthic biodiversity. *BioScience* **41**, 464–469. [8.2.3]

Grassle J.F. & Maciolek N.J. (1992) Deep-sea species richness: regional and local diversity estimates from quantitative bottom samples. *Am. Nat.* **139**, 313–341. [8.2.1]

Greene E. (1987) Sizing up size ratios. *Trends Ecol. Evol.* **2**, 79–81. [8.5.2]

Groombridge B. (ed.) (1992) *Global Biodiversity: status of the Earth's living resources.* Chapman & Hall, London. [8.1] [8.2.2] [8.2.3]

Guillet A. & Crowe T.M. (1986) A preliminary investigation of patterns of distribution and species richness of southern African waterbirds. *S. Afr. J. Wildl. Res.* **16**, 65–81. [8.3.3]

Haffer J. (1990) Avian species richness in tropical South America. *Stud. Neotrop. Fauna Environ.* **25**, 157–183. [8.2.3] [8.3.4]

Harrison S., Ross S.J. & Lawton J.H. (1992) Beta diversity on geographic gradients in Britain. *J. Anim. Ecol.* **61**, 151–158. [8.3.2]

Hawkins B.A. & Compton S.G. (1992) African fig wasp communities: undersaturation and latitudinal gradients in species richness. *J. Anim. Ecol.* **61**, 361–372. [8.5.2]

Henderson A., Churchill S. & Luteyn J. (1991) Neotropical plant diversity. *Nature* **351**, 21–22. [8.2.3]

Holland P.G. (1978) An evolutionary biogeography of the genus *Aloe. J. Biogeog.* **5**, 213–226. [8.3.4]

Howard P.C. (1991) *Nature Conservation in Uganda's Tropical Forest Reserves.* IUCN, Gland, Switzerland. [8.2.4]

Hunter Jr M.L. & Yonzon P. (1993) Altitudinal distributions of birds, mammals, people, forests, and parks in Nepal. *Conserv. Biol.* **7**, 420–423. [8.3.4]

Huston M.A. (1994) *Biological Diversity: the coexistence of species on changing landscapes.* Cambridge University Press. [8.3.4]

ICBP (1992) *Putting Biodiversity on the Map: priority areas for global conservation.* ICBP (BirdLife International), Cambridge. [8.2.4]

Jablonski D. (1993) The tropics as a source of evolutionary novelty through geological time. *Nature* **364**, 142–144. [8.3.1]

Janzen D.H. (1981) The peak in North American ichneumonid species richness lies between 38° and 42°N. *Ecology* **62**, 532–537. [8.3.1]

Järvinen O. (1982) Species-to-genus ratios in biogeography: a historical note. *J. Biogeog.* **9**, 363–370. [8.5.2]

Kiester A.R. (1971) Species density of North American amphibians and reptiles. *Syst. Zool.* **20**, 127–157. [8.3.3] [8.4.2]

Kitching R.L. (1981) The geography of the Australian Papilionoidea. In: *Ecological Biogeography of Australia* (ed. A. Keast), pp. 977–1005. Junk, The Hague. [8.3.1]

Kouki J., Niemelä P. & Viitasaari M. (1994) Reversed latitudinal gradient in species richness of sawflies (Hymenoptera, Symphyta). *Ann. Zool. Fennici* **31**, 83–88. [8.3.1]

Ladiges P.Y. & Humphries C.J. (1986) Relationships in the stringybarks, *Eucalyptus* L'Her. informal subgenus *Monocalyptus* series Capitellatae and Olsenianae: phylogenetic hypotheses, biogeography and classification. *Aust. J. Bot.* **34**, 603–632. [8.3.1]

Ladiges P.Y., Humphries C.J. & Brooker M.I.H. (1983a) Cladistic relationships and biogeographic patterns in the peppermint group of *Eucalyptus* L'Her. (informal subgroup Amydalininae, subgenus *Monocalyptus* Pryor & Johnson) and the description of a new species, *E. willisi*. *Aust. J. Bot.* **31**, 565–584. [8.3.1]

Ladiges P.Y., Humphries C.J. & Brooker M.I.H. (1983b) Cladistic and biogeographic analysis of Western Australian species of *Eucalyptus* L'Her. informal subgenus *Monocalyptus* Pryor & Johnson. *Aust. J. Bot.* **35**, 251–281. [8.3.1]

Ladiges P.Y., Newnham M.R. & Humphries C.J. (1989) Systematics and biogeography of the Australian 'green ash' eucalypts (*Monocalyptus*). *Cladistics* **5**, 345–364. [8.3.1]

Ladiges P.Y., Prober S.M. & Nelson G. (1992) Cladistic and biogeographic analysis of the 'blue ash' eucalypts. *Cladistics* **8**, 103–124. [8.3.1]

Lahti T., Kurtto A. & Väisänen R.A. (1988) Floristic composition and regional species richness of vascular plants in Finland. *Ann. Bot. Fennici* **25**, 281–291. [8.2.4] [8.3.2]

Lambshead P.J.D. (1993) Recent developments in marine benthic biodiversity research. *Océanis* **19**, 5–24. [8.3.1] [8.3.4]

Lawton J.H. (1990) Local and regional species-richness of bracken-feeding insects. In: *Bracken Biology and Management* (eds J.A. Thompson & R.T. Smith), pp. 197–202. Australian Institute of Agricultural Science, Sydney. [8.5.2]

Lawton J.H., MacGarvin M. & Heads P.A. (1987) Effects of altitude on the abundance and species richness of insect herbivores on bracken. *J. Anim. Ecol.* **56**. 147–160. [8.3.4]

Lawton J.H., Prendergast J.R. & Eversham B.C. (1994a) The numbers and spatial distributions of species: analyses of British data. In: *Systematic and Conservation Evaluation* (eds P.L. Forey, C.J. Humphries & R.I. Vane-Wright), pp. 177–195. Clarendon Press, Oxford. [8.3.2]

Lawton J.H., Nee S., Letcher A.J. & Harvey P.H. (1994b) Animal distributions: patterns and processes. In: *Large-scale Ecology and Conservation Biology* (eds P.J. Edwards, R.M. May & N.R. Webb), pp. 41–58. Blackwell Scientific, Oxford. [8.3.2]

Lee K.E. (1991) The diversity of soil organisms. In: *The Biodiversity of Microorganisms and Invertebrates: its role in sustainable agriculture* (ed. D.L. Hawksworth), pp. 73–87. CAB International, Wallingford, Oxon. [8.2.3]

Letcher A.J. & Harvey P.H. (1994) Variation in geographical range size among mammals of the Palearctic. *Am. Nat.* **144**, 30–42. [8.3.2]

Linder H.P. (1991) Environmental correlates of patterns of species richness in the southwestern Cape Province of South Africa. *J. Biogeog.* **18**, 509–518. [8.3.4]

Major J. (1988) Endemism: a botanical perspective. In: *Analytical Biogeography: an integrated approach to the study of animal and plant distributions* (eds A.A. Myers & P.S. Giller), pp. 117–146. Chapman & Hall, London. [8.2.4]

Mares M.A. (1992) Neotropical mammals and the myth of Amazonian biodiversity. *Science* **255**, 976–979.

Marmonier P., Verivier P., Gibert J. & Dole-Olivier M-J. (1993) Biodiversity in ground waters. *Trends Ecol. Evol.* **8**, 392–395. [8.2.3]

Martin J. & Gurrea P. (1990) The peninsular effect in Iberian butterflies (Lepidoptera: Papilionoidea and Hesperioidea). *J. Biogeog.* **17**, 85–96. [8.3.3] [8.3.4]

May R.M. (1992) Biodiversity: bottoms up for the oceans. *Nature* **357**, 278–279. [8.2.1]

May R.M. (1994) Biological diversity: differences between land and sea. *Phil. Trans. R. Soc., Lond. B* **343**, 105–111. [8.2.1]

McAllister D.E., Platania S.P., Schueler F.W., Baldwin M.E. & Lee D.S. (1986) Ichthyofaunal patterns on a geographical grid. In: *Zoogeography of Freshwater Fishes of North America* (eds C.H. Hocutt & E.D. Wiley), pp. 17–51. Wiley, New York. [8.3.3]

McAllister D.E., Schueler F.W., Roberts C.M. & Hawkins J.P. (1994) Mapping and GIS analysis of the global distribution of coral reef fishes on an equal-area grid. In: *Mapping the Diversity of Nature* (ed. R.I. Miller), pp. 155–175. Chapman & Hall, London. [8.2.2] [8.3.3]

McCoy E.D. (1990) The distribution of insects along elevational gradients. *Oikos* **58**, 313–322. [8.3.4]

McCoy E.D. & Connor E.F. (1980) Latitudinal gradients in the species diversity of North American mammals. *Evolution* **34**, 193–203. [8.3.1] [8.3.4]

Macdonald D. (ed.) (1985) *The Encyclopaedia of Mammals* (2 vols). Guild Publishing, London. [Plate 8.3] [8.3.2]

MacKinnon J. & MacKinnon K. (1986) *Review of the Protected Areas System in the Afrotropical Realm*. IUCN and UNEP, Cambridge. [8.3.4] [8.4.1]

McNeely J.A., Miller K.R., Reid W.V., Mittermeier R.A. & Werner T.B. (1990) *Conserving the World's Biodiversity*. IUCN, WRI, CI, WWF and World Bank, Washington DC. [8.2.2]

Macpherson E. & Duarte C.M. (1994) Patterns in species richness, size, and latitudinal range of East Atlantic fishes. *Ecography* **17**, 242–248. [8.3.1] [8.3.2] [8.3.4]

Means D.B. & Simberloff D. (1987) The peninsula effect: habitat-correlated species decline in Florida's herpetofauna. *J. Biogeog.* **14**, 551–568. [8.3.4]

Michener C.D. (1979) Biogeography of the bees. *Ann. Missouri Bot. Gard.* **66**, 277–347. [8.3.1]

Mittermeier R.A. (1988) Primate diversity and the tropical forest: case studies from Brazil and Madagascar and the importance of the megadiversity countries. In: *Biodiversity* (eds E.O. Wilson & F.M. Peters), pp. 145–154. National Academy Press, Washington. [8.2.3]

Mittermeier R.A. & Werner T.B. (1990) Wealth of plants and animals unites 'megadiversity' countries. *Tropicus* **4**, 1, 4–5. [8.2.3]

Myers N. (1988) Threatened biotas: 'hot spots' in tropical forests. *Environmentalist* **8**, 187–208. [8.2.4]

Myers N. (1990) The biodiversity challenge: expanded hot-spots analysis. *Environmentalist* **10**, 243–256. [8.2.4]

Myklestad Å & Birks H.J.B. (1993) A numerical analysis of the distribution patterns of *Salix* L. species in Europe. *J. Biogeog.* **20**, 1–32. [8.3.3]

Nelson B.W., Ferreira C.A.C., da Silva M.F. & Kawasaki M.L. (1990) Endemism centres, refugia and botanical collection density in Brazilian Amazonia. *Nature* **345**, 714–716. [8.1]

O'Brien E.M. (1993) Climatic gradients in woody plant species richness; towards an explanation based on an analysis of southern Africa's woody flora. *J. Biogeog.* **20**, 181–198. [8.3.3] [8.3.4] [8.5.1]

Olmstead K.L. & Wood T.K. (1990) Altitudinal patterns in species richness of Neotropical treehoppers (Homoptera: Membracidae): the role of ants: *Proc. Entomol. Soc. Wash.* **92**, 552–560. [8.3.4]

Owen D.F. & Owen J. (1974) Species diversity in temperate and tropical Ichneumonidae. *Nature* **249**, 583–584. [8.3.1]

Owen J.G. (1989) Patterns of herpetofaunal species richness: relation to temperature, precipitation, and variance in elevation. *J. Biogeog.* **16**, 141–150. [8.3.4]

Pagel M.D., May R.M. & Collie A.R. (1991) Ecological aspects of the geographical distribution and diversity of mammalian species. *Am. Nat.* **137**, 791–815. [8.3.1] [8.3.2] [8.3.3]

Paterson G.L.J. (1993) *Patterns of polychaete assemblage structure from bathymetric transects in the Rockall Trough, NE Atlantic Ocean*. PhD thesis, University of Wales. [8.1]

Patterson B.D. & Atmar W. (1986) Nested subsets and the structure of insular mammalian faunas and archipelagoes. *Biol. J. Linn. Soc.* **28**, 65–82. [8.2.4]

Pianka E.R. (1966) Latitudinal gradients in species diversity: a review of concepts. *Am. Nat.* **100**, 33–46. [8.3.1]

Pianka E.R. & Schall J.J. (1981) Species densities of Australian vertebrates. In: *Ecological Biogeography of Australia* (ed. A. Keast), pp. 1675–1694. Junk, The Hague. [8.3.1] [8.4.1] [8.4.2]

Pineda J. (1993) Boundary effects on the vertical ranges of deep-sea benthic species. *Deep-sea Res.* **40**, 2179–2192. [8.3.4]

Platnick N.I. (1991) Patterns of biodiversity: tropical vs temperate. *J. Nat. Hist.* **25**, 1083–1088. [8.3.1]

Platnick N.I. (1992) Patterns of biodiversity. In: *Systematics, Ecology, and the Biodiversity Crisis* (ed. N. Eldredge), pp. 15–24. Columbia University Press, New York. [8.3.1]

Pomeroy D. (1993) Centers of high biodiversity in Africa. *Conserv. Biol.* **7**, 901–907. [8.2.3] [8.4.2]

Poore G.C.B. & Wilson G.D.F. (1993) Marine species richness. *Nature* **361**, 597–598. [8.3.1]

Prendergast J.R., Wood S.N., Lawton J.H. & Eversham B.C. (1993a) Correcting for variation in recording effort in analyses of diversity hotspots. *Biodiv. Lett.* **1**, 39–53. [8.1] [8.2.4]

Prendergast J.R., Quinn R.M., Lawton J.H., Eversham B.C. & Gibbons D.W. (1993b) Rare species, the coincidence of diversity hotspots and conservation strategies. *Nature* **365**, 335–337. [8.2.4][8.4.1]

Price P.W. (1991) Patterns in communities along latitudinal gradients. In: *Plant–Animal Interactions: evolutionary ecology in tropical and temperate regions* (eds P.W. Price, T.M. Lewinsohn, G.W. Fernandes & W.W. Benson), pp. 51–69. Wiley, New York. [8.3.1]

Rabinovich J.E. & Rapoport E.H. (1975) Geographical variation of diversity in Argentine passerine birds. *J. Biogeog.* **2**, 141–157. [8.1] [8.3.1] [8.3.4]

Rabinowitz D. (1981) Seven forms of rarity. In: *The Biological Aspects of Rare Plant Conservation* (ed. H. Synge), pp. 205–217. Wiley, New York. [8.2.4]

Rapoport E.H. (1982) *Areography: geographical strategies of species.* Pergamon, Oxford. [8.3.2]

Rapoport E.H. (1994) Remarks on marine and continental biogeography: an areographical viewpoint. *Phil. Trans. Roy. Soc., Lond. B* **343**, 71–78. [8.3.4]

Ray G.C. (1991) Coastal zone biodiversity patterns. *BioScience* **41**, 490–498. [8.2.3]

Redford K.H., Taber A. & Simonetti J.A. (1990) There is more to biodiversity than the tropical rain forests. *Conserv. Biol.* **4**, 328–330. [8.2.3]

Rex M.A. (1981) Community structure in the deep-sea benthos. *Annu. Rev. Ecol. Syst.* **12**, 331–354. [8.3.4]

Rex M.A. Stuart C.T., Hessler R.R., Allen J.A., Sanders H.L. & Wilson G.D.F. (1993) Global-scale latitudinal patterns of species diversity in the deep-sea benthos. *Nature* **365**. 636–639. [8.3.1]

Rice J. & Bellard R.J. (1982) A simulation study of moss floras using Jaccard's coefficient of similarity. *J. Biogeog.* **9**, 411–419. [8.5.2]

Rich T.C.G. & Woodruff E.R. (1992) Recording bias in botanical surveys. *Watsonia* **19**, 73–95. [8.1]

Richerson P.J. & Lum K-L. (1980) Patterns of plant species diversity in California: relation to weather and topography. *Am. Nat.* **116**, 504–536. [8.3.4]

Ricklefs R.E. (1987) Community diversity: relative roles of local and regional processes. *Science* **235**, 167–171. [8.1] [8.5.2]

Ricklefs R.E. (1989) Speciation and diversity: integration of local and regional processes. In: *Speciation and its Consequences* (eds D. Otte & J. Endler), pp. 599–622. Sinauer, Sunderland, Massachusetts. [8.5.2]

Ricklefs R.E. & Latham R.E. (1992) Intercontinental correlations of geographical ranges suggests stasis in ecological traits of relict genera of temperate perennial herbs. *Am. Nat.* **139**, 1305–1321. [8.3.2]

Ricklefs R.E. & Schluter D.(eds) (1993) *Species Diversity in Ecological Communities: historical and geographical perspectives.* University of Chicago Press. [8.1]

Rogers J.S. (1976) Species density and taxonomic diversity of Texas amphibians and reptiles. *Syst. Zool.* **25**,26–40. [8.4.2] [8.5.2]

Rohde K. (1992) Latitudinal gradients in species diversity: the search for the primary cause. *Oikos* **65**, 514–527. [8.3.1]

Rohde K., Heap M. & Heap D. (1993) Rapoport's rule does not apply to marine teleosts and cannot explain latitudinal gradients in species richness. *Am. Nat.* **142**, 1–16. [8.3.2]

Rosenzweig M.L. (1992) Species diversity gradients: we know more and less than we thought. *J. Mammal.* **73**, 715–730. [8.2.3] [8.3.1]

Roubik D.W. (1992) Loose niches in tropical communities: why are there so few bees and so many trees? In: *Effects of Resource Distribution on Animal–Plant Interactions* (eds M.D. Hunter, T. Ohgushi & P.W. Price), pp. 327–354. Academic Press, London. [8.3.1]

Roy K., Jablonski D. & Valentine J.W. (1994) Eastern Pacific molluscan provinces and latitudinal diversity gradient: no evidence for Rapoport's rule. *Proc. Nat. Acad. Sci., USA* **91**, 8871–8874. [8.3.2]

Ruggiero A. (1994) Latitudinal correlates of the sizes of mammalian geographical ranges in South America. *J. Biogeog.* **21**, 545–559. [8.3.2]

Ryti R.T. & Gilpin M.E. (1987) The comparative analysis of species occurrence patterns on archipelagos. *Oecologia* **73**, 282–287. [8.2.4]

Samways M.J. (1989) Taxon turnover in Odonata across a 3000 m altitudinal gradient in southern Africa. *Odonatologica* **18**, 263–274. [8.3.4]

Schall J.J. & Pianka E.R. (1977) Species densities of reptiles and amphibians on the Iberian Peninsula. *Acta Vert. Doñana* **4**, 27–34. [8.3.4] [8.4.1] [8.4.2]

Schall J.J. & Pianka E.R. (1978) Geographical trends in numbers of species. *Science* **201**, 679–686. [8.3.4] [8.4.1] [8.4.2] [8.5.1]

Schueler F.W. & McAllister D.E. (1991) Maps of the number of tree species in Canada: a pilot GIS study of tree biodiversity. Part 1. *Can. Biodiv.* **1**, 22–29. [8.2.3]

Schwartz M.W. (1988) Species diversity in woody flora on three North American peninsulas. *J. Biogeog.* **15**, 759–774. [8.3.4]

Shiel R.J. & Williams W.D. (1990) Species richness in tropical fresh waters of Australia. *Hydrobiologia* **202**, 175–183. [8.3.1]

Simpson G.G. (1964) Species density of North American recent mammals. *Syst. Zool.* **13**. 57–63. [8.3.1] [8.3.3] [8.3.4]

Slud P. (1976) Geographic and climatic relationships of avifaunas with special reference to comparative distribution in the neotropics. *Smithson. Contrib. Zool.* **212**, 1–149. [8.1]

Smith F.D.M., May R.M. & Harvey P.H. (1994) Geographical ranges of Australian mammals. *J. Anim. Ecol.* **63**, 441–450. [8.2.4] [8.3.2]

Solem A. (1984) A world model of land snail diversity and abundance. In: *World-wide Snails: biogeographical studies on non-marine Mollusca* (eds A. Solem & A.C. Van Bruggen), pp. 6–22. E.J. Brill/W. Backhuys, Leiden. [8.2.4]

Stehli F.G. (1968) Taxonomic diversity gradients in pole location: the recent model. In: *Evolution and Environment* (ed. E.T. Drake), pp. 163–228. Yale University Press, New Haven. [8.3.1]

Stehli F.G., McAlester A.L. & Helsley C.E. (1967) Taxonomic diversity of recent Bivalves, some implications for geology. *Geol. Soc. Am. Bull.* **78**, 455–466. [8.1] [8.3.1]

Stevens G.C. (1989) The latitudinal gradient in geographical range: how so many species coexist in the tropics. *Am. Nat.* **133**, 240–256. [8.3.1] [8.3.2]

Stevens G. (1992a) Spilling over the competitive limits to species coexistence. In: *Systematics, Ecology, and the Biodiversity Crisis* (ed. N. Eldredge), pp. 40–58. Columbia University Press, New York. [8.3.1] [8.3.2]

Stevens G.C. (1992b) The elevational gradient in altitudinal range: an extension of Rapoport's latitudinal rule to altitude. *Am. Nat.* **140**, 893–911. [8.3.4]

Stuart S.N. & Adams R.J. (1991) *Biodiversity in Sub-Saharan Africa and its Islands.* IUCN, Gland, Switzerland. [8.1]

Sutton S.L. & Collins N.M. (1991) Insects and tropical forest conservation. In: *The Conservation of Insects and their Habitats* (eds N.M. Collins & J.A. Thomas), pp. 405–424. Academic Press, London. [8.2.2]

Swan L.W. (1992) The aeolian biome. *BioScience* **42**, 262–270. [8.2.3]

Taylor J.D. & Taylor C.N. (1977) Latitudinal distribution of predatory gastropods on the eastern Atlantic shelf. *J. Biogeog.* **4**, 73–81. [8.1] [8.3.1]

Tilman D. & Pacala S. (1993) The maintenance of species richness in plant communities. In: *Species Diversity in Ecological Communities: historical and geographical perspectives* (eds R.E. Ricklefs & D. Schluter), pp. 13–25. University of Chicago Press. [8.5.1]

Times Atlas of the World (1987) Comprehensive edn, 7th edn. London. [8.3.3]

Tokeshi M. (1991) Faunal assembly in chironomids (Diptera): generic association and spread. *Biol. J. Linn. Soc.* **44**, 353–367. [8.5.2]

Tonn W.M., Magnuson J.J., Rask M. & Toivonen J. (1990) Intercontinental comparison of small-lake fish assemblages: the balance between local and regional processes. *Am. Nat.* **136**, 345–375. [8.5.2]

Tramer E.J. (1974) On latitudinal gradients in avian diversity. *The Condor* **76**, 123–130. [8.3.1] [8.3.3]

Tsuda S. (1984) *A Distributional List of World Odonata.* Osaka. [8.1]

Turner J.R.G., Gatehouse C.M. & Corey C.A. (1987) Does solar energy control organic diversity? Butterflies, moths and the British climate. *Oikos* **48**, 195–205. [8.1]

Turner J.R.G., Lennon J.J. & Lawrenson J.A. (1988) British bird species distributions and the energy theory. *Nature* **335**, 539–541. [8.1]

Turner J.R.G., Lennon J.J. & Greenwood J.J.D. Does climate cause the global biodiversity gradient? In: *The Genesis and Maintenance of Biological Diversity* (eds M. Hochberg, M.E. Clobert & R. Barbault). Oxford University Press, in press. [8.1]

Usher M.B. (ed.) (1986) *Wildlife Conservation Evaluation.* Chapman & Hall, London. [8.2.4]

Väisänen R. & Heliövaara K. (1994) Hot-spots of insect diversity in northern Europe. *Ann. Zool. Fennici* **31**, 71–81. [8.2.2]

Veron J.E.N. (1993) A biogeographic database of hermatypic corals. *Aust. Instit. Marine Sci. Monogr.* **10**, vii + 433. [8.2.2]

Vincent A. & Clarke A. (1995) Diversity in the marine environment. *Trends Ecol. Evol.* **10**, 55–56. [8.3.1]

Vrba E.S. (1980) Evolution, species and fossils: how does life evolve? *South Afr. J. Sci.* **76**, 61–84. [8.2.1]

Wallen C.C. (ed.) (1970) *Climates of Northern and Western Europe.* Elsevier, Amsterdam. [8.3.3]

Warren P.H. & Gaston K.J. (1992) Predator–prey ratios: a special case of a general pattern? *Phil. Trans. R. Soc., Lond. B* **338**, 113–130. [8.5.2]

West N.E. (1993) Biodiversity of rangelands. *J. Range. Manage.* **46**, 2–13. [8.2.3]

White P.S. & Miller R.I. (1988) Topographic models of vascular plant richness in the southern Appalachian high peaks. *J. Ecol.* **76**, 192–199. [8.3.4]

White P.S., Miller R.I. & Ramseur G.S. (1984) The species–area relationship of the southern Appalachian high peaks: vascular plant richness and rare plant distributions. *Castanea* **49**, 47–61. [8.3.4]

Williams P.H. (1988) Habitat use by bumble bees (*Bombus* spp.) *Ecol. Entomol.* **13**, 223–237. [8.5.2]

Williams P.H. (1991) The bumble bees of the Kashmir Himalaya (Hymenoptera: Apidae, Bombini). *Bull. Br. Mus. nat. Hist.* (Ent.) **60**, 1–204. [8.3.4]

Williams P.H. (1993) Measuring more of biodiversity for choosing conservation areas, using taxonomic relatedness. In: *International Symposium on Biodiversity and Conservation* (ed. T.-Y. Moon), pp. 194–227. Korean Entomological Institute, Korea University, Seoul. [8.2.2] [8.2.4] [Plate 8.1]

Williams P.H. & Gaston K.J. (1994) Measuring more of biodiversity: can higher-taxon richness predict wholesale species richness? *Biol. Conserv.* **67**, 211–217. [8.3.1] [8.4.1] [8.5.2]

Williams P.H., Humphries C.J. & Gaston K.J. (1994) Centres of seed–plant diversity: the family way. *Proc. Roy. Soc. Lond. B* **256**, 67–70. [8.2.4]

Willig M.R. & Sandlin E.A. (1991) Gradients of species density and species turnover in New World bats: a comparison of quadrat and band methodologies. In: *Latin American Mammalogy: history, biodiversity, and conservation* (eds M.A. Mares & D.J. Schmidly), pp. 81–96. University of Oklahoma Press, Norman. [8.3.1] [8.3.2] [8.3.3]

Willig M.R. & Selcer K.W. (1989) Bat species density gradients in the New World: a statistical assessment. *J. Biogeog.* **16**, 189–195. [8.3.1]

Wilson J.W. (1974) Analytical zoogeography of North American mammals. *Evolution* **28**, 124–140. [8.1] [8.3.1]

Wolda H. (1987) Altitude, habitat and tropical insect diversity. *Biol. J. Linn. Soc.* **30**, 313–323. [8.3.4]

Wood T.K. & Olmstead K.L. (1984) Latitudinal effects on treehopper species richness (Homoptera: Membracidae). *Ecol. Entomol.* **9**, 109–115. [8.3.1]

Woodward F.I. (1987) *Climate and Plant Distribution.* Cambridge University Press. [8.3.1]

Wright D.H., Currie D.J. & Maurer B.A. (1993) Energy supply and patterns of species richness on local and regional scales. In: *Species Diversity in Ecological Communities: historical and geographical perspectives* (eds R.E. Ricklefs & D. Schluter), pp. 66–74. Chicago University Press. [8.5.1]

Zug G.R. (1993) *Herpetology: an introductory biology of amphibians and reptiles.* Academic Press, San Diego. [Plate 8.1]

Zwölfer H. & Brandl R. (1989) Niches and size relationships in Coleoptera associated with Cardueae host plants: adaptations to resource gradients. *Oecologia* **78**, 60–68. [8.5.2]

9: Temporal changes in biodiversity: detecting patterns and identifying causes

SEAN NEE, TIMOTHY G. BARRACLOUGH and
PAUL H. HARVEY

9.1 Introduction

The origins of contemporary biodiversity are on the evolutionary time scale of speciation and extinction. As a consequence, questions about the origins and maintenance of biodiversity are often evolutionary questions about the patterns of, and factors responsible for, lineage splitting and lineage extinction. However, inferring the temporal patterns of change in biodiversity has not proved easy because the fossil record is notoriously incomplete. Phylogenetic trees, or phylogenies for short (recognizing that a phylogenetic tree is an estimate of the true phylogeny), are therefore also incomplete. Indeed, the vast majority of independently evolving lineages have arisen and gone extinct without leaving a trace in the fossil record. Nevertheless, a vast amount of research effort has gone into detecting patterns in the fossil record and that material has been reviewed many times elsewhere (e.g. Stanley, 1979; Raup, 1991; Wilson, 1992; Ridley, 1993). Accordingly, it is not our intention to cover that same material again. Instead, we shall discuss a new and profitable approach for detecting long- and short-term patterns in biodiversity which exploits phylogenies produced from genetic differences among extant individuals. As we shall see, patterns in such phylogenies can tell us a lot about the origin and maintenance of biodiversity even in the absence of a fossil record.

Phylogenies based on the degree of genetic differences among extant species are becoming increasingly abundant with the advent of new technologies such as the polymerase chain reaction, which allows small amounts of DNA or RNA to be cloned, and the use of automatic gene sequencing machines. Molecular phylogenies contain much more information than just the relationships amongst taxa. In this chapter we will discuss a series of techniques for extracting information from two contrasting types of phylogeny: those with, and those without, a temporal dimension providing information about the relative temporal orderings of the nodes. Although it is, in principle, possible for some groups to date the nodes from the fossil record, most phylogenies with temporal information are molecular phylogenies which exploit molecular clocks. Sequence divergence provides information about relative times, albeit imperfect, and the phrase 'molecular clock' no longer contains the assumption that lineages are endowed with synchronized metronomes.

The phylogenies we refer to may describe the relationships amongst species or higher taxa, from which one may make inferences about the macroevolution (by which we mean long-term changes in biodiversity) of the group. Or they may be phylogenies of, for example, mitochondria from a single species, in which case they are the genealogies of the mitochondrial lineages that have been sampled. The latter phylogenies can be used to make inferences about the more recent dynamical history of the population. The theoretical tools we will describe have meaningful applications with respect to both sorts of phylogeny, but we will describe them in the context of the sort of phylogeny that was in mind when they were developed.

The main thesis of this chapter, therefore, is that phylogenies can tell us much about the macroevolution of a group that has no fossil record (long-term perspective on biodiversity), and much about the population dynamical history (short-term perspective) of a population which has never been the subject of a census. Our approach to extracting the information contained in molecular phylogenies is to ask what those phylogenies would be expected to look like under the simplest models that we can imagine, and then to ask what sort of processes could have given rise to any deviations that we find from the expected structures.

In the first section of the chapter, we describe methods of making inferences about temporal patterns of change in biodiversity, on both evolutionary and ecological time scales. In the second section, we describe how to identify clades that are more, or less, diverse than one would expect under a null model of equal propensities for diversification. This theory is useful for identifying when and where events which were important for the generation of biodiversity occurred. In the final section we turn to questions of inferring the evolutionary and ecological causes of temporal patterns of change in biodiversity.

9.2 The history of clades

In this section, we are concerned only with phylogenies that have a temporal dimension. One set of data that such phylogenies provide consists of the time intervals between the nodes. For several analyses, we shall represent the data in the form of lineages-through-time plots (Fig. 9.1).

9.2.1 Constant rate models

One of the simplest models of evolution is a single parameter model in which lineages do not go extinct and each existing lineage at each point in time has the same probability of dividing into two. That is, the phylogeny has grown as a constant rate, pure birth process and the number of lineages has

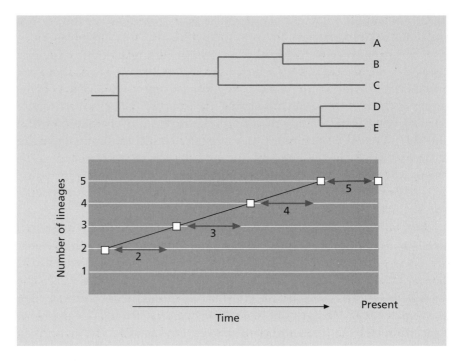

Fig. 9.1 The top of the figure shows a hypothetical phylogeny. The times between the nodes are represented in the form of a lineages-through-time plot (bottom). A dot is plotted at the times the second, third, fourth . . . lineages appear, and we then connect the dots. The patterns that emerge inform us visually of the relationships among the inter-node time intervals, and it is this information that we wish to interpret. The units of the time axis are arbitrary and, so, are unspecified in this and subsequent figures.

increased exponentially with time. Under this model, a semi-logarithmic lineages-through-time plot (with the logarithm of the number of lineages plotted against time) is expected to be a straight line, with stochastic wobbles. If we find deviations from a straight line that cannot be attributed to the stochastic process, then we must consider some other model of evolution.

What if we add a constant rate of extinction (or lineage death) to our model, so that at each point in time each lineage has the same probability of going extinct, with the extinction rate being lower than the birth rate? This is a constant-rates birth–death process model for the growth of the phylogeny. It is convenient to introduce some terminology. The 'actual' phylogeny of a clade is the phylogeny as it would appear in a perfect fossil record, complete with all those lineages which have no descendants at the present day. The 'reconstructed' phylogeny is the one we construct from the information provided by the extant members of the clade, so only lineages with contemporary descendants appear in it. It is the reconstructed phylogenies that we have for analysis. If we consider three intuitively obvious points, we can get a long way to understanding the characteristic footprint left in a phylogeny by a constant-rates birth–death process. First, in the present day, the number of lineages in the actual and the reconstructed phylogenies must be the same. Second, in the past the reconstructed phylogeny will contain fewer lineages than the actual phylogeny, the extent of the deficit being dependent on the extinction rate. Third, once stochastic effects due to small sample sizes are not influencing the picture, the semi-logarithmic lineages-through-time plot for the actual phylogeny (if we had it) would be a straight line with a slope equal to the growth rate (i.e. the birth rate minus the death rate). This all implies, and analytical and simulation studies confirm (Harvey *et al.*, 1994), that under this simple birth–death model, semi-logarithmic lineages-through-time plots of reconstructed phylogenies should have a region of curvature, steepening towards the present (Fig. 9.2).

If we have satisfied ourselves that this model is an appropriate one for the data, we can then estimate birth and death rates from molecular phylogenies. A useful way of portraying a suitable analysis is as a two-dimensional likelihood surface, where one axis is the growth rate and the other is the death rate divided by the birth rate. Figure 9.3 gives an example from Plethodontid salamanders (after Nee *et al.*, 1994b). Inspection of the semi-logarithmic lineages-through-time plot shows the expected linear phase followed by a

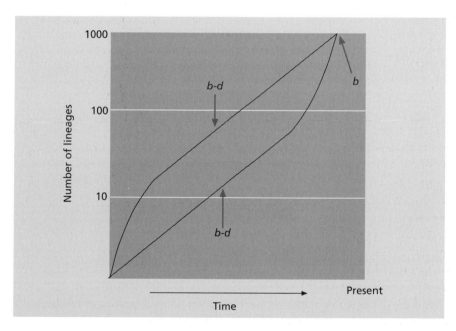

Fig. 9.2 The top curve shows the expected lineages-through-time plot for an actual phylogeny that has been growing according to a constant-rates birth–death process. The bottom curve is the expectation for the reconstructed phylogeny. If the death rate, *d*, was zero, the two curves would be superimposed straight lines with a slope equal to the birth rate. Most of the time, the slopes of the curves are the net growth rate, *b* – *d*, the birth rate minus the death rate. The slope of the reconstructed curve asymptotically approaches the birth rate, *b*, as we approach the present.

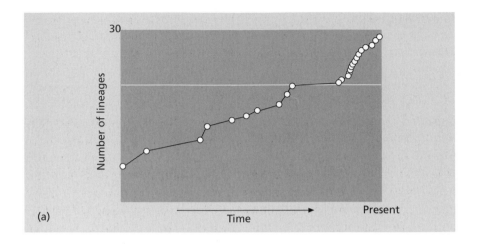

(a)

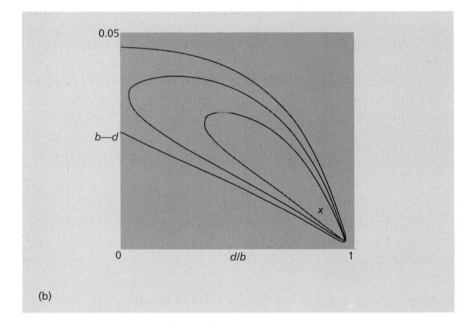

(b)

Fig. 9.3 (a) Lineages-through-time plot for salamanders of the genus *Plethodon*. The data come from a molecular phylogeny constructed by Highton and Larson (1979). Visually satisfied that the constant-rates birth–death process model is not inappropriate for these data, we use the model to estimate the birth and death rates from the data, using the usual techniques of maximum likelihood. (b) Likelihood surface for the composite parameters $b-d$ and d/b. The lines drawn are contour lines (drawn at one, two and three units of support, e.g. Edwards, 1972) and the x indicates the highest point of the surface, the maximum likelihood estimate of the parameters.

steepening towards the present (Fig. 9.3a), which suggests that the constant-rates birth–death model is a reasonable one for these data and a statistical analysis based on that model is therefore appropriate (Fig. 9.3b).

9.2.2 Models in which rates vary

So far, we have discussed how semi-logarithmic lineages-through-time plots which are linear or which steepen towards the present after an original linear phase, are to be expected from constant-rates pure birth or constant rate birth–death models. Deviations from these expectations suggest that constant rate models are not appropriate for the data being considered. For the moment, we continue to assume that each lineage at any particular point in time has the same birth and death parameters as any other lineage, and that it is how these parameters change through time which interests us. If birth rates have been decreasing through time or death rates have been increasing, then we can get a plot which becomes more shallow (rather than steepening) towards the present. For example, analysis of Sibley and Ahlquist's molecular phylogeny of the birds (Sibley & Ahlquist, 1990) revealed just such a case (Nee *et al.*, 1992), which may result from the net rate of lineage production declining as a consequence of density dependent processes such as niche filling. Lineage birth and death rates may also have shown punctuational change in the past, such as mass extinction events (Harvey *et al.*, 1994). Lineage-through-time plots cannot only reveal evidence for mass extinction events, but they also have the potential to inform us of the extent and nature of the event. For example, if 80% of species went extinct, it is of interest to know whether they were concentrated in a particular clade or whether they were a phylogenetic random sample of those then in existence. We could find no evidence of a mass extinction event when we examined Sibley and Ahlquist's phylogeny of the birds (Harvey & Nee, 1994).

9.2.3 When only a few lineages have been sampled

If we imagine some abstract space consisting of all possible hypotheses that we could entertain about the history of a clade, we have now seen that semi-logarithmic lineages-through-time plots can be used to guide us through this space from the simplest evolutionary scenarios to more complex ones. We proceed from simple to more complex until we find a model that is compatible with the data. So far we have assumed that all the extant lineages of a clade are represented in the phylogeny. But what if the data set is incomplete? It is possible to get complete molecular phylogenies of independently evolving sexual lineages by using representatives of all living species of a clade in our sample. For asexual taxa (e.g. many viruses and mitochondria), each individual entity is the equivalent of a species and we could never hope to get molecular sequence data from more than a tiny sample of individuals. Fortunately, we have been able to develop the approach outlined in the previous paragraphs to tackle this problem (Nee *et al.*, 1994b, 1995). We assume that our sample of individuals is random with respect to phylogenetic affinity. When only a small sample of independently evolving lineages is used to reconstruct a phylogeny, recent nodes tend to be disproportionately under-represented. The reason is that there are fewer descendent lineages from each daughter lineage of a recent node than from an ancient node. Accordingly, it is increasingly likely as nodes become more recent that one or both daughter lineages of a node will not be represented in the sample being considered.

The result of the under-representation of more recent nodes is that semi-logarithmic lineages-through-time plots for phylogenies produced under a constant-rates birth–death process, with the birth rate being greater than the death rate, tend to level off towards the present (Fig. 9.4). The upturn towards the present vanishes when only a very small sample of lineages is used. An example of a lineages-through-time plot for such a case is that for the HIV-1 virus (subtype B) which causes AIDS (Fig. 9.5). For this case it has proved possible to estimate the actual number of infected individuals in the United States in the 1980s from the phylogeny produced from a small sample of 24 individuals (Nee *et al.*, 1994b).

Just as the shapes of semi-logarithmic lineages-through-time plots help guide us through hypothesis space when using the full molecular phylogeny, those same plots can provide a useful starting point for similar analyses when

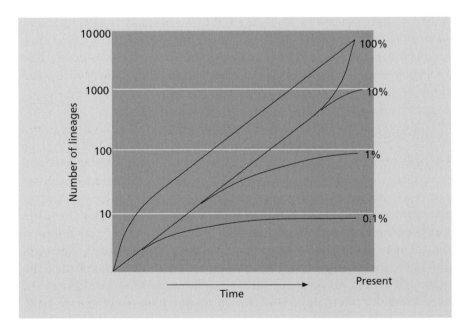

Fig. 9.4 As Fig. 9.2, but now showing the expected lineages-through-time plots when the reconstructed phylogeny is based on successively smaller samples of a clade.

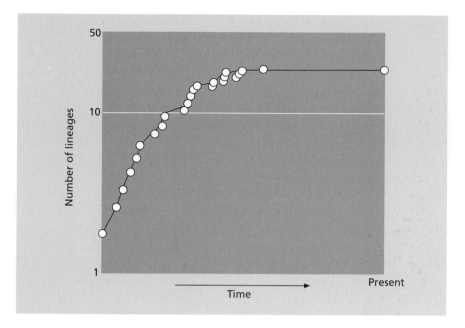

Fig. 9.5 Lineages-through-time plot for a molecular phylogeny of 24 isolates of HIV (all from different patients) taken primarily from North Americans in the mid to late 1980s.

only a small number of lineages has been sampled. We have already seen that for an exponentially growing phylogeny (one in which the birth and death rates have been constant with the birth rate being greater than the death rate) such plots should level off towards the present. In contrast, if the number of lineages has stayed constant through time, then semi-logarithmic lineages-through-time plots are expected to steepen towards the present. This qualitative difference provides a useful starting point for analyses of virus and mitochondrial phylogenies (Fig. 9.6).

If the semi-logarithmic plot levels off towards the present, we know that such a pattern accords with an exponentially growing phylogeny (a phylogeny growing multiplicatively at a constant rate). But that same shape is also characteristic of populations that have been growing exponentially at both an increasing and a decreasing rate. So we now apply a new transformation to the data (different to the logarithmic) which, if the phylogeny has been growing exponentially at a constant rate, produces a linear lineages-through-time plot; if the plot steepens we know the phylogeny has been growing exponentially at an accelerating rate and if it levels off then the population has been growing exponentially at a decelerating rate. We call this the 'epidemic transformation' and the mathematical details can be found in Nee *et al.* (1995).

In contrast, if the semi-logarithmic plot steepens towards the present, we may suspect that the phylogeny has been of constant size, but it might also

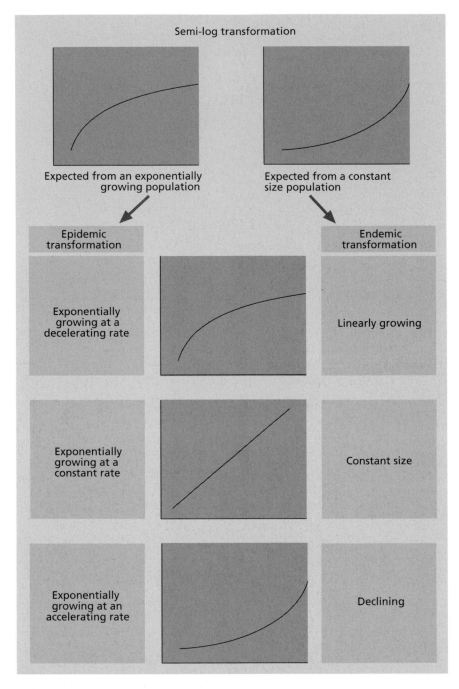

Fig. 9.6 Flow chart of the transformations used for the interpretation of lineages-through-time plots of molecular phylogenies constructed from small samples of clades. Depending on the direction of concavity of the plot under a semi-logarithmic transformation (top), either the epidemic or endemic transformation is then applied to the data. The implications of the various appearances the data may have under the appropriate transformation are described (see text for more detail).

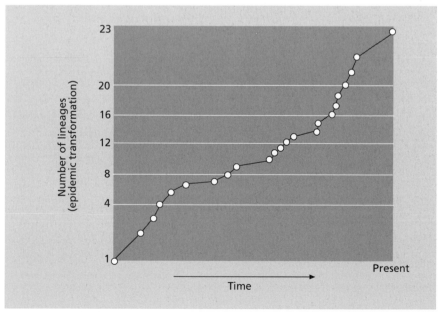

have been growing slowly (say growing arithmetically) or even declining. In order to distinguish among these possibilities, we apply a different transformation, which we call the 'endemic transformation'. If, indeed, the phylogeny has been of constant size, then the lineages-through-time plot is a straight line. However, if the phylogeny has been growing slowly the plot levels off towards the present, and if the phylogeny has been declining the plot steepens towards the present (Nee *et al.*, 1995).

Using this approach, we can examine viral and mitochondrial phylogenies. Several semi-logarithmic plots become shallower towards the present, and we therefore apply the epidemic transformation. Under the epidemic transformation, the HIV 1 (subtype B: Fig. 9.7) and human mitochondrial plots are roughly linear, according with exponentially growing populations. But under the same transformation a Dengue fever virus plot steepens towards the present, suggesting a phylogeny which has been growing exponentially at an increasing rate (Fig. 9.8). In contrast, other semi-logarithmic plots steepen towards the present and so we apply the endemic transformation. Nee *et al.* (1995) present an example of this for humpback whale *Megaptera novaeangliae* mitochondria.

In some circumstances, we can distinguish phases of population history during which different models apply. For example, Hepatitis C virus phylogenies derived from data from different parts of the world always seem to have been characterized by a phase of slow then another phase of fast exponential increase. The change in the epidemic status of the virus was

Fig. 9.7 The HIV data of Fig. 9.5 under the epidemic transformation. The linearity of the plot implies that these data are consistent with a constant-rate exponential growth model of the HIV clade.

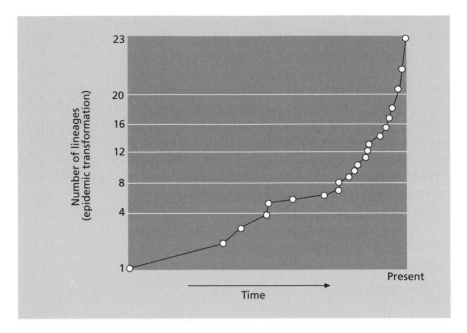

Fig. 9.8 Epidemically transformed data from a molecular phylogeny of Dengue fever virus taken from patients in 13 different countries. The direction of concavity implies that Dengue fever is spreading exponentially at an accelerating rate.

probably associated with various methods of mixing blood, such as transfusion and syringe use, which became more widespread between 30 and 50 years ago.

9.3 Identifying anomalous clades

The theory of the previous section assumed that all lineages are equivalent with respect to features that determine rates of cladogenesis. We now turn to the question of testing that assumption and identifying clades which are unusually large or small, with reference to a null model of cladogenesis.

Consider Fig. 9.9 which is a hypothetical molecular phylogeny. We ask, for each lineage crossing the dotted line (the parent lineage), how many progeny lineages does it have at the present day? This generates a distribution of progeny numbers. We now require a model for this distribution and will exploit a remarkably robust result (Nee *et al.*, 1994a; Nee & Harvey, 1994). Suppose that there are k lineages at the earlier time, a total of n lineages at the present day, and the i'th lineage at the earlier time has n_i progeny (including itself), at the present day. Then, under the null hypothesis that there are no differences between the lineages in features pertaining to their probabilities of diversification, all vectors of progeny number $(n_1, n_2, \ldots, n_k)$, such that the n_i sum to n, are equiprobable. This is MacArthur's broken stick distribution,

240

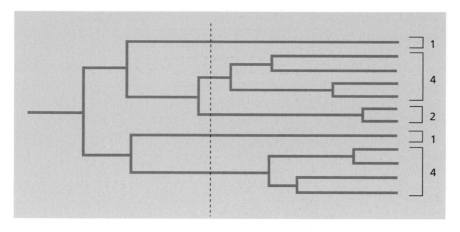

Fig, 9.9 Hypothetical phylogeny. See text for details.

suggested as a model of the relative abundance of species in community ecology (MacArthur, 1960).

This result does not depend on any particular model of diversification. It follows solely from the assumption that the lineages do not differ from each other in their probabilities of cladogenesis at any point in time. So it does not matter if speciation and extinction rates change through time. Nor does it matter if the entire clade has been growing through time or, instead, has always been the same size, so that a species goes extinct whenever a new one arises.

Armed with the broken stick distribution as a null model, it is possible to ask whether or not a particular clade is unusually large, or small, simply by breaking sticks in a computer simulation and observing the resulting statistics. Such an analysis of Sibley and Ahlquist's molecular phylogeny of the birds identified the Passerine and Ciconiiforme radiations as being unusually large, in the statistical sense (Nee *et al.*, 1992). As a new example, we will consider the question posed by Strathman and Slatkin (1983), which is implied by the title of their paper, 'The improbability of animal phyla with few species'. (For this example, we ignore all issues concerning the identification of species and so on.)

Of course, we do not have a molecular phylogeny of the animal kingdom, and the broken stick distribution is only valid for the study of clades which have had the same amount of time to diversify. However, given that most phyla which leave good fossils seem to have been in existence from at least the end of the Cambrian, for the purpose of this example we will make the objectionable assumption that they originated at the same time. Table 9.1 lists the numbers of species in animal phyla, according to Rothschild (1965). Of course, many of the world's species have not been identified and, for

Table 9.1 Numbers of species in the animal phyla. After Rothschild (1965).

Phylum	Species number
Protozoa	30 000
Mesozoa	50
Porifera	4 200
Cnidaria	9 600
Ctenophora	80
Platyhelminthes	15 000
Nemertina	550
Aschelminthes	11 995
Acanthocephala	300
Entoprocta	60
Polyzoa	4 000
Phoronida	15
Brachiopoda	260
Mollusca	100 000
Sipuncula	275
Echiura	80
Annelida	7 000
Arthropoda	766 757
Chaetognatha	50
Pogonophora	43
Echinodermata	5 700
Chordata	44 794

example, smaller bodied species are less likely to have been identified than larger bodied species, introducing biases. However, the theory does not require a complete census, simply one that is random with respect to phylogenetic affinity. We will proceed on this assumption. Simulations of the appropriate stick breaking procedure inform us that the probability of the smallest phylum having less than 100 species is less than 0.005. It follows that to have 5 phyla with less than 100 species is wildly improbable. (We count species rather than, say, genera simply because it is convenient: all the theory described here is relevant in the more general context of questions about 'how many sub-taxa per taxon'. An example of a higher level analysis can be found in Nee *et al.*, 1992.)

Strathman and Slatkin also concluded that small phyla were improbable, but they meant this in a very different sense. They asked: if a phylum has been growing according to a birth–death process, what is the probability that it will survive to the present day and have a small number of species? The answer is that this probability is very small, because if the parameters of the process were such as to generate a small number of species, then it is far more likely that the phylum would be extinct. They concluded that a model of phyletic growth which incorporated some density dependence (i.e. high rates of cladogenesis at low numbers of species, and low rates at higher numbers) was a more suitable model of phyletic growth, given the observation that

242

there are indeed many small phyla. We, on the other hand, are asking about the distribution of species amongst extant phyla, and asking if some phyla are anomalously small under the hypothesis that phyla have no intrinsic differences in their diversification tendencies. Thousands of clades, which we would recognize as distinct phyla, may have gone extinct; it matters not.

Turning from questions of the anomalously small to the perhaps unfeasibly large, we can address the question of whether or not arthropods are unusually speciose with an analytical result based on the broken stick distribution. Given k parents and n progeny in total, the probability that any parent has more than a progeny is given by:

$$1 - \frac{\sum_{v=0}(-1)^v \binom{k}{v}\binom{n-av-1}{k-1}}{\binom{n-1}{k-1}} \tag{9.1}$$

where the summation is for positive $n - av - 1$. This expression is derived from a routine application of covering theorems (e.g. Feller, 1966). The probability of any phylum having 766 757 or more species, is 0.00041, so the arthropods are, indeed, unusual.

In the absence of a molecular phylogeny, the above theory can still be used to compare the sizes of sister taxa which, because they are sister taxa, originated at the same time. All the theory applies, but with the number of parents, k, equal to 2. The theory for this special case was originally derived by others (e.g. Slowinski & Guyer, 1989). As an example of the use of this theory, consider an analysis of angiosperm diversification by Sanderson and Donoghue (1994).

Prompted by notions of 'key innovations' for explaining radiations, Sanderson and Donoghue examine when angiosperm diversification really got off the ground. In particular, they ask two questions. First of all, are the angiosperms, as a whole, significantly more diverse than their sister group, the Gnetales? Second, is there a significant difference in diversity between the two basal sister taxa in the angiosperms? Positive answers to both questions, which is what they obtain, imply that some angiosperms acquired features promoting diversification after angiosperm diversification had already begun. Figure 9.10 illustrates the diversities of the clades under four hypotheses about what the basal node of the angiosperms actually is.

Sanderson and Donoghue addressed the two questions of the previous paragraph by modelling the growth of the clades as a pure birth process and then estimating the parameters of the processes under a variety of hypotheses. The performances of these hypotheses were then statistically compared. A simpler method, which does not make the assumption that plants diversified according to a pure birth process, is to use formula 9.1 with $k = 2$: if the

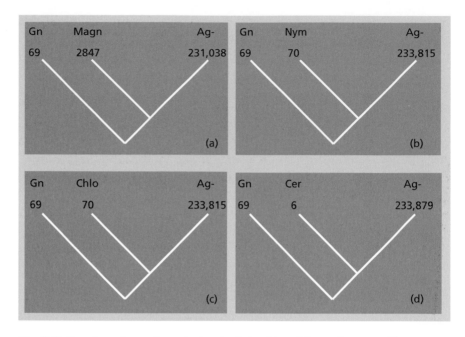

Fig. 9.10 Four hypotheses about the basal relationships of the angiosperms. The angiosperms are divided into their two most basal clades, variously hypothesized to be the Magnoliales (Magn), Nymphaeaceae (Nym), Chloranthaceae (Chlo) or Ceratophyllaceae (Cer) and the rest of the angiosperms (Ag–). The species diversity of these clades, the numbers in the figure, are compared with the diversity of the Gnetales (Gn), the angiosperms' sister group. Adapted from Sanderson and Donoghue (1994).

larger group has r species out of a total of n (so the sister group has $n - r$ species), the probability of an inequality in clade size as large or larger than that observed is $2(n - r)/(n - 1)$.

If we apply this statistic to the data in Fig. 9.10, and simply compare the angiosperm diversity with that of Gnetales (discarding the diversity of the putative basal angiosperm clade, e.g. the Magnoliales in Fig. 9.10a, to get a conservative test), we find that, indeed, angiosperms are significantly more diverse than gnetales (Nee & Harvey, 1994). Comparing the diversity of the two basal angiosperm sister groups, whatever they are, we find a significant inequality (i.e. $P < 0.05$) in these four comparisons as well. Hence, some angiosperms acquired features promoting diversification after angiosperm diversification had already begun. There is one difference between the results of this analysis and the results of Sanderson and Donoghue. This analysis finds the Magnoliales to be significantly less diverse than the rest of the angiosperms, whereas their analysis does not. One possible reason for this discrepancy is that it may be an artefact of the numerical procedures they used in their analysis.

It has often been remarked upon that the distribution of clade sizes is 'skewed'; within a higher taxon most lower taxa have a small number of species while a few have a very large number of species (e.g. Willis, 1922; Dial & Marzluff, 1989; Burlando, 1990; Niklas & Tiffney, 1994). It would be more remarkable if this was not the case. Such skewed distributions are readily generated from the stick breaking procedure discussed above (Raup, 1991; Nee *et al.*, 1992). Somewhat more interesting is the observation that these skewed distributions can be more precisely described by power–law relationships, i.e. a relationship of the form, $G = AN^{-d}$, where G is, say, the number of genera with N species, A is a constant and the exponent d is usually between 1 and 2. A simple explanation for this generalization is as follows. The fundamental probability distribution arising from multiplicative processes (such as the growth of a phylogenetic tree) is the geometric distribution. This distribution has one parameter. Now, if this parameter varies among genera, say, then this generates power–law distributions (Nee *et al.*, 1992). Just as the fact that human heights are normally distributed is not informative about the factors influencing height, neither is the fact that clade sizes are skewed. Skewed distributions are to be expected because of the branching nature of phylogenies.

9.4 Correlates of diversification

We now turn to techniques for testing hypotheses such as, 'phytophagy has promoted cladogenesis in insects' (Mitter *et al.*, 1988). These are hypotheses that take the form, 'a two state trait (e.g. phytophagous/non-phytophagous) influences rates of cladogenesis in a clade'. We will confine ourselves to directional hypotheses, leading to one-tailed tests, for simplicity. Following from the previous section, we focus here on tests using phylogenies that do not have a temporal dimension; tests appropriate for molecular phylogenies with a time dimension have been described elsewhere (Harvey *et al.*, 1991; Nee *et al.*, 1992). Current thinking in comparative biology argues that the relevant data for testing such diversification hypotheses are sister-group (or other statistically independent) comparisons, with the sister groups differing in the state of the trait (e.g. Burt, 1989; Harvey & Pagel, 1991; Slowinski & Guyer, 1993). Mitter *et al.* (1988) were interested in whether phytophagous clades are larger than non-phytophagous clades. In 11 of 13 sister group comparisons this turned out to be the case, which is significant under a one-tailed sign test of the null hypothesis that phagy is irrelevant to biodiversification ($P < 0.05$).

Now consider the imaginary data pertaining to the phytophagy hypothesis described in Table 9.2. There are seven sister-group comparisons, and we see that the phytophagous clade is larger in five of these. This is not statistically significant ($P < 0.23$, one-tailed sign test). However, we notice that in those sister-group comparisons that support the hypothesis, the phytophagous clade is *much* bigger, unlike the situation in those comparisons which do not

Table 9.2 Imaginary diversity data. See text for details.

Non-phytophagous	Phytophagous	Proportion phytophagous
1	99	0.99
1	99	0.99
2	1	1/3
2	1	1/3
1	99	0.99
1	99	0.99
1	99	0.99

support the hypothesis, where there is not much difference in diversity between the groups. One naturally feels that the sign test, which ignores the quantitative differences in diversity, is failing to detect something of biological significance.

This leads one to search for tests of diversification hypotheses which incorporate quantitative information about diversity differences. A recent attempt to do so comes from Slowinski and Guyer (1993). Their test is very simple. For each sister-group comparison, one asks what is the probability of observing an inequality in diversity as large, or larger than that observed, under the null hypothesis that rates of cladogenesis have not been different in the two clades. This probability (independently derived by Slowinski and Guyer) has already been calculated using formula 9.1 in the previous section: one divides that probability by 1/2 to get the one-tailed probability. One then combines the probabilities so calculated for each node using Fisher's combined probability test.

This test is useful in a different context (discussed below) but, unfortunately, cannot be used to test hypotheses about the correlates of diversification. This becomes obvious when one realizes that, using this procedure, it would be possible to establish significance (e.g. phytophagy promotes diversification) with just one sister-group comparison, if there is a large difference in size between the clades. Current understanding of inference from comparative data says that a single sister-group comparison is uninformative, since anything which distinguishes the two clades could be responsible for the difference in their sizes. Furthermore, suppose that we had a million sister-group comparisons in which the phytophagous clade had two species and the non-phytophagous clade had a single species. In this case, it is obvious that phytophagy is implicated in diversity, but the Slowinski–Guyer test could not detect it (since the diversity difference at each node is given a probability of 0.5). In fact, the Slowinski–Guyer test studies whether or not a collection of nodes are anomalous, with respect to the null hypothesis that the observed inequalities in clade size have arisen by chance.

An alternative test can be developed simply as follows, using the logic of randomization tests (Cox & Hinkley, 1992). Quantify the difference in the size of the clades for each sister comparison; for example, calculate the proportion of the species in the two taxa that are in the phytophagous group (Table 9.2). Define a test statistic: we will use the sum of the proportions in the phytophagous group. Under the null hypothesis that phagy is irrelevant, a node that returns the proportion p could have returned the proportion $(1 - p)$ with equal probability (0.5). The distribution of the test statistic can now be determined simply by running a computer program implementing the randomization procedure implied by the null hypothesis (e.g. Siegel, 1956). Applying this test to the imaginary data of Table 9.2, we reject the null hypothesis that phytophagy does not promote cladogenesis ($P < 0.05$).

This indeed seems like an improvement over the sign test. But now consider the imaginary data of Table 9.3. Here phytophagy is significantly implicated as a correlate of diversity under the sign test ($P < 0.02$), but not under the test described in the previous paragraph ($P < 0.07$), and this is because of the magnitude of the difference in clade sizes in the one comparison that does not support the hypothesis. So which test should one use? There is no simple answer to this question. As with all statistical analysis, one's choice of a test statistic depends on what sort of departure from the null hypothesis one wishes to detect. If we are asking the question, 'is phytophagy associated with the inequalities we observe in clade size?', then the sign test is appropriate. If the question is, 'does phytophagy promote large inequalities in clade size, in a specified direction', then the quantitative test is appropriate. A desirable state of affairs is to have very well defined null and alternative hypotheses, in which case it will be clear what significance test to choose. But given that our hypotheses concerning biodiversity are often unlikely to be sufficiently precisely defined to point unambiguously to a particular statistic, it is likely that a number will be investigated, thus blurring the distinction between hypothesis testing and data exploration. We do not think this is a problem.

Table 9.3 Imaginary diversity data. See text for details.

Non-phytophagous	Phytophagous
1	2
1	2
1	2
1	2
1	2
1	2
1	2
1	2
100	1

We quantified the magnitudes of differences in clade size in terms of proportions. Other choices are possible. Wiegmann *et al.* (1993) were interested in the hypothesis that the extraordinary diversity of insects is owing to the accelerated radiation of groups that acquire specialized trophic habits. To explore this idea, they compared the clade sizes of sister-taxa, one of which is parasitic and the other predacious, under the assumption that parasitism is a more specialized habit and, so, the parasitic insect clade should be more diverse. They found no evidence in support of the hypothesis. They carried out the comparison using the Wilcoxon signed rank test. Under this test, the difference in magnitude of the clade sizes in each sister-taxa comparison is quantified as its rank in an ordering of the absolute values of the differences. Then this quantity is given a plus or a minus, depending on the direction of the difference. The distribution of the sum of these quantities is then determined under the null hypothesis that the assignment of pluses and minuses is random (i.e. each has a probability of 0.5). This could, in principle, be determined by randomization on a computer, as above, but exact results are tabulated. (In practice, one need only study the sum of the positive or negative quantities since this automatically determines the total sum.)

One of us (T.G.B.) has used the sign test to explore the hypothesis that intersexual selection promotes diversification, an hypothesis that has been suggested by numerous authors (Lande, 1981; West-Eberhard, 1983; Carson, 1986). It is assumed that intersexual selection is a stronger force in sexually dichromatic passerine birds than in monochromatic species. Using Sibley and Ahlquist's molecular phylogeny of birds (Sibley & Ahlquist, 1990), the proportion of sexually dichromatic species in each tribe was assessed by reference to field guides. Sister tribes are then considered to differ in their level of dichromatism if the proportion of dichromatic species in the tribes differs by more than 0.2. Using this criterion, in 19 out of 25 comparisons between sister tribes, the tribe with a greater proportion of dichromatic species also has more species ($P < 0.01$, one-tailed sign test). The result is robust, remaining significant when the data are analysed in a large number of different ways. For a full account, see Barraclough *et al.* (1995).

We return now to the statistical test of Slowinski and Guyer as a further illustration of the consequences of imprecision in the specification of hypotheses. Suppose that we have a collection of nodes (sister-group comparisons) that we wish to test for imbalance: that is, our hypothesis is that, for some reason, in the collection of sister-group comparisons defined by these nodes, the sister-group diversities are more unequal than we would expect from the null model of diversification which we have used above. Then the Slowinski–Guyer test is a reasonable test to apply.

But let us probe further into what this test actually is. If there are only two nodes, it is easy to discuss with two-dimensional drawings (and the higher dimensional generalizations are obvious). We will discuss the issues in the general context of combined probability tests, rather than unnecessarily adhering to the particular biological context relevant here. The basis of all combined probability tests is the fact that, under the null hypothesis, P-values are uniformly distributed between 0 and 1 (e.g. Cox & Hinkley, 1992). In the one-dimensional case, i.e. the usual case in which we are calculating a P-value for a single experiment, this means that the graphical interpretation of the statement 'the result is significant, $P < 0.05$' is as follows. We imagine that the experiment consists of picking a point at random on a line one unit long, and if the point we pick falls in the interval from 0 to 0.05, which happens with probability 0.05, we call the result significant. If we wish to combine the results of two experiments, each with a reported P-value, P_1 and P_2, say, we imagine that the combined experiment consists of picking a point at random in a square with unit area (Fig. 9.11), each side of the square corresponding to one of the P-values. If the point we pick falls in a region with an area of 0.05, we say the result of the combined probability test is significant. But what is the shape of this region?

Fisher's combined probability test specifies the region in Fig. 9.11a bounded by the curved line. This line is defined by $P_1P_2 = 0.0087$. So we call a result significant if the product of the P-values is less than 0.0087. (This is equivalent to the more conventional description of the test, which is that a result is significant if $-2\ln P_1 -2\ln P_2 > 9.488$, where 9.488 is the critical value of the chi-square distribution with four degrees of freedom). Now, the only virtue in specifying a region of this shape is that the analysis, and its generalizations to the combination of more than two tests, can be performed using tabulated chi-square values. Other choices of region spring to mind. We

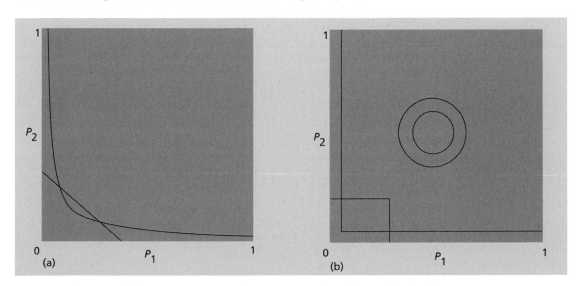

Fig. 9.11 Significance regions for combined probability tests. No attempt has been made to draw these regions quantitatively accurately, i.e. they certainly do not have areas of 0.05, but their shapes are correct.

might want to call a result significant if the sum of the *P*-values, rather than the product, is less than a critical number, i.e. if $P_1 + P_2 < 0.1$, which specifies the triangular region in Fig. 9.11a which has an area of 0.05. More exotic regions are illustrated in Fig. 9.11b. These correspond to combined probability tests which establish significance if the largest *P*-value is less than a critical value (the square region, the critical value being 0.22) or if the smallest is less than a critical value (the L-shaped region, the critical value being approximately 0.025). We will come to the doughnut region below.

To focus the discussion, let us now make up a hypothesis that one might use the Slowinski–Guyer test to explore. The hypothesis is that features which result in non-random diversification arise early in the evolution of a group, so we expect sister-group comparisons at the family level, say, to depart from the null model of diversification, but sister-group comparisons at the level of the tribe to conform with the model. Having constructed a *P*-value for each node according to the Slowinski–Guyer formula, we now have to choose a region for the hypothesis test. We probably do not expect *all* the family level nodes to depart from the null model, which may incline us toward a region shaped more like a boomerang than a square (although data inspection might change our minds). Conversely, we may not expect *all* the lower level comparisons to conform with the null model, inclining us to a region shaped more like a square than a boomerang for these nodes. So we will probably end up using regions of different shapes. Further, the lower level comparisons might, on inspection, look *too* balanced, leading us to choose a region in the top right hand corner of the square instead, to test the hypothesis that taxonomists try, subconsciously, to even up clade sizes. For those so inclined, it is entertaining to try to think of hypotheses that would require rather fanciful shapes to test. In that spirit, we offer a prize of £100 to the first person who can think of a hypothesis, which is meaningful for biodiversity studies, that would be tested with the doughnut region of Fig. 9.11b.

9.5 Conclusion

This chapter is part of the struggle against the pessimism of Joni Mitchell's lyric, 'you don't know what you've got 'til it's gone'. (One wonders if she derives any pleasure from her successful prediction of 'tree museums'.) We have described some tools for the analysis of phylogenies that may be of use in making inferences about the processes that have given rise to what we've got left. These tools are useful for the analysis of phylogenies that cover the very different time scales of macroevolution and population dynamics. In conjunction with other sources of knowledge about the subject of interest, they allow inferences about history in the absence of any direct knowledge of that history.

Acknowledgments

We acknowledge the support of the BBSRC (S.N. & P.H.H.), the NERC (T.G.B.) and The Wellcome Trust (P.H.H.).

References

Barraclough T.G., Harvey P.H. & Nee S. (1995) Sexual selection as a cause of diversity in passerine birds. *Proc. Roy. Soc. Lond., B* **259**, 211–215. [9.4]

Burlando B. (1990) The fractal dimension of taxonomic systems. *J. Theor. Biol.* **146**, 99–114. [9.3]

Burt A. (1989) Comparative methods using phylogenetically independent contrasts. In: *Oxford Surveys in Evolutionary Biology* (eds P.H. Harvey & L. Partridge), pp. 33–53. Oxford University Press. [9.4]

Carson H.L. (1986) Sexual selection and speciation. In: *Evolutionary Processes and Theory* (eds S. Karlin & E. Nevo), pp. 391–409. Academic Press, Orlando. [9.4]

Cox D.R. & Hinkley D.V. (1992) *Theoretical Statistics*. Chapman & Hall, London. [9.4]

Dial K.P. & Marzluff J.M. (1989) Non-random diversification within taxonomic assemblages. *Syst. Zool.* **38**, 26–37. [9.3]

Edwards A.W.F. (1972) *Likelihood*. Cambridge University Press. [9.2.1]

Feller W. (1996) *An Introduction to Probability Theory and its Applications*. Vol. II. John Wiley, New York. [9.3]

Havey P.H. & Nee S. (1994) Comparing real with expected patterns from molecular phylogenies. In: *Phylogenetics and Ecology* (eds P. Eggleton & R.I. Vane-Wright), pp. 219–232. Academic Press, London. [9.2.2]

Harvey P.H. & Pagel M.D. (1991) *The Comparative Method in Evolutionary Biology*. Oxford University Press. [9.4]

Harvey P.H., Nee S., Mooers A.Ø. & Partridge L. (1991) These hierarchical views of life: phylogenies and metapopulations. In: *Genes in Ecology* (eds R.J. Berry & T.J. Crawford), pp. 123–137. Blackwell Scientific, Oxford. [9.4]

Harvey P.H., May R.M. & Nee S. (1994) Phylogenies without fossils: estimating lineage birth and death rates. *Evolution* **48**, 523–529. [9.2.2]

Highton R. & Larson A. (1979) The genetic relationships of the salamanders of the genus *Plethodon*. *Syst. Zool.* **28**, 579–599. [9.2.1]

Lande R. (1981) Models of speciation by sexual selection on polygenic traits. *Proc. Nat. Acad. Sci., USA* **78**, 3721–3725. [9.4]

MacArthur R.H. (1960) On the relative abundance of species. *Am. Nat.* **94**, 25–36. [9.3]

Mitter C., Farrell C.B. & Wiegmann B. (1988) The phylogenetic study of adaptive zones: has phytophagy promoted insect diversification? *Am. Nat.* **132**, 107–128. [9.4]

Nee S. & Harvey P.H. (1994) Getting to the roots of flowering plant diversity. *Science* **264**, 1549–1550. [9.3]

Nee S., Mooers A.Ø. & Harvey P.H. (1992) The tempo and mode of evolution revealed from molecular phylogenies. *Proc. Nat. Acad. Sci., USA* **89**, 8322–8326. [9.2.2] [9.3] [9.4]

Nee S., May R.M. & Harvey P.H. (1994a) The reconstructed evolutionary process. *Phil. Trans. Roy. Soc., Lond. B* **344**, 305–311. [9.3]

Nee S., Holmes E.C., May R.M. & Harvey P.H. (1994b) Estimating extinction from molecular phylogenies. In: *Estimating Extinction Rates* (eds J.L. Lawton & R.M. May). Oxford University Press. [9.2.1] [9.2.3]

Nee S., Holmes E.C., Rambaut A. & Harvey P.H. (1995) Inferring population history from molecular phylogenies. *Phil. Trans. Roy. Soc. Lond., B* **349**, 25–31. [9.2.3]

Niklas K.J. & Tiffney B.H. (1994) The quantification of plant biodiversity through time. *Phil. Trans. Roy. Soc., Lond. B* **345**, 35–44. [9.3]

Raup D.M. (1991) *Extinction: bad genes or bad luck?* Oxford University Press. [9.1] [9.3]

Ridley M. (1993) *Evolution.* Blackwell Scientific, Oxford. [9.1]

Rothschild Lord (1965) *A Classification Of Living Animals.* The University Press, Glasgow. [9.3] [9.2]

Sanderson M.J. & Donoghue M.J. (1994) Shifts in diversification rate with the origin of the angiosperms. *Science* **264**, 1590–1593. [9.3]

Sibley C.G. & Ahlquist, J.E. (1990) *Phylogeny and Classification of Birds.* Yale University Press, New Haven. [9.2.2] [9.4]

Siegel S. (1956) *Nonparametric Statistics.* McGraw-Hill, Tokyo. [9.4]

Slowinski J.B. & Guyer C. (1989) Testing the stochasticity of patterns of organismal diversity: an improved null model. *Am Nat.* **134**, 907–921. [9.3]

Slowinski J.B. & Guyer C. (1993) Testing whether certain traits have caused amplified diversification: an improved method based on a model of random speciation and extinction. *Am. Nat.* **142**, 1019–1024. [9.4]

Stanley S.M. (1979) *Macroevolution: pattern and process.* Freeman, San Francisco. [9.1]

Strathmann R.R. & Slatkin M. (1983) The improbability of animal phyla with few species. *Paleobiology* **9**, 97–106. [9.3]

West-Eberhard M.J. (1983) Sexual selection, social competition, and speciation. *Quart. Rev. Biol.* **58**, 155–183. [9.4]

Wiegmann B.M., Mitter C. & Farrell B. (1993) Diversification of carnivorous parasitic insects: extraordinary radiation or specialized dead end? *Am. Nat.* **142**, 737–754. [9.4]

Willis C.J. (1922) *Age and Area: a study in geographical distribution and origin of species.* Cambridge University Press. [9.3]

Wilson E.O. (1992) *The Diversity of Life.* Penguin, London. [9.1]

10: Spatial and temporal patterns in functional diversity

SCOTT L. COLLINS and TRACY L. BENNING

10.1 Introduction

The challenge at hand, to discuss pattern in the spatial and temporal dynamics of functional diversity, is truly daunting. Three recent volumes (Pimm, 1991; Schulze & Mooney, 1993a; Huston, 1994) devoted a total of approximately 1500 pages of text to components of this topic. Rather than simply begin by proffering excuses to narrow the focus of our chapter in order to hide our ignorance, we will assume the broad objective of linking the dynamics of ecosystem function to three related issues: (i) communities and ecosystems are dynamic in space and time; (ii) individual species affect community and ecosystem dynamics at different spatial and temporal scales; and (iii) species diversity may affect ecosystem structure and function. The first two topics have well developed conceptual frameworks resulting, in part, from a long history of focused research. Analyses related to the third issue have been more theoretical than empirical, but experimental data bearing upon this subject are accumulating. In this chapter we will briefly discuss community dynamics, focusing primarily on our own experience in grassland ecosystems. We will then discuss the extent to which species matter in ecosystems. Next, we will discuss the limited empirical data regarding the relationship between species diversity and ecosystem function and stability. We include a discussion of how the strength of the relationship between species diversity and ecosystem function will vary with scale. These issues have been, and currently are, the focus of considerable attention; however, important questions remain unanswered. In a postscript, we conclude with some perspectives on research that we feel needs to be done.

10.2 Essential definitions

Given that terms such as 'stability' and 'function' mean different things to different people, we begin by providing definitions as we will use them in this chapter. Certainly, one of the most controversial concepts in ecology is stability. Stability is theoretically defined to measure whether or not a system returns to equilibrium following a perturbation (Pimm, 1991). We will use this definition despite the contention engendered by implying that equilibrium conditions exist. Ecosystem function refers to the processing and dynamics of resources (nutrients, organic matter, biomass) and energy through systems. Obviously, this processing is accomplished by a multitude of interacting species which can be classified into functional groups. Functional diversity then refers to the number of such groups in a community (Smith & Huston, 1989), each of which contains one or more species. Species within functional groups may compete strongly, whereas interactions among species in different groups are weaker and may be positive or negative.

Discussions of functional diversity and ecosystem processes inevitably lead to the topic of functional redundancy. Redundancy implies that some species within ecosystems are expendable, because they do not provide a unique contribution to ecosystem function. The issue is controversial because if redundancy does indeed exist, then species could be deemed expendable from the standpoint of conservation of biodiversity. However, it has been hypothesized that many species may have disproportionate and unknown effects on ecosystem function, and thus, may contribute directly to valuable ecosystem services. Such services are assumed to have disproportionate effects on our own well-being (Schulze & Mooney, 1993b), making conservation of species diversity imperative. In this chapter, we concur with Martinez (Chapter 5) that a more realistic approach would be to focus on *functional similarity* rather than redundancy. Similarity can be quantified to varying degrees, whereas redundancy implies an 'all or none' perspective. Given that species may be redundant in some characters but not others, functional similarity can then be compared quantitatively across levels of resolution from physiology to populations.

The potential importance of biodiversity has been advanced recently by the use of technological analogies. For example, Ehrlich (1993) used the rivet-popper hypothesis as a means of relating species diversity and ecosystem function. Ehrlich notes that airplane wings have many rivets, probably more than are necessary to keep them intact. This embodies the notion that individual rivets in a wing are functionally similar. One then asks how many rivets can be popped out of the wing before the number drops below the threshold of structural integrity? From a biological standpoint, the underlying premise is that each species has a small, equal and cumulative effect on ecosystem function. As we gradually eliminate species, we will reach a threshold at which a major ecosystem function will be disrupted, negatively effected, or lost altogether.

Schulze and Mooney (1993b) offer an alternative technological analogy that provides less significance to functional similarity and more emphasis on the interactive functioning of different ecosystem components over time. They note that automobiles contain many essential parts that function when the vehicle is moving, whereas other parts function only occasionally. Bumpers, for example, may decrease the overall functional efficiency of an automobile by adding weight and increasing fuel consumption. Under such circumstances they might be considered expendable. Bumpers do, however, provide valuable services under unusual circumstances, such as low speed collisions. A biological translation would suggest that all species provide crucial ecosystem functions, although at times some species may seem similar in function; their relative importance will vary as environmental conditions change. Given that ecosystems are periodically subjected to natural disturbances, we might conclude that all species effectively perform crucial ecosystem functions within some time frame.

These two analogies do not necessarily lead to mutually exclusive predictions, but the automobile analogy may provide a more realistic argument for the preservation of maximum biodiversity because it incorporates the potential for all species to contribute to ecosystem function. In any case, it is probably best not to gauge the validity of either analogy based upon the annual accident rates of air *versus* ground travel! In addition, there are numerous other reasons to preserve biodiversity regardless of whether or not species are functionally similar (Chapter 11; Ehrlich, 1993).

10.3 Community dynamics

As stated above, our first objective is to briefly describe some of the natural variability exhibited by communities and ecosystems. Our primary focus in this chapter will be on the dynamics of non-successional communities, although we will briefly start with a discussion of temporal dynamics. Temporal dynamics have been well documented in both ecological and evolutionary time (Pickett *et al.*, 1987; Delcourt & Delcourt, 1991). There is ample evidence that many types of communities are subjected to complex disturbance regimes, and that post-disturbance succession is occurring at a variety of spatial and temporal scales (Pickett & White, 1985; Davis, 1986; Glenn-Lewin *et al.*, 1992). The interactive effects of disturbance and succession yield a mosaic of different aged patches in the landscape. A scale may exist at which this mosaic forms a regionally dynamic equilibrium (Bormann & Likens, 1979), yet this scale may be considerably larger than any known preserve (Romme, 1982; Baker, 1989). The maintenance of this dynamic mosaic contributes to landscape heterogeneity which promotes regional biodiversity (Kolasa & Pickett, 1991). Regional biodiversity constrains local biodiversity (Ricklefs, 1987).

Successional dynamics following disturbance are related to resilience, a measure of how quickly a variable returns to equilibrium following a perturbation (Pimm, 1991). Few studies have demonstrated that succession leads to the convergence of vegetation on pre-disturbance conditions or some regional 'norm' (Christensen & Peet, 1984). Thus, species composition in plant communities does not appear to be particularly resilient. However, other non-compositional factors such as nutrient retention (e.g. Bormann & Likens, 1979) might be. Resilience would depend on achieving a certain level of similarity between pre- and post-disturbance vegetation, but establishing objective and non-arbitrary criteria for determining when pre- and post-disturbance conditions are sufficiently similar appears difficult.

One of the more interesting characteristics of species assemblages is their high degree of compositional variation over small spatial scales in the absence of localized, high-intensity disturbances, such as treefall gaps or gopher mounds. This phenomenon has been increasingly documented in grassland vegetation (Glenn & Collins, 1990, 1993; Collins & Glenn, 1991; Herben *et al.*, 1993; van der Maarel & Sykes, 1993). Based on data from Konza Prairie, Kansas, the species richness of mature prairie is highly variable from one year to the next (Glenn & Collins, 1992). On an annually burned site, 37 of 88 species were found in permanent plots (a total of twenty 10 m² plots) in only 5 out of 11 years (Fig. 10.1). On an unburned site, 58 of 113 species occurred in less than 5 of 11 years. This does not necessarily imply that these species became locally extinct. Rather, they often remain present either as viable below-ground tissue and/or in the seed bank. However, their contribution to ecosystem function would certainly vary in concert with their contribution to above-ground production. This also lends support to the notion that a species' contribution to ecosystem function is temporally variable, rendering difficult the assessment of functional similarity when considered over extended time frames.

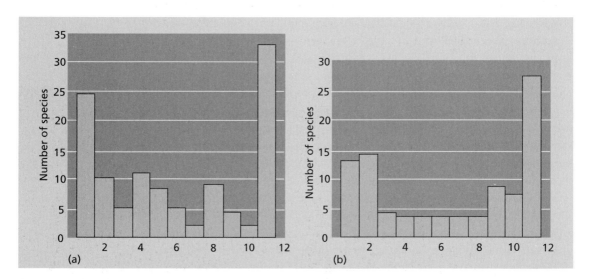

(a)

(b)

Fig. 10.1 The number of years in which a species was present in at least one of twenty permanent 10 m² plots on (a) unburned and (b) annually burned tallgrass prairie over an 11 year period. Data were collected from 1981 to 1992 in twenty 10 m² permanent plots on two burning treatments on the Konza Prairie Long-term Ecological Research site near Manhattan, Kansas. Patterns are bimodal indicating that most species are either present throughout the 11 year period or highly transient. Nevertheless, about 25% of the species appear and disappear over time on each site.

The likelihood that species will disappear locally is somewhat a function of average abundance (Fig. 10.2). Species with high average abundance are typically persistent over time, whereas species with low average abundance may remain persistent or may disappear and reappear. These fluctuations in composition and species abundance produce a constantly changing local assemblage of species (Fig. 10.3). Indeed, on both the annually burned and unburned sites, per cent similarity varies on average about 15% from one year to the next, with occasional changes of up to 25% in response to extreme environmental conditions such as drought (see also Tilman & El Haddi, 1992). This annual variation in composition and abundance produces a consistent pattern of community change over time (Fig. 10.3).

High rates of local species dynamics have been reported for grasslands in Sweden, Hungary, and the Czech Republic, as well as in the United States (Czaran & Bartha, 1992; Glenn & Collins, 1993; Herben *et al.*, 1993; van der Maarel & Sykes, 1993). van der Maarel and Sykes (1993), for example, noted that at very small spatial scales in limestone grassland on the island of Öland, Sweden, species assemblages were unpredictable from one year to the next. Each species would eventually appear and disappear in each of the 3.2 × 3.2 cm sample quadrats. At larger spatial scales, however, the community was floristically stable, in that all species were generally present from one year to the next within the entire 2.5 m² grid of sample quadrats.

Much current ecological research focuses on spatial and temporal heterogeneity and variability. This has been referred to as a new paradigm (Kolasa & Pickett, 1991; Murdoch, 1991; Wu & Loucks, in press), in contrast with the traditional view that communities exist within some equilibrium condition.

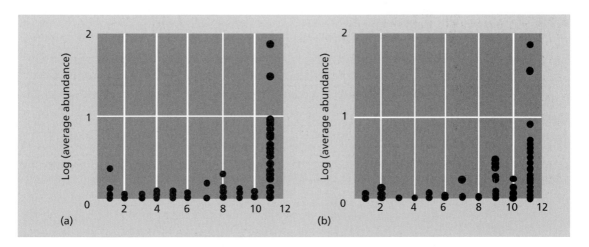

Fig. 10.2 Relationship between number of years in which a species was present on an (a) unburned and (b) annually burned tallgrass prairie, and its log (average abundance). Data from Fig. 10.1. Persistent species are not necessarily abundant, but transient species are not abundant when they occur.

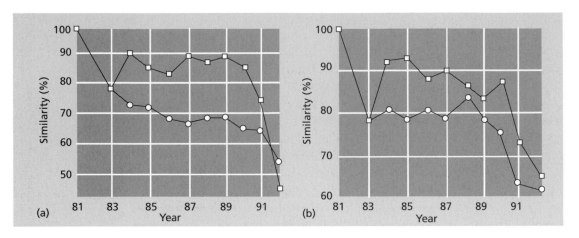

Fig. 10.3 Rate of change in vegetation on an (a) unburned and (b) annually burned tallgrass prairie over an 11 year period. Data from Fig. 10.1. Total change (—□—) represents the quantitative similarity of each sample with the original sample in 1981. Annual rate of change (—○—) represents the quantitative similarity of each sample with the sample from the preceding year.

The new paradigm is based on the notion that disturbance is so frequent that communities do not have time to reach an equilibrium before the next disturbance event (Loucks, 1970). However, as noted above, changes also occur in non-successional communities in the absence of severe disturbances, such as animal diggings or tree-fall gaps (Glenn & Collins, 1993; Herben *et al.*, 1993; van der Maarel & Sykes, 1993). Thus, communities are highly variable and it is this variability that leads to high species diversity (Connell, 1978).

Conversely, there is the notion that diversity begets stability (May, 1973; Pimm, 1991; Tilman & Downing, 1994). Many theoretical and some empirical data support this argument (Dodd *et al.*, 1994; Tilman & Downing, 1994). Hence, we are confronted with the contradiction that communities are both dynamic and stable, and that diversity is somehow related to both dynamics and stability. To some extent, this dichotomy has developed because of differences in spatial and temporal scales of resolution (Collins, 1995). This begs the question, 'at what scale would diversity effect ecosystem stability?' A 5 × 5 cm quadrat is a phenomenally large and heterogeneous area for soil microbes. Yet, at that scale only a very small number of plants can occur at any one time (Watkins & Wilson, 1994). Therefore, above-ground diversity is highly constrained at the spatial scales that might reasonably apply to the soil biota that are driving ecosystem function. In addition, the temporal dynamics at this scale are quite large, so how can diversity affect ecosystem stability when diversity itself is so dynamic? Before we address this topic, we will turn our attention to consider the extent to which species matter in ecosystems.

10.4 Species effects

Population and ecosystem ecology have developed along divergent paths (McIntosh, 1985), yet recently this divergence has narrowed both conceptually and experimentally. Top-down versus bottom-up controls and the keystone species concept are only two examples of conceptual frameworks that directly link species to community and ecosystem function. To better address this relationship, questions regarding species–ecosystem interactions should include an explicit context that frames spatial and temporal scales. This should acknowledge the well known variability in the distribution, abundance and dynamics of species within and among ecosystems. In this section, we briefly review selected examples of species effects on ecosystem processes at different spatial and temporal scales. We draw our examples from terrestrial and aquatic systems, and from microbes to moose.

The notion that species affect ecosystems is hardly new. One of the founding conceptual frameworks in North American plant ecology is Clement's (1916) succession-to-climax paradigm. The primary mechanism driving succession in this model is *reaction*, a condition by which species in one successional stage modify the environment such that it is no longer suitable for their own existence, but is more suitable for species in the next successional stage. Thus, reaction leads to facilitation, the driving force in Clement's (1916) formulation of succession. In modern manifestations of reaction, this mechanism has been developed into a process-oriented feedback system between organisms and environment which leads vegetation on a dynamic trajectory through environmental space over time (Roberts, 1987). An alternative concept to reaction is the notion of 'switches' (Wilson & Agnew, 1992), in which positive or negative feedbacks lead to distinctly different plant communities that establish under identical environmental conditions. In this case, the ecosystem divergence results from dynamic feedbacks between vegetation and environment in which environmental starting conditions are similar, but species composition varies among sites. This differs from Clement's reaction mechanism in which different starting conditions, both environment and vegetation, converge to the regional climax conditions over time. In addition to reaction, Connell and Slatyer (1977) identified tolerance and inhibition as alternative species-based general mechanisms that also may drive community dynamics.

The keystone species concept (Paine, 1969; Menge *et al.*, 1994) provides another organizational framework for synthesizing research on the extent to which species matter in ecosystems. By definition, keystone species affect community or ecosystem processes to a degree that appears greater than their relative abundance in a community would indicate (Bond, 1993). Menge *et al.* (1994) documented that keystone predation, one 'function' of keystone species, occurred in some terrestrial and aquatic situations, yet they were

unable to specify conclusively the conditions under which keystone predation was likely to occur. Their analyses suggested that the occurrence of
keystone predation was not consistently related to specific environmental
conditions, intensity of predation, or degree of dominance. Some evidence
did suggest that keystone predation was related to high rates of resource
supply, which ultimately could be a function of primary production.

Recently, Jones *et al.* (1994) described a related concept, ecological engineering, in which organisms directly modulate the availability of resources by
causing physical state changes in biotic and abiotic materials. Ecological
engineering does not include direct biotic interactions, such as predation or
competition. Like keystone species, ecological engineers can have direct and
dramatic effects on ecosystem dynamics. In other contexts, ecological engineers may also function as keystone species. As an example, we briefly
describe the impact of a 'keystone engineer', the North American bison, *Bison
bison*, on plant community and ecosystem dynamics in the Great Plains
grasslands of North America.

Historical estimates indicate that up to 60 000 000 bison roamed the Great
Plains of North America prior to European settlement (England & DeVos,
1969). Bison have a considerable and direct impact on plant community
structure through their grazing, trampling, and wallowing behaviours. In
addition, their excrement has a significant impact on above-ground and
below-ground production, small-scale patch dynamics, and the location of
future grazing lawns (*sensu* McNaughton, 1984). Bison can be considered to
be keystone herbivores because their preferred forage in the mesic tallgrass
prairie is the dominant C_4 grass, big bluestem *Andropogon gerardii*. Experimental evidence suggests that plant species richness is significantly higher on
grazed compared to ungrazed tallgrass prairie (Table 10.1); these effects are
greater on burned compared to unburned grassland (Collins, 1987). This
differential effect results because growth and palatability of big bluestem are
enhanced on burned compared to unburned prairies. In the absence of
grazing, plant species richness decreases and dominance increases with
greater frequency of burning (Fig. 10.4; Collins, 1992; Collins *et al.*, 1995).

Table 10.1 Average plant species richness (n = 3) on four experimental treatments in
tallgrass prairie: control (−G−B), burned only (−G+B), grazed only (+G−B), and burned
plus grazed (+G+B). Values in each row with similar superscripts were not significantly
different based on Kruskal–Wallis one-way analysis of variance. Data from Collins
(1987).

Year	Treatment			
	−G−B	+G−B	−G+B	+G+B
1985	26.7[a]	30.7[b]	31.1[b]	32.7[b]
1986	24.3[a]	24.7[a]	27.7[a,b]	29.0[b]

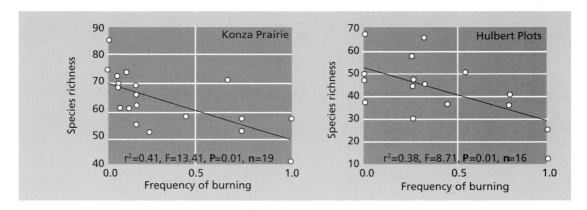

Fig. 10.4 Relationship between frequency of burning and species richness on two experimental data sets. The Konza Prairie data are from watershed-scale management units subjected to burning in April every 1, 2, 4, or 20 years. The Hulbert Plots are 5 × 10 m experimental plots subjected to the same burning frequencies as the Konza Prairie management units. From Collins *et al.* (1995).

The increase in dominance is a function of enhanced growth of the C_4 grasses following fire (Knapp & Seastedt, 1986). In addition, soil nitrogen mineralization rates increase immediately after a burn which leads to higher leaf nitrogen content and lower leaf carbon:nitrogen ratios (Seastedt, 1988). This flush of nutritious growth is then preferentially grazed by bison. In effect, the disturbance (fire) enhances growth of the competitive dominant, big bluestem, yet that dominant becomes more susceptible to grazing by bison. Grazing intensity increases on burned areas which ultimately leads to significantly higher species diversity compared to sites that are only either grazed or burned.

From an engineering standpoint, bison have a tremendous impact on the physical environment through wallowing, a dust-bathing behaviour in which a bison repeatedly rolls back and forth on its side. Wallowing creates an ≅1.5 m diameter area of bare, compacted soil (Fig. 10.5). Vegetation and soil nutrients in wallows are significantly different from those of adjacent areas (Polley & Collins, 1984). What is more impressive, however, is that wallows can still be found in areas that have not been grazed by bison for nearly 100 years. The impact of wallowing behaviour is both severe and long-lasting. Vegetation in these antique wallows is still significantly different from that in neighbouring areas outside wallows (Polley & Collins, 1984).

Ungulates also engineer nitrogen retention in tallgrass prairies at the landscape level. Frequent spring burning in grasslands leads to a reduction in total nitrogen retention because fire volatilizes nitrogen stored in aboveground tissue (Schimel *et al.*, 1991). Nevertheless, with average to above average precipitation, production is generally higher on burned compared to unburned grasslands (Hulbert, 1988; Seastedt *et al.*, 1991). Thus, more carbon is fixed per gram of nitrogen after fire on frequently burned areas, which

(a)

(b)

Fig. 10.5 Evidence of ecological engineering. Bison wallows are formed by bison as they roll on the ground as a threat display, during the rutting season, or just because it feels good. Wallows may be depressions of bare soil which may be colonized by annuals and weedy species. During spring rains, wallows serve as vernal pools supporting aquatic

(c)

(d)

Fig. 10.5 (*continued*) hydrophytes and zooplankton. These areas of higher soil moisture may also serve as refugia from fire. This unburned wallow occurs in a prairie that has not been grazed by bison in over 100 years. Plant species composition in these ancient wallows is different from vegetation in surrounding prairie.

263

increases the carbon:nitrogen ratio in leaf tissue. Tissue with a high carbon:nitrogen ratio decays slowly, creating a feedback loop that further decreases nitrogen availability in the soil. Hobbs *et al.* (1991) demonstrated experimentally, however, that combustion losses of nitrogen on plots grazed by cattle were half those of ungrazed plots following spring burning. Therefore, grazing by cattle (and probably by bison, as well) effectively conserves nitrogen at the landscape scale by subjecting nitrogen stored in above-ground tissues to alternative recycling pathways rather than volatilization from burning.

Bison are also keystone engineers because of the complicated impacts of their faeces and urine. Coffin and Lauenroth (1988) have shown that faecal deposition is lethal to the dominant C_4 bunchgrass of shortgrass prairie, *Bouteloua gracilis*. In this way, faecal deposition creates patches of nutrient rich soil on which new individuals of *B. gracilis* and other species become established. In mesic tallgrass and mixed-grass prairies, bison urine has been shown to alter plant species composition, above- and below-ground production, clonal growth, and tissue nitrogen concentrations of plants growing on urine patches (Day & Detling, 1990; Jaramillo & Detling, 1992; Steinauer & Collins, 1995). These patches of enhanced tissue nitrogen concentration are subsequently grazed by bison and other herbivores, such as grasshoppers. Small urine patches (generally $0.4 m^2$) serve as foci for grazing events that initiate larger patches of heavily grazed vegetation (Steinauer, 1994).

There are numerous other examples of species effects on ecosystem structure, function and dynamics (Jones & Lawton, 1995). We present the following examples primarily because they illustrate the need to consider spatial constraints when judging whether or not a given species affects ecosystem structure and function. Continuing with grasslands, Vinton and Burke (1995) have shown the microbial biomass carbon and soil carbon mineralization rates were significantly different in the root zones of six common plant species in the shortgrass steppe of Colorado. Wedin and Tilman (1990) reported similar findings for soil nitrogen beneath different species from tallgrass prairie. However, when scaling up to the ecosystem in the shortgrass steppe, species effects on soil carbon, carbon mineralization rate, and microbial biomass carbon were less important than simple comparisons between root zones and bare areas (Vinton & Burke, 1995). Thus, at larger spatial scales in the shortgrass steppe, species effects gave way to 'vegetation' effects (Table 10.2).

In aquatic systems, consumers adversely affect primary producers by grazing. However, herbivores also alter phytoplankton community structure by altering supply rates, amounts, and ratios of essential resources, such as carbon, phosphorus, and nitrogen (Sterner *et al.*, 1992; Elser & Hassett, 1994; Vanni, in press). The movement of benthic-feeding fish transports nutrients which subsequently are released by excretion into pelagic zones. Vanni

Table 10.2 Estimates of total soil carbon (kg/ha), carbon mineralization (kg/ha/day) and microbial biomass carbon (kg/ha) on four experimental treatment plots: control (−N−W), nitrogen added (+N−W), water added (−N+W), and nitrogen plus water added (+N+W). Measurements compare plants versus bare soil, and plants weighted by species abundances. At the hectare scale, the main differences occur between plants and bare soil, whereas calculations taking into account the relative abundances of species add little further resolution. Statistical differences (superscripts) among estimates for each variable based on analysis of variance averaging across all treatments. Data from Vinton and Burke (1995).

Variable	Estimate	Treatment			
		−N−W	+N−W	+N+W	−N+W
Total soil carbon	Bare soil[a]	5350	6259	7973	7615
	Plant[a]	6560	7502	7804	7119
	Species[a]	6369	7646	7930	6996
Carbon mineralization	Bare soil[a]	4.7	9.5	10.8	7.4
	Plant[b]	11.4	15.8	13.5	10.6
	Species[b]	11.5	15.8	13.3	10.5
Microbial biomass carbon	Bare soil[a]	78	61	89	40
	Plant[a]	85	74	79	55
	Species[a]	80	81	83	55

(1995) has calculated that a population of gizzard shad *Dorosoma cepedianum*, a primarily benthic-feeding fish, released 0.44–1.34 µg of nitrogen and 0.07–0.24 µg of phosphorus per litre of water per day in Acton Lake, Ohio. There are scaling issues involved, however. Based on principles of mass balance (Sterner, 1990), the carbon, nitrogen, and phosphorus content of excretion is a function of the materials being consumed by the fish, time since feeding, the age, and the nutrient balance of the fish (Sterner & Hessen, 1994; Vanni, 1995). Phosphorus excretion rates of gizzard shad declined to baseline about 12 hours after feeding, whereas nitrogen excretion rates remain high for more than 24 hours. Thus, over the course of a day, gizzard shad are releasing variable nitrogen:phosphorus ratios. It has been demonstrated that variable nutrient supply rates and ratios directly affect phytoplankton assemblage structure and dynamics (Tilman, 1982; Sommer, 1989). Thus, movement and feeding patterns of gizzard shad impact significantly on the composition and dynamics of phytoplankton communities. Carpenter *et al.* (1985) provide a more detailed introduction to the impacts of these 'trophic cascades' on aquatic community and ecosystem dynamics.

Another potential keystone engineer is the moose *Alces alces*. Although herbivory does not constitute engineering as defined by Jones *et al.* (1994), the secondary consequences of moose herbivory dramatically change plant community dynamics (McClaren & Peterson, 1994) which is a form of ecological engineering. Essentially, selective foraging by moose on hardwoods and avoidance of conifers alters plant community composition and structure, which, in turn, affects nutrient cycling and productivity. High rates of moose browsing depress nitrogen mineralization rates and net primary production through indirect effects on recruitment of tree seedlings and saplings, and subsequent depression of nitrogen release from decomposing leaf litter. These effects are then amplified by positive feedbacks between plant litter and soil nutrient availability. Because moose prefer hardwoods over conifers, browsing favours dominance by conifers in northern forests. Conifers provide a low quality source of soil nitrogen because they have high leaf retention rates, and these leaves contain large amounts of lignin and secondary metabolites, both of which contribute to slow decomposition rates. Poor nutrient cycling renders fewer resources available, which further reduces hardwood growth relative to conifers (Pastor & Naiman, 1992; Pastor *et al.*, 1993).

Ultimately, species effects on nutrient cycling in terrestrial systems are the direct result of microbial activity in the soil. Differences in the chemical composition of leaf litter along with environmental factors, such as moisture and temperature, have direct effects on microbial community structure and function. At the same time, microbial species are primarily responsible for ecosystem processes. Perhaps the search for functional similarity is better directed towards microbial processing than to energy-fixing organisms. The diversity, composition, and community structure of microbes is poorly known, yet it is through microbial processes that functional similarity may be widely detectable at the ecosystem level (Zak *et al.*, 1994; Schimel, 1995; Ward *et al.*, 1995). Microorganisms occupy a greater range of ecological niches than macroorganisms (Price, 1988) and some groups were diverse as long as three and a half billion years ago (Schopf, 1993). They occur in all niches where life is thermodynamically possible, thus, the chances of functional similarity or even redundancy seem quite high. Within the wide range of aerobic heterotrophs, the importance of substrate functional diversity is frequently cited for carbon metabolism during the decomposition of plant material (e.g. specialists on lignin, cellulose, pectins; Meyer, 1993; Schimel, 1995). However, even within the group that specializes on lignin, metabolism of this recalcitrant plant compound is carried out by hundreds of species of fungi and bacteria at a given location and by thousands of species world-wide (Meyer, 1993). Thus, there appears to be good potential for functional similarity in ecosystem processes at the microbial level. The importance of these functional groups is widely acknowledged, and the loss of an entire group would lead to the loss of a crucial ecosystem process. Overall, the risk from biodiversity loss may be low if many different microorganisms are able to carry out ecosystem processes. As with macroorganisms, however, the extent to which within-group diversity of microbes results in process heterogeneity or buffering against environmental fluctuations has yet to be determined.

Microbial ecologists have recently made significant progress in peering

266

into the microbial black box at the mysteries of functional processing within ecosystems (Woese, 1994). Zak *et al.* (1994) used Biolog®, a commercially available suite of 128 carbon compounds, to measure differences in functional diversity of soil microbes in six community types in the northern Chihuahuan Desert of New Mexico. The number of substrates utilized, and the rate of reactions differed among the six communities (Fig. 10.6), suggesting that plant communities support different functional assemblages of microbes. Unfortunately, Biolog® plating methods do not provide information on species richness, and physiological activity may be biased by effects of dominant microbial species. Nevertheless, such analyses provide the first step towards establishing that functional similarity may occur in microbial processes.

These patterns also indicate that a relationship exists between aboveground plant community composition and microbial community structure and function. These patterns have been investigated extensively with regard to mycorrhizal effects on plant species interactions (Allen, 1991). However, not all soil microbial-plant feedbacks will be favourable. For example, Bever (1994) demonstrated a negative feedback between species and their own soil communities. That is, species grown with soil innocula derived from soils beneath conspecifics exhibited reduced survival rates compared to soils with microbial communities derived from different species. Such interactions may help to drive small-scale temporal dynamics in non-successional communities.

Given the direct relationship between microbial community structure and ecosystem processes, there is a fundamental need to decipher the mysteries of

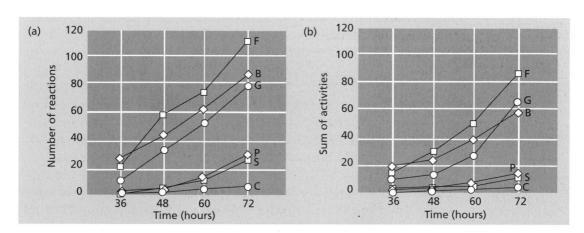

Fig. 10.6 Biolog® substrate utilization patterns by functional groups of soil microbes in six communities along the Jornada Long-term Ecological Research site near Las Cruces, New Mexico. (a) Number of substrates used by soil microbes, and (b) activity levels for each community. F, mesquite-playa fringe community; B, herbaceous bajada; G, black grama grasslands; C, *Larrea tridentata* dominated bajada; P, playa grassland; and S, *Sporobolus* grassland. From Zak *et al.* (1994).

267

microbial community composition and diversity. Previously, microbial communities were identified through rather coarse filters including morphological differences, fatty-acid methyl esters, or plate culture techniques. Ward *et al.* (1990) concluded that these techniques grossly underestimated microbial diversity in samples from natural ecosystems. Using cloned rcDNA from 16 S ribosomal RNA subunits, samples from microbial mats in hot springs in Yellowstone National Park, Wyoming, indicated that microbial diversity may be six to seven times greater than estimated by traditional methodologies. The rcDNA technique may also provide a glimpse into the functional relationships of these previously undescribed microbes. Most of the species identified by this method have not been described morphologically or physiologically; however, the genetic similarity of unknown to known species may indicate aspects of their function. That is, several unknown species may yield rcDNA that is highly similar to known species of, say, green non-sulphur eubacteria (Ward *et al.*, in press), offering a potential glimpse of the degree of functional similarity in microbial assemblages.

With regard to decomposition, Schimel (1995) noted that at small spatial scales, rate of litter decomposition was sensitive to microbial community structure and environmental conditions. These factors reflect physiologically narrow processes at small scales in which microbial community structure seems to relate directly to ecosystem function. However, ecosystem process models often generate meaningful results, even though microbial community structure is not incorporated into them (Schimel, 1995). In addition, Sugai and Schimel (1993) found no differences in functional processing by microbes among successional stages in Alaskan taiga. Thus, at the ecosystem scale, microbial community structure may not be important. Species are either functionally similar or ubiquitous, and effects of community structure are averaged out over larger spatial scales (Schimel, 1995).

10.5 Species diversity and ecosystem dynamics

Thus far, we have directed our discussion to two well-documented sets of phenomena: communities and ecosystems are dynamic; and species (or functional groups in the case of microbes) have significant impacts on community and ecosystem dynamics. Questions regarding our third area of interest, the specific effects of species diversity on ecosystem dynamics, have received far more attention from theoreticians than from experimentalists (e.g. May, 1973; Pimm, 1991), but experimental analyses are increasing and data on this important topic are accumulating.

Given that the measurement of stability, in itself, requires long-term data, only a few field-oriented experiments exist that can directly address the relationship between species diversity and ecosystem function. Perhaps the longest permanent plot field experiment on non-successional vegetation is the Park Grass Experiment at Rothamsted Experiment Station. Beginning in

1856, a series of plots was established to assess the effects of different fertilizer regimes on hay yield in permanent grasslands (Lawes *et al.*, 1882). Biomass on these plots is estimated each year by clipping, and periodically these biomass samples are sorted to species yielding a measure of species composition and structure. Thus, the Park Grass Experiment provides an exceptionally long record of biomass production and compositional change. Although biomass has been shown to fluctuate from year to year on these plots in response to changes in annual precipitation, there has been no overall trend in biomass production during the past 100 years (Silvertown, 1980). Thus, biomass production appears to be stable. Recently, Dodd *et al.* (1994) and Silvertown *et al.* (1994) used these same data to address the relationship between species richness and biomass production, the latter of which is considered to be one measure of ecosystem function. Although changes in biomass were related to precipitation, changes in species composition were not (Silvertown *et al.*, 1994). Thus, ecosystem function was not tied directly to species composition. In addition, this research group also noted that in some years biomass variability was lowest on species-rich plots compared to species-poor plots (Dodd *et al.*, 1994). This would suggest that species diversity enhances ecosystem stability. However, this relationship was inconsistent and often weak over the long history of the Park Grass Experiment.

Tilman and Downing (1994) recently provided an experimental analysis relating directly to the diversity–stability debate. Using long-term data from fertilization experiments in a series of different aged old-fields in Minnesota, Tilman and Downing found that ecosystem productivity, measured as aboveground plant biomass, in high diversity plots was more resistant and resilient to drought than in low diversity plots. The data suggested an incremental effect of species richness on production up to an apparent plateau, suggesting that redundancy did not occur in this system. Instead, each species contributed to total ecosystem production. In this case, the diversity effects were played out through the greater likelihood of finding drought tolerant species in more species-rich plots.

The effect of species richness on biomass stability noted by Tilman and Downing (1994) is not consistent among grasslands. In our analysis of long-term data from mature tallgrass prairie vegetation in Kansas, biomass production on high diversity plots was more resistant to drought only in lowland prairie, but there was no such relationship on upland sites (Fig. 10.7). Both sites supported herbaceous communities with considerable overlap in species composition. Upland soils retain less soil moisture, in general, than lowland soils, thus, during drought, the upland soils are more uniformly dry than lowland soils. Greater spatial variation in moisture retention in lowlands compared to uplands may lead to higher overall biomass production where species richness is highest. The relationship between stability and spatial variation in resource supply needs further investigation.

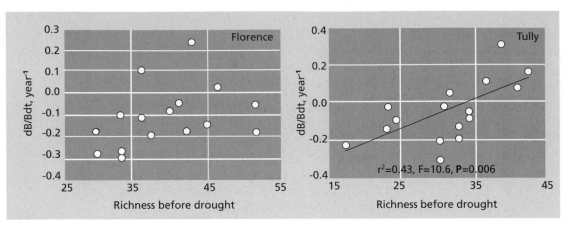

Fig. 10.7 Relationship between drought resistance of upland (Florence soils) and lowland (Tully soils) grasslands and plant species richness preceding a severe drought. Resistance was measured following Tilman and Downing (1994) as 0.5 (ln[biomass$_{1988}$/biomass$_{1987}$]) where 1988 was a drought year and 1987 was a year of above average precipitation. The relationship between diversity and stability was positive and significant only on the Tully soils suggesting topography and soil type may constrain this relationship.

An important experiment by Naeem *et al.* (1994), using the Ecotron at Silwood Park in England, directly addressed the effects of species diversity on ecosystem function. Naeem *et al.* created experimental communities of high, medium, and low diversity (Fig. 10.8) in which the lower diversity communities were subsets of the high diversity communities. This experiment was unique in that it included several trophic levels (decomposers, primary producers, primary consumers, and secondary consumers) in the experimental design. Results indicated that the high diversity community fixed more carbon and had higher cover, hence more light interception, than the low diversity community (Fig. 10.8). Other ecosystem functions, such as nutrient retention, water retention, respiration, and decomposition, varied among treatments but did not show any pattern with respect to species diversity.

Herbaceous communities continue to provide additional examples that confound attempts to seek generality in the diversity–stability debate. Frank and McNaughton (1991) reported that species diversity enhanced compositional stability in grasslands at Yellowstone National Park, Wyoming. Rodriguez *et al.* (1994) reported the opposite. Compositional changes were greatest, hence stability was lowest, on species-rich grasslands in Spain. Overall, results from these analyses demonstrate that species diversity, *per se*, has impacts on some ecosystem functions, but not necessarily on community or ecosystem stability. The relationship between diversity and stability is still unclear and difficult to generalize. Clearly, a consistent set of variables needs

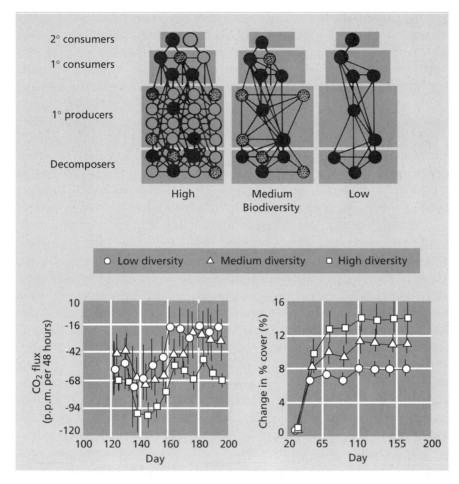

Fig. 10.8 (a) Experimental design and (b) effects of species diversity on ecosystem function. Experimental microcosms were constructed with several trophic levels and three levels of biodiversity. Each lower level of diversity was a subset of the next higher level. Treatments with higher diversity fixed more carbon and had higher overall cover than low diversity treatments suggesting that species diversity directly affects ecosystem function. From Naeem *et al.* (1994).

to be measured across several systems to quantitatively determine the spatial and temporal domains within which species diversity may enhance ecosystem stability.

10.6 Linking species diversity, functional similarity and ecosystem processes

Thus far, we have attempted to demonstrate with a few examples that communities are highly variable in space and time, and that certain species in communities can have considerable impacts on ecosystem structure and function. These impacts then produce feedback loops which may further perpetuate community dynamics. In addition, evidence is accumulating which suggests that species diversity may affect ecosystem structure and function. Again, this is not a particularly new topic, yet direct experimental tests of explicit hypotheses derived from diversity–stability theory are only beginning to accumulate. Essentially, we note that a few species can either increase or decrease community and ecosystem stability over short time-frames. The degree of stability or variability will be a function of the level of resolution under investigation. Perhaps species composition changes considerably from year to year, but biomass production is less variable. On the other hand, biomass production may fluctuate independently of species richness over the long term. Thus, the relationship between species composition, stability, and ecosystem function needs to be investigated over a variety of spatial and temporal scales.

The need for manipulative experiments to address questions regarding the effects of diversity on ecosystem processes is paramount to answering questions regarding functional similarity among species. Given that different species may provide a variable suite of characteristics that affect ecosystem function, functional similarity may occur through diversity because the mechanisms associated with ecosystem processes will be represented some-where within a given collection of species. For example, two equally diverse communities with different sets of species may produce the same impacts on ecosystem function even though individual species in each community do not contain the same combination of characters (Table 10.3). At this scale, species are functionally similar, but the assemblages are functionally redundant. Indeed, the likelihood that all ecosystem processes will be represented increases as diversity increases. Eventually, a point may be reached where increases in species richness will not add any new processes into an ecosystem. The implication would be that functional saturation occurs above a certain level of species richness (Tilman & Downing, 1994). This cannot be the case, however, if functional similarity among species is explicitly defined in the context of a suite of physiological characteristics. For example, two legume species may be functionally similar because both contain bacterial symbionts for nitrogen fixation. However, if the species differ in drought tolerance, the contributions of each species to nitrogen fixation may vary, yet the collective contribution to ecosystem function will remain stable in the face of a perturbation, such as drought. In this case, redundancy cannot exist even though a certain process is represented by more than one species because these species are not similar in other characteristics.

Unfortunately, much of the redundancy debate has not been couched within a clear spatial and temporal framework, and data to address the controversy are often of limited scope. Indeed, we believe that issues of redundancy and functional similarity have been improperly focused at the

Table 10.3 Communities with equal species richness in which each community contains seven 'functions' (A–G). In community 1, redundancy occurs for functions B–F, but not for A and G. The same seven functions occur in community 2, but exist in different combinations among the species. Thus, functional diversity is the same between the two communities, but redundancy differs because E and F are not redundant in community 2. Also, this illustrates the principle that species may be redundant in one function but not another.

Species							
Community 1				Community 2			
I	II	III	IV	I	II	III	IV
A	B	D	E	A	B	G	B
B	C	E	F	C	D	A	G
C	D	F	G	F	E	C	D

species level. Rather, functional similarity should be focused at the process level and determined by estimating the number of individuals, regardless of taxonomic affiliation, in an assemblage that contributes to that function. If more than one individual contributes to a function, then functional similarity may exist at the community or ecosystem level. Overall, we argue that functional similarity must be defined objectively on a case by case basis, and that the spatial and temporal domain of the function being addressed must be specified. For example, a given plant species will make multiple contributions to ecosystem function, including a symbiotic relationship with nitrogen-fixing bacteria, offering floral rewards to pollinators, serving as a resource base for insect larvae, fixing carbon, hosting mycorrhizal fungi, producing shade, supporting bird nests, contributing to soil organic matter, adding organic debris to stream ecosystems, etc. Therefore, similarity in one characteristic does not lead to the conclusion that an individual is expendable, because it may provide a crucial contribution to a different but equally important ecological process.

We have attempted to illustrate the well-known fact that communities and ecosystems are highly variable in space and time. This variability is directly related to the non-equilibrium paradigm of communities (Murdoch, 1991; Wu & Loucks, in press). In many cases, community dynamics can be related to certain species either because they are highly abundant, or because some of their activities have disproportionate impacts on structure and function. In the latter case, such keystone species or ecological engineers are, by definition, not redundant. However, ecological engineers and keystone species do not appear to be the norm (Jones *et al.*, 1994; Menge *et al.*, 1994). Most communities comprise a collection of species that differ in abundance, yet all contribute in one way or another to ecosystem structure and function. For example, all producers contribute to biomass production, which is considered to be a primary ecosystem function. In such a case, all producer species are truly redundant. At the ecosystem level then, the question becomes, 'How many producer functional groups are necessary to maintain this ecosystem function in a dynamic equilibrium?'

If functional redundancy is to occur, several conditions should be defined and measured explicitly: the number of individuals in the area of interest; the variable combination of functional characteristics within these individuals; the effects of environmental variation on those characteristics; and the spatial and temporal scales of reference. Functional similarity will be a consequence of the collective functional traits possessed among the different individuals in a defined area. Individuals possess a combination of several characteristics (e.g. carbon fixation, biomass production, nitrogen storage, drought tolerance) that link community structure to ecosystem function. In total, functional similarity is a product of multiple individuals that together combine traits into related ecosystem functions. There is a need to re-focus research on processes and the individuals that represent them. The distribution and abundance of individuals contributing to processes need to be measured. Thus, studies quantifying functional similarity need to incorporate measurement of density, spatial pattern, physiological ecology and population dynamics at spatial scales that relate to ecosystem function. In addition, the mechanistic details of ecosystem function, particularly for microbial systems, need to be carefully addressed.

All of these factors must be tempered by an additional question regarding scale. Can a boundary be placed around an ecosystem such that the species assemblage and its collective traits will be quantifiable and functional similarity can be determined? At small scales where distinct boundaries can be arbitrarily imposed (e.g. 2×2 m plots), diversity of plants and animals is constrained by the scale of measurement (Collins, 1995). On the other hand, 2×2 m is a tremendously large and unmanageable area for studying microbial processes. Hierarchical and nested sampling schemes may help to alleviate some of this scale incompatibility. Once boundaries are established, we then must ask whether or not such small plots would ever be considered stable under the current definition. That is, can stability and functional similarity among groups and processes be measured at the same small scale? Given the high degree of annual turnover exhibited on small plots, compositional stability seems unlikely, but functional stability may not be linked to compositional stability. Biological stability, in true terms, is really a 'meta-stability' or dynamic equilibrium with oscillations around a so-called equilibrium point. If the amplitude and degree of periodicity of oscillations can be characterized statistically, then the meta-stability can be defined explicitly. Instability occurs when an environmental change or perturbation drives the system outside these statistical limits. Resilience measures if and when the system falls back into the previous realm of variation, or if it is reset

274

to a new long-term dynamic (temporary instability). If no new statistically predictable regime of oscillations develops, the system remains unstable. Answers to these issues will only be derived by long-term data from which trends in community and ecosystem dynamics, and their associated statistical properties, can be estimated. Only when temporal variability and scale are considered explicitly can we fully evaluate the relationship between diversity, stability, and functional similarity. Theoretical analyses and empirical data that directly address these issues are only just beginning to accumulate.

10.7 Postscript

As noted earlier, the topic of functional redundancy is highly controversial. However, with a clear focus on the issue, and an explicit statement of the functions being considered, experiments to determine the existence and importance of functional similarity among species seem relatively straightforward. For example, Lawton and Brown (1993) posed the question: do similar species, occupying similar trophic positions, have similar effects on community organization? Clearly, it is often difficult to determine 'similarity'. Undaunted by this challenge, Morin (1995) directly addressed this question. He demonstrated experimentally that two predator species, adult spotted newts *Notophthalmus viridescens* and larval marbled salamanders *Ambystoma opacum* were functionally similar with regard to their effects on the abundances of three prey species of tadpoles. Prey species abundance was the same when the two predators occurred singly or in combination. Thus, in this clearly defined scenario, these predators are functionally similar. More experimental tests of this nature using different predator and prey species, and different trophic levels are necessary to determine the generality of functional similarity in communities. In addition, such experiments would benefit by the inclusion of environmental heterogeneity because variable environments may alter the patterns found under more stable ecological conditions.

Microbial communities would seem to be ideal systems in which to address functional similarity in ecosystem-level processes. There is a clear need for microbial ecology to expand beyond the bioremediation arena, to which it has made outstanding contributions, in order to address questions regarding microbial diversity and ecological processes in natural systems (Olsen *et al.*, 1994). Given that microbial diversity is relatively unexplored and difficult to measure (Ward *et al.*, 1990), studies that link microbial community structure and ecosystem processes along spatial and temporal gradients would be of immense value. Within the realm of microbial processing, additional research that addresses functional diversity would be relevant (e.g. Zak *et al.*, 1994). Rather than using the medically oriented substrates in Biolog®, however, future studies might employ a suite of substrates commonly found within ecological systems to determine if variation in rates and degree of microbial processing exists among different habitats or along gradients.

More experiments that directly address functional redundancy, biodiversity, and their relationship to ecosystem function could be conducted in controlled conditions, such as the Ecotron (Lawton *et al.*, 1993). For example, it is important to tease apart the role of functional diversity versus species diversity in ecosystems. A simple experiment using six plant species could address this issue. If diversity, *per se*, affects stability, then experimental plantings of, say, six grass species should produce the same 'stability' as experimental plantings of three grasses and three forbs under increasing drought conditions.

Such an experiment would only address the role of one component of diversity, species richness. In most studies thus far, diversity has referred to the number of species per unit area. However, the distribution of dominance among species could also be experimentally addressed in relation to ecological stability. For example, experiments could be conducted in which richness is held constant but dominance is varied, as another means to directly address the effects of diversity on ecosystem processes and stability.

In most cases, the experiments we envision are designed to fine-tune the diversity–stability relationship, as well as link these studies to the issue of functional similarity. As data accumulate, we hypothesize that the focus of stability and diversity will shift from the level of species to other levels of resolution, such as individuals, populations, or functional groups, and from scales ranging from patches to ecosystems. Furthermore, parameters related to ecosystem-level processes should also be measured, such as net primary production, nutrient cycling and carbon evolution, to define and relate the role of species assemblages to ecosystem functions. To us, the linkage of diversity, functional similarity, and ecosystem processes seems like an ideal conceptual framework that can further the cooperative research efforts among those whose traditional research focus has been at a single level of ecological organization. This conceptual framework is clearly an exciting and important arena for further inquiry.

Acknowledgments

We would like to thank Neo Martinez, Peter Vitousek and two anonymous reviewers for many helpful discussions and comments on earlier versions of the manuscript.

References

Allen M.F. (1991) *The Ecology of Mycorrhizae*. Cambridge University Press. [10.4]

Baker W.L. (1989) Landscape ecology and nature reserve design in the Boundary Waters Canoe Area, Minnesota. *Ecology* **70**, 23–35. [10.3]

Bever J.D. (1994) Feedback between plants and their soil communities in an old field community. *Ecology* **75**, 1965–1977. [10.4]

Bond W.J. (1993) Keystone species. In: *Biodiversity and Ecosystem Function* (eds E.-D. Schulze & H.A. Mooney), pp. 237–253. Springer-Verlag, Berlin. [10.4]

Bormann F. & Likens G. (1979) *Pattern and Process in a Forested Ecosystem.* Springer-Verlag, Berlin. [10.3]

Carpenter S.R., Kitchell J.F. & Hodgson J.R. (1985) Cascading trophic interactions and lake productivity. *BioScience* **35**, 634–639. [10.4]

Christensen N.L. & Peet R.K. (1984) Convergence during secondary forest succession. *J.Ecol.* **72**, 25–36. [10.3]

Clements F.E. (1916) *Plant Succession.* Carnegie Institute of Washington Publication No. 242. [10.4]

Coffin D.P. & Lauenroth W.K. (1988) The effects of disturbance size and frequency on a shortgrass prairie plant community. *Ecology* **69**, 1609–1617. [10.4]

Collins S.L. (1987) Interaction of disturbances in tallgrass prairie: a field experiment. *Ecology* **68**, 1243–1250. [10.4]

Collins S.L. (1992) Fire frequency and community heterogeneity in tallgrass prairie vegetation. *Ecology* **73**, 2001–2006. [10.4]

Collins S.L. (1995) The measurement of stability in grasslands. *Trends Ecol. Evol.* **10**, 95–96. [10.6]

Collins S.L. & Glenn S.M. (1991) Importance of spatial and temporal dynamics in species regional abundance and distribution. *Ecology* **72**, 654–664. [10.3]

Collins S.L., Glenn S.M. & Gibson D.J. (1995) Experimental analysis of intermediate disturbance and initial floristic composition: decoupling cause and effect. *Ecology* **76**, 486–492. [10.3] [10.4]

Connell J.H. (1978) Diversity in tropical rainforests and coral reefs. *Science* **199**, 1302–1310. [10.3]

Connell J.H. & Slatyer R.O. (1977) Mechanisms of succession in natural communities and their roles in community stability and organization. *Am. Nat.* **111**, 1119–1144. [10.4]

Czaran T. & Bartha S. (1992) Spatio-temporal models of plant populations and communities. *Trends Ecol. Evol.* **7**, 38–42. [10.3]

Davis M.B. (1986) Climatic instability, time lags, and community disequilibrium. In: *Community Ecology* (eds J. Diamond & T. Case), pp. 269–284. Harper & Row, New York. [10.3]

Day T. & Detling J.K. (1990) Grassland patch dynamics and herbivore grazing preference following urine deposition. *Ecology* **71**, 180–188. [10.4]

Delcourt H.R. & Delcourt P.A. (1991) *Quaternary Ecology.* Chapman & Hall, London. [10.3]

Dodd M.E., Silvertown J., McConway K., Potts J. & Crawley M. (1994) Stability in the plant communities of the Park Grass Experiment: the relationships between species richness, soil pH, and biomass variability. *Phil. Trans. Roy. Soc., Lond. B* **346**, 185–193. [10.3] [10.5]

Ehrlich P.R. (1993) Forward. Biodiversity and ecosystem function: need we know more? In: *Biodiversity and Ecosystem Function* (eds E.-D. Schulze & H.A. Mooney), pp. vii–xi. Springer-Verlag, Berlin. [10.2]

Elser J.J. & Hassett R.P. (1994) A stoichiometric analysis of the zooplankton–phytoplankton interaction in marine and freshwater ecosystems. *Nature* **370**, 211–213. [10.4]

England R.E. & DeVos A. (1969) Influence of animals on pristine conditions on the Canadian grasslands. *J. Range Manage.* **22**, 87–94. [10.4]

Frank D.A. & McNaughton S.J. (1991) Stability increases with diversity in plant communities: empirical evidence from the 1988 Yellowstone drought. *Oikos* **62**, 360–362. [10.5]

Glenn S.M. & Collins S.L. (1990) Patch structure in tallgrass prairies: dynamics of satellite species. *Oikos* **57**, 229–236. [10.3]

Glenn S.M. & Collins S.L. (1992) Effects of scale and disturbance on rates of immigration and extinction in prairies. *Oikos* **63**, 273–280. [10.3]

Glenn S.M. & Collins S.L. (1993) Experimental analysis of patch dynamics in tallgrass prairie plant communities. *J. Veg. Sci.* **4**, 157–162. [10.3]

Glenn-Lewin D.C., Peet R.K. & Veblen T.T. (1992) *Plant Succession: theory and application.* Chapman & Hall, London. [10.3]

Herben T., Krahulec F., Hadinocova V. & Skalova H. (1993) Small-scale variability as a mechanism for large-scale stability in mountain grasslands. *J. Veg. Sci.* **4**, 163–170. [10.3]

Hobbs N.T., Schimel D.S., Owensby C.E. & Ojima D.S. (1991) Fire and grazing in the tallgrass prairie: contingent effects on nitrogen budgets. *Ecology* **72**, 1374–1382. [10.4]

Hulbert L.C. (1988) Causes of fire effects on tallgrass prairie. *Ecology* **69**, 46–58. [10.4]

Huston M. (1994) *Biological Diversity.* Cambridge University Press. [10.1]

Jaramillo V.J. & Detling J.K. (1992) Small-scale grazing in semi-arid North American grassland. II. Cattle grazing of simulated urine patches. *J. Appl. Ecol.* **29**, 9–13. [10.4]

Jones C.G. & Lawton J.H. (eds) (1995) *Linking Species and Ecosystems.* Chapman & Hall, London. [10.4]

Jones C.G., Lawton J.H. & Shachak M. (1994) Organisms as ecological engineers. *Oikos* **69**, 373–386. [10.4] [10.6]

Knapp A.K. & Seastedt T.R. (1986) Detritus accumulation limits productivity of tallgrass prairie. *BioScience* **36**, 662–668. [10.4]

Kolasa J. & Pickett S.T.A. (1991) *Ecological Heterogeneity.* Springer-Verlag, Berlin. [10.3]

Lawes J.B., Gilbert J.H. & Masters M.T. (1882) Agricultural, botanical and chemical results of experiments on the mixed herbage of permanent meadow, conducted for more than twenty years in succession on the same land. Part II. The botanical results. *Phil. Trans. Roy. Soc. A & B* **173**, 1181–1413. [10.5]

Lawton J.H. & Brown V.K. (1993) Redundancy in ecosystems. In: *Biodiversity and Ecosystem Function* (eds E.-D. Schulze & H.A. Mooney), pp. 255–270. Springer-Verlag, Berlin. [10.7]

Lawton J.H., Naeem S., Woodfin R.M., Brown V.K., Grange A., Godfray H.J.C., Heads P.A., Lawler S., Magda D., Thomas C.D., Thompson L.J. & Young S. (1993) The Ecotron: a controlled environmental facility for the investigation of population and ecosystem processes. *Phil. Trans. Roy. Soc., Lond. B* **341**, 181–194. [10.7]

Loucks O.L. (1970) Evolution of diversity, efficiency and community stability. *Am. Zool.* **10**, 17–25. [10.3]

May R.M. (1973) *Stability and Complexity in Model Ecosystems.* Princeton University Press, New Jersey. [10.3] [10.5]

McClaren B.E. & Peterson R.O. (1994) Wolves, moose, and tree rings on Isle Royale. *Science* **266**, 1555–1558. [10.4]

McIntosh R.P. (1985) *The Background of Ecology.* Cambridge University Press. [10.4]

McNaughton S.J. (1984) Grazing lawns: animals in herds, plant form, and coevolution. *Am. Nat.* **124**, 863–886. [10.4]

Menge B.A., Berlow E.L., Blanchette C.A., Navarrete S.A. & Yamada S.B. (1994) The keystone species concept: variation in interaction strength in a rocky intertidal habitat. *Ecol. Monogr.* **64**, 249–286. [10.6]

Meyer O. (1993) Functional groups of microorganisms. In: *Biodiversity and Ecosystem Function* (eds E.-D. Schulze & H.A. Mooney), pp. 67–96. Springer-Verlag, Berlin. [10.4]

Morin P.J. (1995) Functional redundancy, non-additive interactions, and supply-side dynamics in experimental pond communities. *Ecology* **76**, 133–149. [10.7]

Murdoch W.W. (1991) The shift from an equilibrium to a non-equilibrium paradigm in ecology. *Bull. Ecol. Soc. Am.* **72**, 49–51. [10.3] [10.6]

Naeem S., Thompson L.J., Lawler S.P., Lawton J.H. & Woodfin R.M. (1994) Declining biodiversity can alter the performance of ecosystems. *Nature* **368**, 734–737. [10.5]

Olsen G.J., Woese C.R. & Overbeek R. (1994) The winds of (evolutionary) change: breathing new life into microbiology. *J. Bacteriol.* **176**, 1–6. [10.7]

Paine R.T. (1969) A note on trophic complexity and community stability. *Am. Nat.* **103**, 91–93. [10.4]

Pastor J. & Naiman R.J. (1992) Selective foraging and ecosystem processes in boreal forests. *Am. Nat.* **139**, 690–705. [10.4]

Pastor J., Dewey B., Naiman R.J., McInnes P.F. & Cohen Y. (1993) Moose browsing and soil fertility in the boreal forests of Isle Royale National Park. *Ecology* **74**, 467–480. [10.4]

Pickett S.T.A. & White P.S (1985) *The Ecology of Natural Disturbance and Patch Dynamics.* Academic Press, New York. [10.3]

Pickett S.T.A., Collins S.L. & Armesto J.J. (1987) Models mechanisms and pathways of succession. *Bot. Rev.* **53**, 335–371. [10.3]

Pimm S.L. (1991) *The Balance of Nature?* University of Chicago Press. [10.1] [10.2] [10.3] [10.5]

Polley H.W. & Collins S.L. (1984) Relationships of vegetation and environment in buffalo wallows. *Am. Midl. Nat.* **112**, 178–186. [10.4]

Price D. (1988) An overview of organismal interactions in ecosystems in evolutionary and ecological time. *Agric. Ecosyst. Environ.* **24**, 369–377. [10.4]

Ricklefs R.E. (1987) Community diversity: relative roles of local and regional processes. *Science* **235**, 167–171. [10.3]

Roberts D.W. (1987) A dynamical systems perspective on vegetation theory. *Vegetatio* **69**, 27–34. [10.4]

Rodriguez M.A. & Gomez-Sal A. (1994) Stability may decrease with diversity in grassland communities: empirical evidence from the 1986 Cantabrian Mountains (Spain) drought. *Oikos* **71**, 177–180. [10.5]

Romme W.H. (1982) Fire and landscape diversity in subalpine forests of Yellowstone National Park. *Ecol. Monogr.* **52**, 199–221. [10.3]

Schimel D.S., Kittel T.G.F., Knapp A.K., Seastedt T.R., Parton W.J. & Brown V.B. (1991) Physiological interactions along resource gradients in a tallgrass prairie. *Ecology* **72**, 672–684. [10.4]

Schimel J.P. (1995) Ecosystem consequences of microbial diversity and community structure. In: *The Role of Biodiversity in Tundra Ecosystems* (eds F.S. Chapin & C. Korner), pp. 239–254. Springer-Verlag, Berlin. [10.4]

Schopf J.W. (1993) Microfossils of the early archaean apex chert: new evidence of the antiquity of life. *Science* **260**, 640–646. [10.4]

Schulze E.-D. & Mooney H.A. (eds) (1993a) *Biodiversity and Ecosystem Function.* Springer-Verlag, Berlin. [10.1]

Schulze E.-D. & Mooney H.A. (1993b) Ecosystem function of biodiversity: a summary. In: *Biodiversity and Ecosystem Function* (eds E.-D. Schulze & H.A. Mooney), pp. 497–510. Springer-Verlag, Berlin. [10.2]

Seastedt T.R. (1988) Mass, nitrogen, and phosphorus dynamics in foliage and root detritus of tallgrass prairie. *Ecology* **69**, 59–65. [10.4]

Seastedt T.R., Briggs J.B. & Gibson D.J. (1991) Controls of nitrogen limitation in tallgrass prairie. *Oecologia* **87**, 72–79. [10.4]

Silvertown J. (1980) The dynamics of a grassland ecosystem: botanical equilibrium in the Park Grass experiment. *J. Appl. Ecol.* **17**. 491–504. [10.5]

Silvertown J., Dodd M.E., McConway K., Potts J. & Crawley M. (1994) Rainfall, biomass variation, and community composition in the Park Grass Experiment. *Ecology* **75**, 2430–2437. [10.5]

Smith T. & Huston M. (1989) A theory of the spatial and temporal dynamics of plant communities. *Vegetatio* **83**, 49–69. [10.2]

Sommer U. (1989) The role of competition for resources in phytoplankton succession. In: *Plankton Ecology: succession in plankton communities* (ed. U. Sommer), pp. 57–106. Springer-Verlag, Berlin. [10.4]

Steinauer E.M. (1994) 'Effects of Urine Deposition on Small-scale Patch Structure and

Vegetative Patterns in Tallgrass and Sandhills Prairies.' Ph.D. Dissertation, University of Oklahoma, Norman. [10.4]

Steinauer E.M. & Collins S.L. (1995) Effects of urine deposition on small-scale patch structure in prairie vegetation. *Ecology* **76**, 1195–1205. [10.4]

Sterner R.W. (1990) N:P resupply by herbivores: zooplankton and the algal competitive arena. *Am. Nat.* **136**, 209–229. [10.4]

Sterner R.W. & Hessen D.O. (1994) Algal nutrient limitation and the nutrition of aquatic herbivores. *Annu. Rev. Ecol. Syst.* **25**, 1–29. [10.4]

Sterner R.W., Elser J.J. & Hessen D.O. (1992) Stoichiometric relationships among producers, consumers and nutrient cycling in pelagic ecosystems. *Biogeochemistry* **17**, 49–67. [10.4]

Sugai S.F. & Schimel J.P. (1993) Decomposition and biomass incorporation of ^{14}C labeled glucose and phenolics in taiga forest floor: effect of substrate quality, successional state, and season. *Soil Biol. Biochem.* **25**, 1379–1389. [10.4]

Tilman D. (1982) *Resource Competition and Community Structure*. Princeton University Press, New Jersey. [10.4]

Tilman D. & Downing J.A. (1994) Biodiversity and stability in grasslands. *Nature* **367**, 363–365. [10.3] [10.5] [10.6]

Tilman D. & El Haddi A. (1992) Drought and biodiversity in grasslands. *Oecologia* **89**, 257–264. [10.3]

van der Maarel E. & Sykes M.T. (1993) Small-scale plant species turnover in a limestone grassland: the carousel model and some comments on the niche concept. *J. Veg. Sci.* **4**, 179–188. [10.3]

Vanni M.J. (1995) Nutrient transport and recycling by consumers in lake food webs: implications for algal communities. In: *Food Webs: integration of patterns and dynamics* (eds G.A. Polis & K.O. Winemiller). Chapman & Hall, London. in press. [10.4]

Vinton M.A. & Burke I.C. (1995) Interactions between individual plant species and soil nutrient status in shortgrass-steppe. *Ecology* **76**, 1116–1133. [10.4]

Ward D.M., Weller R. & Bateson M.M. (1990) 16S rRNA sequences reveal numerous uncultured microorganisms in a natural community. *Nature* **344**, 63–65. [10.4] [10.7]

Ward D.M., Ferris M.J., Nold S.C., Bateson M.M., Kopczynski E.D. & Ruff-Roberts A.L. (1995) Species diversity in hot spring microbial mats as revealed by both molecular and enrichment culture approaches — relationship between biodiversity and community structure. In: *Microbial Mats: structure, development and environmental significance* (eds L. Stal & P. Coumette), NATO/ASI Series, in press. [10.4]

Watkins A.J. & Wilson J.B. (1994) Plant community structure, and its relation to the vertical complexity of communities: dominance/diversity and spatial rank consistency. *Oikos* **70**, 91–98. [10.3]

Wedin D.A. & Tilman D. (1990) Species effects on nitrogen cycling: a test with perennial grasses. *Oecologia* **84**, 433–441. [10.4]

Wilson J.B. & Agnew A.D.Q. (1992) Positive-feedback switches in plant communities. *Adv. Ecol. Res.* **23**, 263–336, [10.4]

Woese C.R. (1994) There must be a prokaryote somewhere: microbiology's search for itself. *Microbiol. Rev.* **58**, 1–9. [10.4]

Wu J. & Loucks O.L. (1995) From the balance of nature to hierarchical patch dynamics: a paradigm shift in ecology. *Quart. Rev. Biol*, in press. [10.6]

Zak J.C., Willig M.R., Moorehead D.L. & Wildman H.G. (1994) Functional diversity of microbial communities: a quantitative approach. *Soil Biol. Biochem.* **26**, 1101–1108. [10.4] [10.7]

PART 3: CONSERVATION AND MANAGEMENT

11: Does biodiversity matter?
Evaluating the case for conserving species

WILLIAM E. KUNIN and JOHN H. LAWTON

How different would the world be without the fly to help in the decomposition of wastes and carcasses? Without the fly as an experimental animal, how much less would we know about population cycles, about nervous function, about heredity? What is a fly worth, an oak tree, a crow, a wisp of thistledown? By how much would life be diminished if Shelley had not written his *Ode to a Skylark*, if Emerson had not penned *The Rhodora* or Lanier *The Marshes of Glynn*? How many persons would not be alive today if we had not discovered penicillin, the improbable product of a lowly green mold? If it is true that half our new drugs are being produced from botanical sources, how can we afford to neglect or destroy any portion of the earth's green mantle? Who can say what obscure plant or animal may someday be precious to us? Are not all precious, since in fact we understand so little about the interdependence of living things, since life itself is the most precious of all? The earth has spawned such a diversity of remarkable creatures that I sometimes wonder why we do not all live in a state of perpetual awe and astonishment.

> H.E. Evans (1966); *Life on a Little-Known Planet*.

11.1 Introduction

11.1.1 On the valuation of species

In a world of scarcity, it is not enough to decry the continual erosion of the earth's biotic riches. If we expect the majority of people to care, we must have some way to evaluate the loss. Just what is a species worth? The subject has been addressed at length by economists, biologists, and philosophers (e.g. Oldfield, 1984; Ehrlich, 1985, 1988; Norse, 1985; Callicot, 1986; Randall, 1986; Norton, 1986, 1987; Brown, 1990; Orians & Kunin, 1990) and we can do little more than summarize what we believe to be some of the key points in the space of this chapter.

What we can add, if anything, is an attempt at objectivity. The literature on biological diversity too often seems polarized between enthusiasts and sceptics; crusaders devoted either to selling the cause of conservation or to sinking it. One side seems prone to parade each speculative notion as established and general fact, the other side opposes any projection beyond that which is known with certainty and measurable in cash receipts. The valuation of biodiversity is concerned largely with unpriced goods and services, with predicting and appropriately discounting future developments, and with deep and difficult arguments about the sustainability of the human enterprise. As such, we are dealing with an inherently uncertain business, which can complicate any attempt at a crisp cost-benefit analysis (Ehrlich & Daily, 1993). But policy questions of all sorts are commonly fraught with great uncertainty, and if no policy were made until all the facts were known, then nothing would ever be done about anything.

The case for conserving species can be gathered under several headings (Lawton, 1991; Ehrlich & Ehrlich, 1992), which we list here together with some obvious rejoinders.

1 Humankind has moral and ethical responsibilities to care for life on earth. This argument cuts little ice with barbarians.

2 Many organisms — flowers, birds, butterflies — bring pleasure to countless people and enrich our lives, as do medieval cathedrals, Mozart concertos and Monet paintings. This is an important and honest argument, but it does little to help the losers in life's beauty contest — slugs, sea-scorpions, or slime moulds. It is also easily taken hostage by determined, myopic accountants and economists. How much is a puffin worth?

3 Species can be useful, as treasure-troves of drugs, new food-stuffs, genes etc. sometimes, but unfortunately, most species will probably never be useful, and we currently have no certain way of knowing which species may turn out to be so.

4 Organisms provide essential 'ecosystem services', maintaining the life-support systems of the planet. Ecosystems impoverished in species, so the argument goes, perform these functions less well, or not at all. Possibly, but the jury is still out.

5 Species are the touchstone of whether we are, or are not, using the planet sustainably. If we cannot maintain biodiversity in the short- to medium-term, it is doubtful whether civilized human activities and a decent quality of life are sustainable in the longer term. Sceptics, unconvinced by arguments **1–4**, are unlikely to find this a compelling argument either.

We explore these ideas in the sections that follow, although not in this order, and not with equal weight; we have neither the expertise (moral philosophy is not one of our strong points, for example), nor the space to treat each equally. Rather we concentrate on those parts of the arguments (especially **3** and **4**) that appear to us to be most compelling, or for which there is new evidence.

11.2 Species as sources of marketable commodities

11.2.1 Food

The most fundamental benefit we derive from the other species with whom we share the planet is in our food. Through most of human history we have depended on wild animal and plant populations to feed us, and wild sources still play a significant role in meeting the nutritional needs of people in many of the world's poorest nations (Groombridge, 1992). Even in the 'developed' world, where most of the foods we eat come from domesticated species, our food supplies are critically dependent on wild populations, and a significant percentage continues, in fact, to be foraged from the wild. Despite the rise of aquaculture, more than 85% of the fish and marine invertebrates we eat are captured from wild stocks.

In subsistence economies, the range of foods harvested from the wild is much wider. Scoones *et al.* (1992) provide a comprehensive review of the literature. For instance, in the apparently maize dominated agricultural system of Bungomo in Kenya, people consume at least 100 different species of wild fruits and vegetables in 70 genera and 35 families. In all parts of the world, wild foods are not the exclusive preserve of 'hunter-gatherers.' Rather, they are an essential part of the diets of many people, particularly during certain seasons of the year, and during major periods of stress such as droughts. They are particularly important for women, children and the poor. Because wild foods are often collected from 'common land', secure access to such areas is vital for those dependent upon them. The problem, of course, is that 'development', frequently involving changes in land use (e.g. from forest to pasture), can eliminate many species and deprive local people of essential resources. For them, loss of biodiversity is not an abstract, philosophical or moral dilemma. Some will go hungry. Yet few studies have addressed the degree of dependency on these food sources or their economic value to local people (Scoones *et al.*, 1992). Because development, by and large, ignores problems caused by the loss of wild foods, biodiversity is severely undervalued.

One recent study of a hectare of forest in the Peruvian Amazon is instructive (Peters *et al.*, 1989). Cut over for timber, the patch was worth $1000. This net payment is a one-off. But the patch also contained plants that could be harvested sustainably for fruit, as well as non-food products such as latex. Peters *et al.* argued that the net value of these products, when sold in the nearest town, Iquitos, was $422 per annum ($400 for fruit and $22 for latex). How best to equate the instantaneous value of the forest for timber, with its value as a sustainable source of food and other non-timber forest products, depends upon a series of assumptions about interest rates, inflation, the discount rate, future markets for the products, what happens to the timber etc. By their calculations, harvested sustainably, the net present value of the forest was $6330, about 13 times its value as timber. One can argue with the detailed economic calculations, but not with the biological lunacy of obliterating a resource that yields (as a one-off payment) only twice what it can generate sustainably each year, particularly if, as seems likely, the local people benefit not one iota from the timber harvest. As these authors note: 'Tropical timber is sold in international markets and generates substantial amounts of foreign exchange . . . controlled by the government and supported by large federal expenditures. Non-wood resources, on the other hand, are collected and sold in local markets by an incalculable number of subsistence farmers, forest collectors, middlemen and shop owners. These decentralized trade networks are extremely hard to monitor and easy to ignore in national accounting schemes.' But they are very real, nonetheless.

Human beings, however, also have a depressing propensity to utilize wild foods in an unsustainable manner. The full extent of the havoc wrought to bird populations by Polynesian colonists in the Pacific (Pimm *et al.*, 1994; Steadman, 1995) is only now becoming apparent. Many of the larger vertebrates on endangered species lists are threatened precisely because they are of value to us as food resources. The great auk, passenger pigeon, dodo, and Steller's sea cow were all driven to extinction because of their value as human food. Many species of whales, sea turtles, fish and ungulates may soon join the list: potentially renewable food resources that will be lost to future generations because of mismanagement. As Oldfield (1984) points out, it is bitterly ironic that the natural resources we most commonly destroy irretrievably are the biotic ones usually deemed to be 'renewable'. Economically speaking, this depressing aspect of human behaviour may even be termed rational (Clark, 1981; Lande *et al.*, 1994); it makes economic sense to hunt a species to extinction if the resulting money reproduces more quickly in the bank than the organisms themselves reproduce in nature.

Whereas the direct contribution of wild animals and plants to the human diet is gradually diminishing, their indirect role seems likely to grow. In part, this may come about as a result of the growing shortage of arable land, creating incentives for the development of new crop and livestock species with unconventional environmental tolerances. Wild ungulate species, for example, show promise as subjects for ranching in areas of Africa where tsetse flies (and the sleeping sickness they spread) make cattle ranching impractical (Coe, 1980; Kyle, 1987). Similarly, a variety of promising potential crop plants suitable for the moist tropics (e.g. the palms *Guilielma gasipaes* and *Jessenia polycarpa*) and arid lands (e.g. the legumes *Phaseolis acutifolius* and *Tylosema esculentum*) have received little agronomical research to date (Vietmeyer, 1986). Of the 300 000 or so flowering plant species, about 20 000 are known to be edible and perhaps 3000 have a significant history of human utilization. Yet most of our food comes from about two dozen widely grown

286

species. The value of preserving the remainder is perhaps best demonstrated by the recent history of soybean, oil palm, safflower, and sunflower, each of which has grown from a minor regional speciality to a major global crop species in the last century (Langer & Hill, 1982). The extinction or genetic impoverishment of species with similar but as of yet unrealized economic potential forecloses the agricultural options of future generations.

If adding new species to the agricultural roster is of potentially great value, at least as great is the potential genetic contribution that wild species may make to increasing the yields, expanding the tolerances, and improving the disease resistance of currently domesticated species. The past track record is impressive: in rice, for example, resistance to grassy stunt virus was incorporated from the wild *Oryza nivara*; *Fusarium* wilt resistance in tomatoes comes from the wild *Lycopersicon pimpinellifolium*; and potato blight resistance genes were taken from the wild *Solanum demissum* (Oldfield, 1984). Agricultural pests commonly evolve means of circumventing the defences of resistant strains over time, so the need for such genetic inputs from wild populations seems likely to continue indefinitely. Until recently, only species closely related to domesticated species were likely to have value in crop and livestock breeding programmes. With improvements in genetic technology, however, the circle of species of potential value as agricultural gene sources is growing progressively wider (Goodman *et al.*, 1987; Hilder *et al.*, 1987). A good example is the Gram-positive bacterium *Bacillus thuringiensis*, a fatal pathogen of lepidopteran caterpillars. Modified versions of the key bacterial gene, known as bt2, responsible for the production of a toxic protein, can now be engineered into higher plants; transgenic plants synthesize the insecticidal protein and kill caterpillars attempting to eat their leaves (Vaeck *et al.*, 1987). As time and technology progress, practically any organism with a novel arsenal of chemical defences or weapons could become a genetic library of potentially great agricultural value. Before long, breeders aiming to improve the resistance of wheat (for instance) to herbivores or pathogens might be as likely to rely on genes for the anti-predator defensive chemicals of South American rain forest flatworms or the allelopathic toxins used by some Southeast Asian mould against its competitors, rather than turning to wild crop relatives as they have in the past. The list of species of potential genetic value to us is thus expanding quickly.

If genes from some wild species have helped eliminate certain agricultural pests, other wild species have participated in pest control more actively. Biological pest control enlists the services of the natural enemies of species we find harmful. The most valuable species in this regard are not the cuddly vertebrates that attract most of the attention in conservation, but rather insects, nematodes, predatory mites, and pathogenic microorganisms (Wood & Way, 1988; Waage, 1991). The parade of stars is instructive if not particularly charismatic: the moth *Cactoblastis cactorum* (responsible for controlling the *Opuntia* cactus outbreak in Australia), the fly *Ptychomyia remota* (which

287

ended the infestation of coconut moths in Fiji), the *Chrysolina* beetles (responsible for controlling Klamath weed in California), the mirid bug *Cyrtorhinus mundulus* (which saved the Hawaiian sugarcane industry from leafhopper infestation), and the *Aphytis* wasps (responsible for controlling citrus scale outbreaks world-wide). In most cases, biocontrol agents are recruited from the native range of the introduced weed or insect pest. Thus, perversely, the native populations of species that have expanded their ranges and become pests (along with the natural enemies they harbour) may be particularly high priorities for conservation.

The scale of biological pest control world-wide is impressive, but less well-known than it ought to be. Julien (1992) provides a comprehensive, global review of biological weed control to the end of 1991. Biocontrol programmes have been attempted against nearly 200 species of weeds, using over 310 species of potential control agents. A similarly up-to-date, comprehensive catalogue of biocontrol programmes against insect pests is not available, but the scale of activity is even larger; Waage and Greathead (1988) estimate that over 560 species of insect natural enemies have been established for control of nearly 300 exotic insect pests in the last 100 years. Not all attempted control programmes work (although the average success rate is orders of magnitude higher than the search for new chemical pesticides); approximately 40% of insect biocontrol programmes, and 30% of weed control programmes are successful (Waage & Greathead, 1988; Lawton, 1990).

The beauty of classical biological control is that once an agent is established, and the pest brought under control, no further expenditure of time, effort or money is necessary, unless something happens to disrupt the new order. Paradoxically, this most basic advantage of biological control may be one of the reasons why it does not enjoy the public status it deserves. It is literally a case of 'out of sight, out of mind'! Yet the economic returns are impressive.

Total costs of biocontrol of the appropriately named, exotic mealybug *Rastrococcus invadens* in Togo by an introduced parasitoid *Gyranusoides tebygi* were US$172 thousand; the *annual* benefits in increased fruit production are ten times that figure (Waage, 1991). Biocontrol of the exotic waterweed *Salvinia molesta* in Sri Lanka provides a particularly well-studied example of the economics of biological control (Doelman, 1989; Room, 1990). *Salvinia* interfered with irrigation and drainage of rice paddies (at an estimated annual median cost, prior to biocontrol, of 16.1 million rupees in lost harvest), it reduced fish catches (median cost 6.1 million rupees), and it posed a health hazard by providing additional breeding sites for mosquitoes (median cost 8.7 million rupees). Abatement costs, largely mechanical clearance, were of the order of 9.5 million rupees a year. Successful control by the weevil *Cyrtobagus salviniae*, imported from Brazil (see below), is estimated to be worth 433 million rupees (A$16 million) over a 25-year time horizon at

1987 prices (Doelman, 1989) — a return of 53:1 on the costs invested in the programme. Similar successful control of *Salvinia* by *Cyrtobagus* in Australia, Papua New Guinea, India and Namibia has not been valued in detail (Room, 1990), but the benefits must run to hundreds of millions of dollars. Multiply such benefits across hundreds of successful weed and insect biocontrol programmes world-wide, and the scale of the enterprise is mind-boggling.

What has all this got to do with biodiversity? Biological control agents are biodiversity. Or rather, they are a particular sample of it. Agents have to be very carefully matched to the target for host specificity, for example, and climatically matched to the proposed area of introduction. They are usually collected from the country of origin of the exotic pest, but may only be found in a small area. The parasitoid *Epidinocarsis lopezi*, responsible for spectacular biological control of cassava mealybug in Africa (Neuenschwander & Herren, 1988) was eventually collected from a small number of localities in Brazil and Bolivia, after extensive searches throughout Central and South America. Other natural enemies from a very closely related, but more widespread mealybug, proved to be totally ineffective as control agents. The *Cyrtobagus–Salvinia* story is very similar. The effective control agent, *C. salviniae*, was eventually collected from *S. molesta* in southeast Brazil; a close relative (*C. singularis*), which was the original focus of the search for biocontrol agents of *Salvinia*, had no impact on the weed (Room, 1990).

Natural and semi-natural habitats, and indigenous agricultural lands hold the world's stock of potential biological control agents. Their erosion compromises the ability of future generations to reap the benefits of new biological control opportunities, and is clearly not compatible with sustainable development.

11.2.2 Medicine

A significant fraction of all drugs are derived either directly or indirectly from biological sources (e.g. Fig. 11.1). Many plants, marine invertebrates, fungi, microorganisms, reptiles and amphibians have evolved chemical defences to protect them from their natural enemies or to subdue their prey. These compounds are effective because they are biologically active, with properties that disrupt the physiology of their target organisms. Those same properties often prove useful in medicine, either as defences, against human pathogens and parasites or to influence human physiology towards some desired objective. Indeed, according to Caporale (1995), all of the drugs discovered at the Merck research laboratories in North America that have become available to patients in the last decade emerged from programmes that benefited from knowledge of biological diversity.

World-wide, many drugs are derived directly from plants, either cultivated or (in species that resist cultivation) from the wild. Examples include

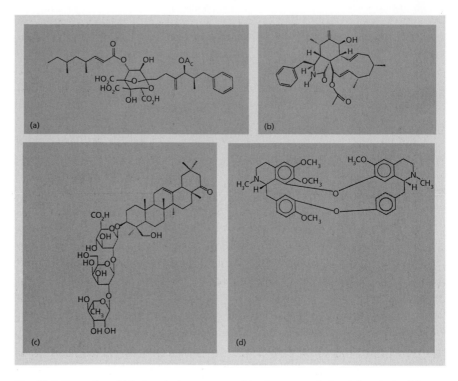

Fig. 11.1 Examples of biochemical structures with pharmaceutical activity, derived from living organisms (Caporale, 1995). (a) Zaragozic acid A, an inhibitor of mammalian and fungal sterol synthesis, obtained from fungi; it is a broad spectrum fungicide and can lower mammalian cholesterol levels. (b) L-696474, an inhibitor of HIV protease, also obtained from fungi. (c) Dehydrosoyasaponin I, an agonist of the calcium-activated potassium channel, derived from a Ghanaian medicinal plant (*Desmodium adscendens*) used to treat asthma. (d) Tetrandrine, an inhibitor of L-type calcium channels, extracted from a Chinese medicinal herb (*Stepania tetranda*).

Reserpine, a tranquillizer and anti-schizophrenic drug derived from the tropical shrub *Rauvolfia serpentina*, and quinodine, a drug used against cardiac arrhythmia and derived from the plant, *Cinchona calisaya*, that also produces quinine (Oldfield, 1984). Altogether, about 90 plant species world-wide are used as sources of medicinal drugs (Groombridge, 1992), with some industrial countries (e.g. Australia—Collins *et al.*, 1990) investing considerable effort in the search for new compounds. Recent work on Indonesian rain forest plants, for instance, has led to the discovery of promising compounds for use against HIV and cancer (Soejarto, 1993; Fuller *et al.*, 1994). Estimates of the economic benefits of plant-based drugs range from US$34–US$300 billion per year (in 1984 dollars) in the United States alone, depending largely on the value placed on a human life (Groombridge, 1992).

Animals, too, have yielded promising medicinal compounds. Promising anticancer drugs have been isolated from the wings of Asian sulphur butterflies (*Catopsilia crocale*) and from the legs of Taiwanese stag beetles (*Allomyrina dichotomus*); the venoms of snakes, frogs and toads have yielded a variety of nerve and muscle drugs. The Australian biologist Andrew Beattie has been particularly innovative in searching for new medicines derived from living organisms, both animals and plants (Beattie, 1992, 1994; see below). As Beattie points out (1994), 'many invertebrates live in environments where microorganisms may be expected to be particularly active. As invertebrate immune systems are often non-existent or relatively simple, the production of antibiotics would be expected in these organisms.' Consistent with this prediction, biocidal compounds have been isolated from a variety of beetles, flies, millipedes, cockroaches, snails and nematodes. Ants, in particular, seem to offer considerable potential as sources of antibiotics, a belief that is backed by financial support for pioneering survey work from the pharmaceutical company Ciba-Geigy (Beattie, 1992, 1994).

The objective of such surveys is not necessarily to harvest populations of wild species for new drugs, but rather to use them as a 'library of patents' to discover novel classes of compounds that can then be synthesized by biochemists (e.g. Caporale, 1995). Indeed, many drugs already in use and produced synthetically were first discovered, as close analogues or in their pharmacologically active form, in wild species. The synthetic anti-leukaemia drug cytosine arabinoside (Ara-C), for example, was first isolated from sponges (*Cryototethya crypta*) from the Caribbean. We do not yet understand the biochemistry of life well enough to design drugs for our needs from first principles — we need to rely on the accumulated experimentation and experience of the millions of species that share the planet with us to recommend interesting solutions. The value of this information is non-trivial; one estimate based on pharmaceutical sales (Principe, 1991) places the expected value in terms of medicinal potential of *each* unexplored species of plant at approximately US$1.6 million. This is very probably an overestimate, because it assumes that the species studied to date for pharmacological activity are a random sample from all plants, but still it gives an indication of the rough scale of the figures involved.

Some species are more likely than others to harbour potentially useful drugs. Plants from tropical countries, for example, are much more likely to harbour defensive alkaloids than are their temperate counterparts (Levin, 1976) and the alkaloids they bear tend to be more toxic (Levin & York, 1978). Among tropical plants, fast-growing sun-loving species show lower levels of a variety of anti-herbivore defences than do shade-tolerant species (Coley, 1988; Coley & Aide, 1991). Similar patterns show up in microorganisms as well; a survey of *Penicillium* moulds from soil samples taken over a wide latitudinal range showed an unusually high incidence of antibiotic activity amongst lowland tropical species (Oldfield, 1984). All of this suggests that species from mature tropical forests may be much more valuable to us, on average, than species from the temperate zone, or the weedy tropical gap species that thrive after deforestation. No one knows how many potentially useful drugs have been lost due to the rapid destruction of much of the earth's primary tropical forest, and the consequent extinctions of a dizzying array of undescribed species within them.

Biological diversity has a second medical benefit. Medical research commonly depends on animal models to test potentially useful therapeutic drugs or techniques. To do so requires laboratory animals that are similar to humans in relevant physiological details; species susceptible to human diseases or especially vulnerable to particular chemical or physical conditions. Most notably, a wide array of primate species have proved valuable, presumably because of their close evolutionary relationship to humans. Yet many primate populations are extremely vulnerable, as they often depend on large tracts of vanishing primary tropical habitats. Medical research values are not, however, restricted to our closest relatives: armadillos (*Dasypus novemcinctus*) have served as valuable clinical model organisms in studies of leprosy (Job *et at.*, 1993). Even more remarkably, horseshoe crabs (*Limulus polyphemus*) have proved important both in the study of vision and in the testing of human blood. Their blood has the peculiar property of clotting extremely rapidly when exposed to Gram-negative bacteria, and so serves as a reliable assay to check for disease contamination in human blood supplies (Groombridge, 1992). Such medical model organisms have played a significant role in human health research, and may not be easily replaced by other species.

A third category of medicinal interest covers those species valued as ingredients in traditional folk medical practices. A wide variety of species of both plants and animals are used in various medical traditions, and their economic value, as expressed by those who buy them, is considerable. From personal experience, we know that local people gather, use, and sell a host of medicines from the 9000 ha Mbalmayo forest in the Cameroon, although their value to the local economy has not been quantified (Holland *et al.*, 1992). The annual global trade in medicinal products derived from deer (antlers, tendons and musk, primarily) is estimated at some US$30 million (Luxmoore, 1989). Even though the value of cures composed of rhinoceros horns or tiger bones is not widely acknowledged in the West, there are clearly enough people who *do* believe in their efficacy to pose a significant danger to the species involved. It seems only fair to count these interests in among the commercial values of conserving the relevant animal stocks. Many traditional cures, however, may turn out to be of profound medicinal value once they are investigated; Cameroonian colleagues, for example, are convinced of the efficacy of medicines prepared from the bark of some forest trees, not least in their ability to reduce malarial fevers (Marc Geroe, pers. comm. to J.H.L., 1994). Indeed, the investigation of folk medicine has proven a valuable tool in the developing art of bioprospecting for pharmaceutical compounds (Mahyar *et al.*, 1991; Cordell *et al.*, 1993; Caporale, 1995). Thousands of years of human ingenuity and experience (mixed, admittedly, with a fair amount of magical nonsense) are tied up in the practices of traditional healers. This para-medical casebook is quickly being erased by the onslaught of western culture. One of the last surviving practitioners of traditional Hawaiian medicine recommended that a local seaworm species, *Lanice conchilega*, be tested for anticancer activity. Crude tentacle extracts were found to reduce tumour growth in mice by 60–100% (Oldfield, 1984). Species involved in traditional remedies should be among the highest priorities for conservation.

11.2.3 Industrial values

Biological diversity also has immense industrial value. Wood is the most obvious example. More than 3.8 billion cubic metres of wood are harvested annually world-wide, for fuel, timber, and pulp. Different tree species have different potential uses, and the loss of tree species diversity brings with it the possibility that species with commercially valuable properties may be lost. Woods differ in their density, colour, workability, and susceptibility to fungi or termites, and the trees involved have differing growth rates and habitat tolerances. The most valuable timber species are generally the first to be harvested, and the pressure is so intense on some that they have been effectively eliminated from much of their range (Groombridge, 1992). The same economic logic that has doomed whales and other organisms that reproduce more slowly than money virtually guarantees the destruction of precisely the tree species we find most useful.

Gradually, as natural woodlands are depleted, forestry in many parts of the world is turning to forest plantations for continuing wood supplies. In so doing, it has created a new reason for preserving wild species. Plantation forestry is after all a form of agriculture; the arguments presented above concerning the potential value of wild species for crop development and improvement hold for tree crops as well (indeed, given how poorly developed agro-forestry is as a science in most parts of the world, the potential value of new tree crop species may be immense). A number of promising tree species with fast growth, disease resistance, useful wood characteristics, and (in some leguminous species) nitrogen-fixing abilities have come to light in recent years (e.g. *Leucaena leucocephala*, *Albizia falcataria*; Oldfield, 1984). There is no way of knowing how many other species that would have shown tremendous economic potential have been, or are being lost, but the lost opportunities are no less real just because they go unrecorded.

There are a wide array of other industrial products that depend directly or indirectly on the maintenance of biological diversity. We do not have the space or the expertise to review them all here (excellent reviews have already been written by Oldfield, 1984; Beattie, 1992, 1994; and Groombridge, 1992); instead we will simply list some of the categories to give a taste of the diversity involved. A wide variety of plants are valuable as sources of fibres. Rattan palms, 90% of which are still harvested from the wild, are a major export from many Southeast Asian nations. Of the 104 species found in peninsular Malaysia, all but six are now categorized as vulnerable or endangered (Kiew & Dransfield, 1987). Natural rubber was originally collected from several plant species, but has come to depend almost exclusively on the Para' rubber tree *Hevea brasiliensis*. Despite the advent of cheaper synthetic rubbers, natural rubber has superior durability and so remains invaluable for many purposes, accounting for more than 30% of world rubber sales. A range of plant species is also being investigated as potential sources of industrial oils and waxes, with jojoba (*Simmondsia chinensis*) gradually gaining economic significance as a replacement source for industrial lubricants formerly derived from sperm whale (*Physeter catodon*) oil, required for many lubrication applications under extremely high pressure. Other premium waxes are collected from Brazilian carnauba palms (*Copernicia cerifera*) and Mexican candelilla shrubs (*Euphorbia antisyphilitica* and *Pedilanthus pavonis*). Plants are also being investigated as potentially valuable replacements for fossil fuels; the tissues of some *Euphorbia* species are nearly 10% oil and could yield 5–25 barrels of oil per hectare annually in cultivation.

The list of actual and potential industrial product values extends, however, well beyond these well-known examples. Many agricultural chemicals (herbicides, fungicides, and insecticides) are derived from natural products and, as mentioned above for medicinal chemicals, many others are now synthesized using natural chemicals as templates. Other chemicals have found uses because of their effects on humans; among them are several promising artificial sweeteners that have recently been identified from rain forest plants of Southeast Asia (e.g. Baek *et al.*, 1993). Perfume manufacturers are very secretive about the ingredients they use, but many are reputed to stem from wild sources, mostly in the tropics. The use of fur, feathers, and reptile skins by the fashion industry has often led to severe pressures on wild animal populations; used sustainably, the species involved are demonstrably valuable to many consumers. Beattie (1992, 1994) lists a fascinating array of actual and potential industrial products derived from obscure animals, including instant adhesives (produced by onychophorans, spiders and centipedes to capture prey); corrosion-resistant glues (from barnacles); sunscreens (from the mucous of corals); heat-stable enzymes (from the bacteria of natural hot springs); and ideas for the design of new materials, including flexible concrete (from molluscan shells).

In other words, as with food crops, the contributions of wild species to industry are both direct and indirect. Many are harvested directly from the wild; others are now grown commercially; still others are a potential source of genes to improve commercial crops; and a final group provide inspiration and blueprints for human engineers and designers.

11.2.4 Recreational harvesting

Whereas commercial fishing and hunting in subsistence economies are important food sources, in affluent economies both hunting and fishing have significant recreational value. These can be measured, at least in part, by the significant sums that people are willing to spend on equipment, transportation, and time to pursue them. A wide variety of vertebrate species — including fish of many families, ducks and geese, gallinaceous birds, ungulates, and many carnivores — are valued for their recreational uses as our prey. An even wider array of species are used recreationally by fanciers of ornamental plants. Some of these species are still gathered from the wild, and horticultural plant collecting has sometimes led to the depletion or extinction of wild stocks of certain ornamental species (e.g. the Chilean bulb *Tecophilaea cyanocrocus*, now feared to be extinct in the wild). In gardening, the very novelty of a rare plant adds to its value, and so the range of species with significant commercial value is quite extensive. In other words, the commercial value of conserving rare tropical orchids or other attractive species may be considerable. Similar arguments apply to certain groups of animals (e.g. parrots, reptiles) that are in significant demand for the pet trade.

There are also a number of non-harvest-related recreational values to wild species; birdwatching and ecotourism, for example, generate significant economic activity. The line between the market and non-market categories is blurred here, but we chose to treat the subject below, under aesthetic values.

11.3 Non-market goods and services

The economic value of a species is not restricted to the value of its fur or fruits on the open market. Many valuable functions are served by wild organisms 'for free', and while these roles may be price-less, they are often far from worth-less. It remains a difficult matter for economists to determine the value of things for which no one need ever pay; they often resort to hypothetical questions of how much one would be willing to pay for the existence of pandas, for the sight of a rare wildflower, or for a clear lake filtered of algae by zooplankton. We share with Ehrlich and Daily (1993) the view that some of these attempts to assign monetary values to species or ecosystems are an exercise in 'crackpot rigor' (detailed quantitative analysis of an intractable problem), or 'suboptimization' (doing the very best way something that should not be done at all). Hence we content ourselves here with cataloguing the varieties of such non-priced roles, roughly ordered from the most concrete to the most abstract.

11.3.1 Environmental modulation

Organisms often have profound influences on the environment in which they live. On a small scale, for instance, a tree shades the ground beneath it and draws up water from deep in the soil, creating a cool, moist microhabitat which other plants and animals may require (and which humans and their livestock may enjoy). On a somewhat bigger scale, a local forest of such trees will affect the local hydrology. Deforestation of a watershed can turn formerly steady streams into seasonal torrents which flood in the rainy season and dry up completely in the dry season, disrupting potential uses of stream water for irrigation, milling, hydroelectric power, or transportation. On an even larger scale, extensive forest cover can affect regional or even global climates. Thus, trees (and many other organisms) can be thought of as 'ecosystem engineers', which Jones *et al.* (1994) define as species that directly or indirectly modulate the availability of resources (other than themselves) to other species, by causing physical state changes in biotic or abiotic materials. In doing so they modify, maintain and/or create habitats. Beaver (*Castor canadensis*) are classical ecosystem engineers. Their dams 'retain sediments and organic matter in the channel, . . . modify nutrient cycling and decomposition dynamics, modify the structure and dynamics of the riparian zone, influence the character of water and materials transported downstream, and ultimately influence plant and animal community composition and diversity' in entire watersheds (Naiman *et al.*, 1988).

There are probably no ecosystems on earth that are not engineered in some way by one or more species. However, species clearly differ markedly in their capacity to act as ecosystem engineers. Most probably never perform this function; a few have major impacts, and qualify for the description 'keystone species'. Although this term is overworked and has been used in ways that extend far beyond its original meaning (Mills *et al.*, 1993), it is nonetheless useful shorthand for species whose removal has big effects on many other taxa and ecosystem functions (Paine, 1969; Daily *et al.*, 1993). Most ecologists probably regard trophic links as the main mechanism by which keystone species act (e.g. Paine, 1966; Estes *et al.*, 1978, see also below); but engineering is arguably even more important (Jones *et al.*, 1994). Whatever the mechanism(s) by which keystone species (broadly defined) impose their effects, their existence means that ecosystem processes may respond dramatically and suddenly to the loss of just one species.

11.3.2 Ecosystem functions

Humanity cannot live in isolation. Other organisms play vital roles in photosynthesizing, forming and holding soils, pumping water, fixing nutrients, transforming gases, and regulating the climate of the world we inhabit (Ehrlich & Ehrlich, 1992). While it is clear that we need some other species to share the planet with us if we are to survive, it is far less clear just how many species we need keep to serve those functions adequately. To what extent do total photosynthesis, nutrient flows, water flows, and other vital processes depend on diversity? Conceptually, it is useful to distinguish between responses measured under average or normal environmental conditions, and responses measured under rare, extreme events.

Under average environmental conditions ecosystem processes might respond in one of three ways to reductions in species richness (Lawton, 1994; Naeem *et al.*, 1995). (Obviously, if we were to remove all the species in an entire functional group—all the plants, or all the termites, for example—ecosystem processes would be drastically impaired; *reductio ad absurdum*, the hypotheses are not interesting). The first, the redundant species hypothesis, suggests that there is a minimal diversity necessary for proper ecosystem functioning, but beyond this minimum, most species are redundant in their roles (Walker, 1992; Lawton & Brown, 1993); above the minimum, adding or deleting species has no detectable effect on the process or processes in question. A second, contrasting view suggests that all species make a contribution to ecosystem processes, so that the functioning of these processes declines progressively as species are lost (Ehrlich & Ehrlich, 1981). For convenience we refer to this as the rivet hypothesis, because Ehrlich and Ehrlich liken species to rivets holding together a complex machine, and postulate that functioning will be impaired as rivets (species) fall out, leading ultimately to the complete collapse of the system. A third view, the idiosyncratic response hypothesis, suggests that it is not the number of species in a community *per se* that is important for ecosystem function, but rather the particulars of the species themselves. Various ecosystem functions may change when diversity changes, but the magnitude and direction of the changes is unpredictable, because the roles of individual species are complex and varied (see Lawton, 1994 and Vitousek & Hooper, 1994 for further discussion of these ideas).

Naeem *et al.* (1994, 1995) provided the first direct experimental test of these hypotheses, using small, artificially constructed terrestrial ecosystems in a controlled environment facility, the Ecotron. Most processes (decomposition rates, nutrient retention etc.) varied significantly but idiosyncratically with species richness. Uptake of carbon dioxide and plant productivity, however, both declined as species richness declined, as predicted by the rivet hypothesis. In general, we suspect that species-loss will generally produce either no, or idiosyncratic, changes in ecosystem function. The fundamental problem is that at the moment we are unable to predict the consequences of most species-deletions. Arguably, the most worrying losses are where essential ecosystem 'services' decline as species disappear, conforming to the rivet hypothesis. For example, as Naeem *et al.* (1994, 1995) point out, if reductions in plant species richness generally result in a reduction in the ability of the terrestrial biosphere to fix CO_2, the consequences for global environmental change are enormous. Biodiversity matters!

But not all biodiversity matters equally. Most studies to date have been carried out on relatively simple communities, and even then there has generally been evidence of decelerating effects of species diversity on ecosystem function. The emerging result seems to be that the value of each additional species declines after the first few (Vitousek & Hopper, 1994). Beyond a dozen or so species the addition of another may result in no measurable improvement in performance. Nonetheless, individual species may have relatively large impacts if they have unusual characteristics (e.g. species with unusual stature or growth forms, nitrogen fixers, or ones with unusual habitat requirements or tolerances).

All three models of diversity effects on ecosystem function are right: in general more species perform better than fewer, some perform much more important roles than others; and yet, on average, each additional species added to the system is increasingly likely to be functionally redundant. This is reminiscent of Ehrlich and Ehrlich's original presentation of their 'rivet' metaphor. An airplane's wing is held on by a large number of rivets. Some of them are much more important structurally than others, and (because of the safety margins inherent in good aircraft design) many of the rivets could be thought of as redundant so long as certain others remain in place. As each additional rivet is removed from the airplane's wing, however, there's an increasing probability of the next removal having disastrous consequences. If the ecosystem effects of adding species tend to follow decelerating functions, the effects of subtracting species (which is, after all, what we're doing) must be accelerating.

Even if high species richness does not always play a significant role in maintaining ecosystem processes under average, or benign, environmental conditions, it may nevertheless be important when conditions change. If dominant species are sensitive to the perturbation, rare species with different environmental tolerances may show compensatory changes in abundance, and maintain ecosystem processes (Walker, 1992; Lawton & Brown, 1993; Coleman *et al.*, 1994; Frost *et al.*, 1994). There is growing evidence for such effects in the field, both in aquatic (Frost *et al.*, 1994) and terrestrial (Fig. 11.2; Tilman & Downing, 1994) ecosystems. Bongers (1990) has argued that some nematode taxa may increase rapidly in numbers when environments are disturbed or polluted. In other words, 'species redundancy' under normal environmental conditions, manifest by suites of species with apparently similar roles, many of them rare, may be critical for the maintenance of ecosystem processes in the face of unusual, extreme events.

11.3.3 Ecological roles

Species may have impacts not only on their (and our) environment, but also on the other species with which they interact. Each species fills a number of different ecological roles: most species eat some number of other species and

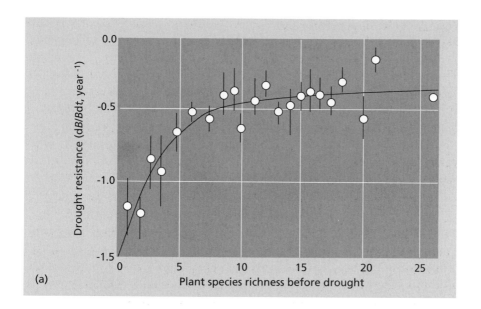

(a)

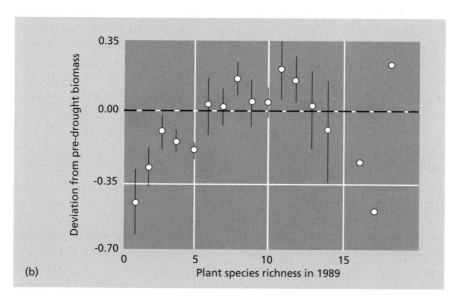

(b)

Fig. 11.2 The effects of plant species richness on the ability of grassland ecosystems to resist, and recover from, drought (Tilman & Downing, 1994). (a) Rate of reduction in plant biomass (drought resistance) as a function of plant species richness in plots; plots that were more species rich in 1986 – the year before the drought – suffered less reduction in biomass during a severe drought that reached its peak in 1988, when these measures were taken. (b) By 1992, more species-rich plots had also recovered more quickly after the drought (recovery is measured as 1992 biomass/average pre-drought biomass).

are themselves food or hosts to a number of others; still others compete with them or are dependent upon them in commensal or mutualistic interactions. The circle of less directly affected organisms linked to them through this first circle is wider still. What we do not know is to what extent species interactions are 'replaceable'. Under what conditions will the extinction of one species cause the subsequent extinction of others, in a sort of 'extinction multiplier effect?' Earlier we considered the case of species that serve 'keystone' roles in communities by virtue of their engineering prowess or as dominant predators, but others may play more subtle, but no less powerful, roles. For instance, a few species of figs and palms in Amazonian Peru provide the only fruit resources available to a host of frugivorous mammals and birds during periods of scarcity (Terborgh, 1986). If we value the preservation of monkeys and toucans (whether for their commercial value or because of our sentimental attachment to them), then we must value these otherwise commercially unimportant plant species as well. Overall, our ignorance of the processes that control the distribution and abundance of the millions of species that share our planet is almost total. This suggests that we ought to be cautious in deciding that any particular species serves no vital role. Ehrlich (1985) cites as an example the *Cactoblastis* moths which were introduced into Australia to act as biocontrol agents (see above) for *Opuntia* cacti. To someone unfamiliar with the history of the system, it would no longer be easy to ascertain the vital role played by these moths. Having brought *Opuntia* under control, *Cactoblastis* is now relatively rare, as is the cactus it consumes. If it were to disappear, however, the now innocent-seeming *Opuntia* populations would presumably explode again. Several authors suggest this is a potent argument for conservation: even if only a few species have critical roles, we may be unable to guess which they are.

11.3.4 Knowledge

Which leads us, naturally enough, to the pursuit of knowledge. The loss of a species represents the loss of information. Who knows what Steller's sea cow might have told us about deep-diving in cold Arctic waters, or whether dodos were essential for the dispersal and germination of fleshy-fruited *Calvaria* trees on Mauritius (Temple, 1977; Owadally, 1979; Witmer & Cheke, 1991). Not only species, but also species interactions and intact ecosystems are important subjects of research. Each species lost from a community represents dozens of lost interactions, leaving a potentially gaping hole in the ecological web. As human domination of the planet grows more complete, there are progressively fewer places where biologists can study natural systems at something resembling evolutionary equilibrium (if that is not an oxymoron). Do interacting species coevolve? It is hard to say unless we can observe assemblages with a long history of coexistence. Have human interventions had significant effects on major ecosystem functions, on

300

hydrology, or on other important natural processes? It is impossible to know unless there are at least a few 'control plots' of unmolested communities that we can use for comparison. This may be dismissed as special pleading; we are both, after all, field ecologists. Fair enough. And yet the pursuit of knowledge, even for its own sake, is something that western society has historically agreed is valuable. Beyond that, a well-developed science of ecology would have practical implications — if we understood more deeply the rules of the game that determines the distributions and abundances of species, we would be able to turn that knowledge to greater advantage, improving agricultural productivity, controlling pests, and learning how to live sustainably on this planet. The maintenance of natural systems, and the information they represent, helps to make that possible.

11.3.5 Aesthetic values

Economists argue that to the extent that people derive pleasure from their experience of wild animals or plants, or even from the knowledge that they exist out there somewhere, those organisms provide a service which should, in theory, be taken into account in decision making. These services are provided free, and so it is often a difficult (if not a downright impossible) matter determining the monetary value people would be willing to pay for them. Consequently, as a practical matter, they are often neglected. It is nonetheless true that a potent reason why people wish to conserve species is aesthetic (Lawton, 1991); we wish to do it because they enrich our lives, just as Mozart concertos, Monet paintings and medieval cathedrals give meaning, shape, colour and hope to the lives of millions of people. Aesthetic values depend on the intrinsic aesthetic qualities of a species (how pleasant we find it to interact with them), but also on the quantity of such interactions (how often humans are likely to interact with them). An unfortunate consequence of this is that the rarer a species is, the lower its total aesthetic value is likely to be. On the other hand, the per capita aesthetic value of the few remaining individuals of a rare species may be enhanced, especially in taxa (e.g. birds and butterflies) where hobbyists have developed a taste for 'twitching'. The enormous amounts of time, energy and money spent by birdwatchers in the pursuit of sightings of very rare birds suggests that aesthetic values can occasionally be quite significant. Indeed, the economic payoffs from hobbyists and from the growing industry of 'ecotourism' may be considerable, often dwarfing some of the more traditional categories of market values listed above. Gordon Orians (pers. comm., 1989) has pointed out that non-consumptive (e.g. aesthetic) uses of wildlife often compete favourably with consumptive uses (e.g. hunting) in terms of benefits conferred: an elk can be shot only once but it can be admired and/or photographed countless times. Moreover, hunting may result in changes in the distribution or behaviour of aesthetically valuable species (i.e. eliminating them or causing them to avoid

301

areas frequented by humans) which make non-consumptive uses, even of the surviving individuals, more difficult. As human populations increase and the availability of natural areas decreases, there should come a point where non-consumptive uses of certain wildlife species outweigh their consumptive uses.

11.3.6 Existence values

The 'existence value' of a species is even harder to quantify, but it accrues whether or not a person directly experiences the organism in question. For a few, highly charismatic or symbolic species (giant pandas, quetzals, and blue whales, for example) these values may be enormous. People give willingly to conservation organizations to aid in the conservation of populations that most of the donors will never have a chance to see. It is no wonder that fund-raising campaigns generally focus on charismatic vertebrate species — it is from them that the public derives this most disembodied of species values. The hard fact of the matter, however, is that the vast majority of species are not valued (neither aesthetically nor for existence) by any but a very few people. The two of us may be concerned about the continued existence of a rare desert mustard (W.E.K.) or a specialist bracken herbivore (J.H.L.), but few others will lose much sleep (or donate much money) over the matter. As public education and communications technology improve, existence and aesthetic values for many species will probably increase. Even so, the vast majority will probably be left out in the cold.

11.4 Intrinsic value

All of the above has assumed that value is measured exclusively in terms of increases in human welfare and happiness. But many people feel that the preservation of a species has value independent of any utilitarian consequences (e.g. Callicot, 1986; Ehrlich & Ehrlich, 1992). The reasons given are as varied as the philosophies of the proponents. Some argue that all species are the product of Divine creation and so must have a role to play, even if we do not perceive it. Those who come from western religious backgrounds may cite the biblical figure Noah as an example to follow. Others suggest that diversity is an intrinsic good for its own sake. Some (including many animal liberationists) see the prevention of pain in all creatures to be the source of moral imperative, and so argue not so much for the value of non-human species but rather for the individual organisms that compose them. Still others argue that morality is based, ultimately, on empathy, and that the extension of human concern to non-human species is but a further elaboration upon it. The arguments presented for each notion are impassioned and sometimes persuasive. Indeed, many who argue for conservation on other grounds probably take the actions they do based

302

partially on personal beliefs concerning the intrinsic value of species. We include ourselves in this category. Unfortunately (or, perhaps, fortunately), such fundamental principles carry little weight in the formation of policy. Opponents of conservation are unlikely to share the values of conservationists. The danger of allowing such personal values to enter the political arena is that they may turn other arguments into apologetics.

11.5 In summary

There is a great deal of fuzzy talk circulating about the value of biodiversity. It is not enough to demonstrate that we benefit from the existence of some other species. Of course we do. But it is exceedingly unlikely that the cow, wheat, or the Hevea rubber tree will become extinct, so long as we continue to find them useful. The real issue is: what becomes of the rest of the 10 million or so species of life on earth? Most of them are not particularly prepossessing: insects, fungi, nematodes, marine invertebrates, microorganisms of various kinds, shrubby plants; and very few of them have any significant track record of known service to humanity. To save them will require real sacrifices by people who have little spare to sacrifice, or significant investments by those who have many other demands on their resources. Can such expenditures be justified?

The arguments for conservation reduce to three basic points:

1 Some of the species at risk *are* of value to us, indeed, some are endangered precisely *because* they are of value to us. Others are known to be valuable, but may be accidentally lost due to our carelessness or the allocation of benefits (that is, the costs of conservation are not borne by those who benefit by it). Still others are hunted to extinction because they breed less quickly than money does when invested. To allow over-exploitation or neglect to destroy a demonstrably useful species is foolish mismanagement: killing the species of goose that lays golden eggs. These are the easiest cases to argue, but they represent only a small fraction of the species in danger.

2 Many species may have great value to us now or in the future, but we do not know which they are. Our uncertainty stems from the fact that the majority of the earth's species have not yet been named, much less studied intensively, from our ignorance of the mechanics of the ecological systems upon which we depend, and from our inability to predict what our needs are likely to be in the future. Because of this uncertainty, a much wider array of species are covered by the argument, but the expected value of each is fairly small—the true value of the relatively few successes must be discounted for its timing (in the mid- to long-term future) and spread over the greater number of species that may never prove valuable. To the extent that we can refine our predictive ability (whether by developing taxonomic or biogeographic generalizations based on past experience, or by the application of evolutionary principles, e.g. Beattie, 1992, 1994) we can increase the expected value of species but decrease the number of species covered.

3 The diversity of life on earth is intrinsically valuable, not for what it contributes to human welfare, but for itself. This argument covers all species, but its persuasive power is limited. Its strength depends on the personal beliefs of its proponents, and so may bear little political weight with those who do not share their philosophy.

Ultimately, the question of how much biodiversity is worth may be at best academic, and at worst an exercise in 'crackpot rigor'. If we are unable to act effectively to protect those species of proven value (category 1 above), is there any point in devoting great energies to quantifying just how much the species in the remaining categories are likely to be worth?

In any case, the exercise of assigning market values to individual species is irrelevant to conservation as it is usually practised. If the great mass of biological diversity on this planet is to be saved, it will not be because of species-centred conservation attempts and their attendant individual valuations, but rather by habitat conservation, with whatever species happen to come along with the territory. Thus, with all but the few established stars (e.g. primates and the wild relatives of agricultural species), if species values are to be taken into account in the policy arena, they will have to be considered *en masse*. Conservation of species via their habitats is a logarithmic process; as a biogeographical rule of thumb, for each 10-fold reduction in habitat area, the number of species retained is approximately halved. Thus, as remaining stands of natural habitat decline in a region, the conservation value (in terms of the mean number of species lost per unit habitat lost) rises exponentially. Multiplying those species numbers by any reasonable positive value, there must come a point where the conservation value of retaining preserves outweighs the value of the land for other uses. If we are to prioritize, it should not be species which we evaluate but sites; the greater the number, size, and (most crucially) variety of sites we can salvage, the more successfully we will preserve the planet's biological heritage for future generations. In Beattie's (1994) words, 'it is currently inappropriate to rule out any habitat type as being not worth saving'.

The open question is not whether species ought to be preserved, but rather whether we as a species will have the sense and the self-control to carry out the task, the resources to do it properly, and the global political willpower and machinery to deliver the goods in the face of a lethal cocktail of collective global madness and indifference.

References

Baek N.I., Chung M.S., Shamon L., Kardono K.B.S., Tsauri S., Padmawinata K., Pezzuto J.M., Soejarto D.D. & Kinghorn A.D. (1993) Potential sweetening agents of plant origin 28. Studies on Indonesian medicinal plants 5. Selligueain-A, a novel highly sweet proanthocyanidin from the rhizomes of *Selliguea feei. J. Nat. Prod.—Lloydia* **56**, 1532–1538. [11.2.3]

Beattie A.J. (1992) Discovering new biological resources—chance or reason? *BioScience* **42**, 290–292. [11.2.2] [11.2.3] [11.5]

Beattie A.J. (1994) Conservation, evolutionary biology and the discovery of future biological resources. In: *Conservation Biology in Australia and Oceania* (eds C. Moritz, J. Kikkawa & D. Doley), pp. 305–312. Surrey Beatty, Sydney. [11.2.2] [11.2.3] [11.5]

Bongers T. (1990) The maturity index: an ecological measure of environmental disturbance based on nematode species composition. *Oecologia* **83**, 14–19. [11.3.2]

Brown G.M.J. (1990) Valuation of genetic resources. In: *The Preservation and Valuation of Biological Resources* (eds G.H. Orians, G.M.J. Brown, W.E. Kunin & J.E. Swierzbinski), pp. 203–245. University of Washington Press, Seattle. [11.1.1]

Callicot J.B. (1986) On the intrinsic value of nonhuman species. In: *The Preservation of Species* (ed. B.G. Norton), pp. 138–172. Princeton University Press, New Jersey. [11.1.1] [11.4]

Caporale L.H. (1995) Chemical ecology: a view from the pharmaceutical industry. *Proc. Nat. Acad. Sci., USA* **92**, 75–82. [11.2.2]

Clark C.W. (1981) Bioeconomics. In: *Theoretical Ecology: principles and applications*, 2nd edn (ed. R.M. May), pp. 387–418. Blackwell Scientific, Oxford. [11.2.1]

Coe M. (1980) African wildlife resources. In: *Conservation Biology: an evolutionary–ecological perspective* (eds M.E. Soulé & B.A. Wilcox), pp. 273–302. Sinauer, Sunderland, Massachusetts. [11.2.1]

Coleman D.C., Dighton J., Ritz K. & Giller K. (1994) Perspectives on the compositional and functional analysis of soil communities. In: *Beyond the Biomass: compositional and functional analysis of microbial communities* (eds K. Ritz, J. Dighton & K. Giller), pp. 261–271. John Wiley, Chichester. [11.3.2]

Coley P.D. (1988) Effects of plant growth rate and leaf lifetime on the amount and type of anti-herbivore defense. *Oecologia* **74**, 531–536. [11.2.2]

Coley P.D. & Aide T.M. (1991) Comparison of herbivory and plant defenses in temperate and tropical broad-leaved forests. In: *Plant–animal Interactions: evolutionary ecology in tropical and temperate regions* (eds P.W. Price, T.M. Lewinsohn, G.W. Fernandes & W.W. Benson), pp. 25–49. Wiley-Interscience, New York. [11.2.2]

Collins D.J., Culvenor C.C.J., Lamberton J.A., Loder J.W. & Price J.R. (1990) *Plants for Medicine: a chemical and pharmacological survey of plants in the Australian region.* CSIRO, Melbourne. [11.2.2]

Cordell G.A., Farnsworth N.R., Beecher C.W.W., Soejarto D.D., Kinghorn A.D., Pezzuto J.M., Wall M.E., Wani M.C., Brown D.M., O'Neill M.J., Lewis J.A., Tait R.M. & Harris T.J.R. (1993) Novel strategies for the discovery of plant-derived anti-cancer agents. *ACS Symp. Series* **534**, 191–204. [11.2.2]

Daily G.C., Ehrlich P.R. & Haddad N.M. (1993) Double keystone bird in a keystone species complex. *Proc. Nat. Acad. Sci. USA* **90**, 592–594. [11.3.1]

Doeleman J.A. (1989) Biological control of *Salvinia molesta* in Sri Lanka: an assessment of costs and benefits. *ACIAR Technical Reports* **12**, 4–14. [11.2.1]

Ehrlich P.R. (1985) Extinctions and ecosystem functions: implications for humankind. In: *Animal Extinctions: what everyone should know* (ed. R.J. Hoage), pp. 159–174. Smithsonian Institution Press, Washington DC. [11.1.1] [11.3.3]

Ehrlich P.R. (1988) The loss of diversity: causes and consequences. In: *Biodiversity* (ed. E.O. Wilson & F.M. Peter), pp. 21–27. National Academy Press, Washington DC. [11.1.1]

Ehrlich P.R. & Daily G.C. (1993) Population extinction and saving biodiversity. *Ambio* **22**, 64–68. [11.1.1] [11.3]

Ehrlich P.R. & Ehrlich A.H. (1981) *Extinction: the causes and consequences of the disappearance of species.* Random House, New York. [11.3.2]

Ehrlich P.R. & Ehrlich A.H. (1992) The value of biodiversity. *Ambio* **21**, 219–226. [11.1.1] [11.4]

Estes J.A., Smith N.S. & Palmisano J.F. (1978) Sea otter predation and community organization in the western Aleutian islands, Alaska. *Ecology* **59**, 822–833. [11.3.1]

Evans H.E. (1966) *Life on a Little-known Planet*. University of Chicago Press. [11]

Frost T.M., Carpenter S.R., Ives A.R. & Kratz T.K. (1994) Species compensation and complementarity in ecosystem function. In: *Linking Species and Ecosystems* (eds C.G. Jones & J.H. Lawton), pp. 224–239. Chapman & Hall, New York. [11.3.2]

Fuller R.W., Bokesch H.R., Gustafson K.R., McKee T.C., Carbellina J.H., McMahon J.B., Cragg G.M., Soejarto D.D. & Boyd M.R. (1994) HIV-inhibiting coumarins from latex of the tropical rainforest tree *Calophyllum teysmannii* var *inophylloide*. *Bioorganic Medicinal Chem. Lett.* **4**, 1961–1964. [11.2.2]

Goodman R., Hauptli H., Crossway A. & Knauff V.C. (1987) Gene transfer in crop improvement. *Science* **236**, 8–14. [11.2.1]

Groombridge B. (ed.) (1992) *Global Biodiversity: status of the earth's living resources*. Chapman & Hall, London. [11.2.1] [11.2.2] [11.2.3]

Hilder V.A., Gatehouse A.M.R., Sheerman S.E., Barker R.F. & Boulter D. (1984) A novel mechanism of insect resistance engineered into tobacco. *Nature* **330**, 160–163. [11.2.1]

Holland M.D., Allen R.K.G., Campbell K., Grimble R.J. & Stickings J.C. (1992) *Natural and Human Resource Studies and Land Use Options*. Department of Nyong and So'o Cameroon. Natural Resources Institute, Chatham. [11.2.2]

Job C.K., Sanchez R.M., Hunt R., Truman R.W. & Hastings R.C. (1993) Armadillos (*Dasypus novemcinctus*) as a model to test antileprosy vaccines – a preliminary report. *Int. J. Leprosy* **61**, 394–397. [11.2.2]

Jones C.G., Lawton J.H. & Shachak M. (1994) Organisms as ecosystem engineers. *Oikos* **69**, 373–386. [11.3.1]

Julien M.H. (ed.) (1992) *Biological Control of Weeds: a world catalogue of agents and their target weeds*, 3rd edn. CAB International/Australian Centre for International Agricultural Research, Wallingford & Canberra. [11.2.1]

Kiew R. & Dransfield J. (1987) The conservation of palms in Malaysia. *Malay. Nat.* **41**, 24–31. [11.2.3]

Kyle R. (1987) *A Feast in the Wild*. Kudu, Oxford. [11.2.1]

Lande R., Engen S. & Saether B.-E. (1994) Optimal harvesting, economic discounting and extinction risk in fluctuating populations. *Nature* **372**, 88–90. [11.2.1]

Langer R.H.M. & Hill G.D. (1982) *Agricultural Plants*. Cambridge University Press. [11.2.1]

Lawton J.H. (1990) Biological control of plants: a review of generalisations, rules and principles using insects as agents. In: *Alternatives to Chemical Control of Weeds. Proceedings of an International Conference, Rotorua, New Zealand* (eds C. Bassett, L.J. Whitehouse & J.A. Zabkiewicz), pp. 3–17. Ministry of Forestry, Forest Research Institute Bulletin 155, Rotorua. [11.2.1]

Lawton J.H. (1991) Are species useful? *Oikos* **62**, 3–4. [11.1.1] [11.3.5]

Lawton J.H. (1994) What do species do in ecosystems? *Oikos* **71**, 1–8. [11.3.2]

Lawton J.H. & Brown V.K. (1993) Redundancy in ecosystems. In: *Biodiversity and Ecosystem Function* (eds E.-D. Schulze & H.A. Mooney), pp. 255–270. Springer-Verlag, Berlin. [11.3.2]

Levin D.A. (1976) Alkaloid-bearing plants: an ecogeographic perspective. *Am. Nat.* **110**, 261–284. [11.2.2]

Levin D.A. & York B.M. Jr (1978) The toxicity of plant alkaloids: an ecogeographic perspective. *Biochem. Syst. Ecol.* **6**, 61–76. [11.2.2]

Luxmoore R.A. (1989) International trade. In: *Wildlife Production Systems: economic utilisation of wild ungulates* (eds R.J. Hudson, K.R. Drew & L.M. Baskin), pp. 28–49. Cambridge University Press. [11.2.2]

Mahyar U.W., Burley J.S., Gyllenhaal C. & Soejarto D.D. (1991) Medicinal plants of Seberida (Riau Province, Sumatra, Indonesia). *J. Ethnopharm.* **31**, 217–237. [11.2.2]

Mills L.S.M., Soulé M.E. & Doak D.F. (1993) The keystone-species concept in ecology and conservation. *BioScience* **43**, 219–224. [11.3.1]

Naeem S., Thompson L.J., Lawler S.P., Lawton J.H. & Woodfin R.M. (1994) Declining biodiversity can alter the performance of ecosystems. *Nature* **368**, 734–737. [11.3.2]

Naeem S., Thompson L.J., Lawler S.P., Lawton J.H. & Woodfin R.M. (1995) Empirical evidence that declining biodiversity may alter the performance of terrestrial ecosystems. *Phil. Trans. R. Soc. Lond. B* **347**, 249–262. [11.3.2]

Naiman R.J., Johnston C.A. & Kelley J.C. (1988) Alteration of North American streams by beaver. *BioScience* **38**, 753–762. [11.3.1]

Neueschwander P. & Herren H.R. (1988) Biological control of cassava mealybug, *Phenacoccus manihoti*, by the exotic parasitoid *Epidinocarsis lopezi* in Africa. *Phil. Trans. Roy. Soc., Lond. B* **318**, 319–333. [11.2.1]

Norse E.A. (1985) The value of animal and plant species for agriculture, medicine, and industry. In: *Animal Extinctions: what everyone should know* (ed. R.J. Hoage), pp. 59–70. Smithsonian Institution Press, Washington DC. [11.1.1]

Norton B.G. (1986) On the inherent danger of undervaluing species. In: *The Preservation of Species* (ed. B.G. Norton), pp. 110–137. Princeton University Press, New Jersey. [11.1.1]

Norton B.G. (1987) *Why Preserve Natural Variety?* Princeton University Press, New Jersey. [11.1.1]

Oldfield M.L. (1984) *The Value of Conserving Genetic Resources*. US Department of Interior, Washington DC. [11.1.1] [11.2.1] [11.2.2] [11.2.3]

Orians G.H. & Kunin W.E. (1990) Ecological uniqueness and loss of species. In: *The Preservation and Valuation of Biological Resources* (eds G.H. Orians, G.M.J. Brown, W.E. Kunin & J.E. Swierzbinski), pp. 146–184. University of Washington Press, Seattle. [11.1.1]

Owadally A.W. (1979) The dodo and the tambalacoque tree. *Science* **203**, 1363–1364. [11.3.4]

Paine R.T. (1966) Food web complexity and species diversity. *Am. Nat.* **100**, 65–75. [11.3.1]

Paine R.T. (1969) A note on trophic complexity and community stability. *Am. Nat.* **103**, 91–93. [11.3.1]

Peters C.M., Gentry A.H. & Mendelsohn R.O. (1989) Valuation of an Amazonian rainforest. *Nature* **339**, 655–666. [11.2.1]

Pimm S.L., Moulton M.P. & Justice L.J. (1994) Bird extinctions in the central Pacific. *Phil. Trans. Roy. Soc. Lond. B* **334**, 27–33. [11.2.1]

Principe P.P. (1991) Valuing the biodiversity of medicinal plants. In: *The Conservation of Medicinal Plants* (eds O. Akerele, V. Heywood & H. Synge), pp. 79–124. Cambridge University Press. [11.2.2]

Randall A. (1986) Human preferences, economics, and the preservation of species. In: *The Preservation of Species* (ed. B.G. Norton), pp. 79–109. Princeton University Press, New Jersey. [11.1.1]

Room P.M. (1990) Ecology of a simple plant–herbivore system: biological control of *Salvinia*. *Trends Ecol. Evol.* **5**, 74–79. [11.2.1]

Scoones I., Melnyk M. & Pretty J.N. (eds) (1992) *The Hidden Harvest: wild foods and agricultural systems. A literature review and annotated bibliography*. International Institute for Environment and Development, London. [11.2.1]

Soejarto D.D. (1993) Logistics and politics in plant drug discovery—the other end of the spectrum. *ACS Symp. Series* **534**, 96–111. [11.2.2]

Steadman D.W. (1995) Prehistoric extinctions of Pacific island birds: biodiversity meets zooarchaeology. *Science* **267**, 1123–1131. [11.2.1]

Temple S.A. (1977) Plant-animal mutualism: coevolution with dodos leads to near extinction of plant. *Science* **197**, 885–886. [11.3.4]

Terborgh J. (1986) Keystone plant resources in the tropical forest. In: *Conservation Biology: science of scarcity and diversity* (ed. M.E. Soulé), pp. 300–344. Sinauer, Sunderland, Massachusetts. [11.3.3]

Tilman D. & Downing J.A. (1994) Biodiversity and stability in grasslands. *Nature* **367**, 363–365. [11.3.2]

Vaeck M., Reynaerts A., Höfte H., Jansens S., De Beuckeleer M., Dean C., Zabeau M., Van Montagu M. & Leemans J. (1987) Transgenic plants protected from insect attack. *Nature* **328**, 33–37. [11.2.1]

Vietmeyer N.D. (1986) Lesser-known plants of potential use in agriculture and forestry. *Science* **232**, 1379–1384. [11.2.1]

Vitousek P.M & Hooper D.U. (1994) Biological diversity and terrestrial ecosystem biogeochemistry. In: *Biodiversity and Ecosystem Function* (eds E.-D. Schulze & H.A. Mooney), pp. 3–14. Springer-Verlag, Berlin. [11.3.2]

Waage J.K. (1991) Biodiversity as a resource for biological control. In: *The Biodiversity of Microorganisms and Invertebrates: its role in sustainable agriculture* (ed. D.L. Hawksworth), pp. 149–163. CAB International, Wallingford. [11.2.1]

Waage J.K. & Greathead D.J. (1988) Biological control: challenges and opportunities. *Phil. Trans. R. Soc., Lond. B* **318**, 111–128. [11.2.1]

Walker B.H. (1992) Biodiversity and ecological redundancy. *Conserv. Biol.* **6**, 18–23. [11.3.2]

Witmer M.C. & Cheke A.S. (1991) The dodo and the tambalacoque tree: an obligate mutualism reconsidered. *Oikos* **61**, 133–137. [11.3.4]

Wood R.K.S. & Way M.J. (eds) (1988) *Biological Control of Pests, Pathogens and Weeds: developments and prospects*. Royal Society, London. [11.2.1]

12: Identifying priorities for the conservation of biodiversity: systematic biological criteria within a socio-political framework

R.I. VANE-WRIGHT

12.1 Objectives and values

Conservation of biological diversity depends on respect for people, including their needs and aspirations, as well as for nature. One of the tasks of conservation biology is to determine appropriate biological priorities for action (Spellerberg, 1992). However, only if such priorities are socially acceptable can progress be expected. Furthermore, priorities are relative, not absolute. Doing more of one thing, or doing it with greater urgency, should not mean doing nothing about everything else — only less, or later.

The values of biological diversity for humanity are many (Chapter 11), including material resources (such as food, oils, drugs, genes for biotechnology, building materials), services (fertile soils, clean air, fresh water, renewable energy) and existence needs (spiritual well-being, knowledge, ethics). Reid (1994) has argued that the primary goal of ecological management should be to maximize human capacity to respond to changing ecological conditions — for which maintenance of biodiversity is a prerequisite. These ideas are encapsulated within the idea of option value (Section 12.2.2; Chapter 3), a concept covering the full range of values, from basic material resources, through human adaptability and the needs of future generations, to ethical and spiritual concerns for other organisms.

Given that biological diversity has so many differing values, it is not surprising that difficulties arise over its non-sustainable use or transformation. Biodiversity is inextricably bound to local geography on the one hand, and to complex transboundary interactions and cycles on the other. What the United Kingdom chooses to do about air pollution affects not only British wildlife, but also animals and plants in other parts of Europe. Trade in tiger bones in China damages conservation and ecotourism in India, and threatens the rights of all future generations to know the 'spirit of the jungle'. An international ban on the sale of ivory can bring about destruction of an African farmer's crops, without legitimate possibility for alternative income from trade in elephant products. Triumph for one human interest group can lead to deprivation or even extermination for another. So, if conservation priorities based on biological factors are to have any impact on human affairs, they must be determined with a view to implementation within a human socio-political context. We need a framework for understanding what biological conservation priorities are, and how they can be related to human self-interests.

As identified by Gadgil (Chapter 13), biodiversity is complex and cannot be conserved adequately simply by attention to special areas or reserves (cf. Westman, 1990). There is a growing realization, based on metapopulation biology, that a far wider range of species than those traditionally considered vulnerable may be threatened by habitat fragmentation (Kareiva & Wennergren, 1995). If areas chosen as priority sites become disconnected patches surrounded by a hostile matrix, then sooner or later much of the diversity for which they were set aside will be lost. Cousins (1994) suggests that selection of hyper-diverse areas involves 'cheating' the species–area curve, so that if only such reserves are managed they will steadily lose diversity until they come back into line with the expected point on the graph. Equally, only a fraction of genes and species can be conserved by *ex situ* means (and functional ecosystems probably not at all). If we conclude, with Gadgil and others, that all of nature must be managed for biodiversity, are priorities meaningless?

Priorities only make sense in relation to responsibilities and obligations. Human affairs are organized hierarchically (family, community, state, federation, global league). We can apply all sorts of measures and analyses to identify biological singularities, maxima or complements, but these can only be translated into action at one or more appropriate levels within the socio-political hierarchy. Moreover, priorities are not absolute but relative, and the relativities must be taken into account at every level if all needs, objections and opportunities for conservation are to be satisfied. This requires harmonization (conflict resolution), which in turn implies information flow and trade-off between levels, if practical successes are to be achieved. Recognition of political reality is essential – otherwise, setting priorities will achieve no purpose whatsoever.

Conservation is, in a sense, at the mercy of a fully democratized political process, in which all legitimate levels of responsibility should be heeded, but in which *rights, responsibilities and obligations of individuals and local communities are paramount*. Rights without obligations, and obligations without rights, are insufficient (Ehrenfeld, 1993; cf. Gadgil *et al.*, 1993). Within human society, those in authority often seek to reserve power (rights) for themselves (e.g. the right to decide), but tend to devolve responsibility downward – demanding 'loyalty' from subordinates whilst frequently eschewing that moral responsibility themselves. Solution of the biodiversity conservation problem cannot be found without addressing the political problems of democracy, and this applies *a fortiori* to the problem of priority setting. In this chapter I will consider what, from a biological perspective, priorities might be – and then try to place such priorities in a suitable framework for action.

12.2 Prerequisites

A change of focus has occurred within conservation, from saving particular charismatic animals from extinction, or special wilderness areas from destruction, to concern for an ever-wider range of threatened species and habitats (Myers, 1979). Over the last two decades this has come to encompass the whole legacy of evolution – the literally countless genes and species that exist, functionally related within a great variety of assemblages and ecosystems. Signature of the 'Rio' global Convention on Biological Diversity in June 1992, and its entry into force on 29th December 1993 (Glowka *et al.*, 1994), underline how far and how rapidly this change has progressed.

Given that biodiversity protection will largely depend on *in situ* measures, two principal scientific issues have emerged: how to conserve it, and where to conserve it. The first, involving applied ecology and population biology, has been dominant because of a direct relationship to managing endangered species and local ecosystems. The central problems of conservation management are caused by the sheer complexity of life processes: if a piece of land, a lake, river or part of the ocean is designated as a biodiversity priority area, how is that system to be managed to preserve the biodiversity found there, or for which it was set aside?

The second issue, where conservation action is most needed, is now coming to the fore (e.g. Austin & Margules, 1986; Margules & Austin, 1991; Spellerberg, 1992; Forey *et al.*, 1994; Humphries *et al.*, in prep.). It involves dealing with the problems caused by the complex geographical patterns, found at varying spatial scales, in the distribution of all living organisms. The practitioners of conservation evaluation acknowledge that resources for conservation, especially land and people needed to manage it, will always be limited. They are primarily concerned with criteria for choosing amongst conservation alternatives to satisfy clearly stated goals.

In choosing between alternative projects, those that fail to take heed of conservation management issues (including the appropriate use of indigenous knowledge: Gadgil *et al.*, 1993) are unlikely to be effective. Research undertaken to improve successful management of specific ecosystems, or particular elements of biodiversity, should be encouraged. However, the 'where?' issue is critical in terms of setting priorities (especially in the shorter term, while some real options for action still remain open). Although scientific reasons alone will never be the only factors governing choice between projects, as conservation evaluation is now concerned not only with sustainability, but also efficiency and value-for-money, it follows that favouring non-biological criteria will normally result in less effective use of available resources in relation to protecting biodiversity *per se*.

12.2.1 Goals

A general idea of representation has existed in the literature for decades, but only now is it emerging as a general principle for goal-setting (Margules & Usher, 1981; Austin & Margules, 1986). Representation is about sampling. Imagine a set of Scrabble letters. All letters (genes or characters) of the alphabet are represented, some several times, some only once. The object of the game is to subsample this pool of letters in a series of attempts to create whole words (species), until all the letters are used. The words have to fit together according to the assembly rules of a crossword, within the confines of the Scrabble board (ecosystem). Scoring depends on individual letter values, their combination in each word, and the position of words on the board. Scrabble is difficult because players subsample the letter pool at random. Very often the subsamples only permit construction of low-value words, or even no words at all — random permutations of letters rarely make words, just as random combinations of genes will not make organisms. Very often the players can make good words, but can find no place to fit them on the board — in the same way that, although a huge variety of different combinations of animals and plants can coexist within different ecosystems, a random combination of species would rarely be stable.

Analogies are limited, but I will push this example one more step. Suppose you had to make Scrabble sets and needed, among other things, to keep a set of letters as patterns (units of replication). Suppose also that you did not have room in your workshop for all 152 letters that make a full set. The players sample the letters at random — would this be good enough for you? Would you be content to put your hand into the bag, pull out a handful, and say 'that's the best I can do!' Obviously not. You want, at a minimum, one copy of every letter in the alphabet (together, ideally, with a formula that says make 9 As, 2 Bs, 2 Cs . . . and 1 Z per set). Every word can be made up from copies of the full set — leave out even one letter-pattern, and a huge slice of the dictionary is lost.

In general, subsamples made from a multi-copy set of heterogeneous entities (like a Scrabble set, or species distributed over a complex landscape) will fall into three sorts: random, hypodiverse, and hyperdiverse. If a subsample is small and chosen at random, then it will usually fail to include representatives of all the different entities. If not random, then subsamples can either be hypodiverse (fewer entities than expected by chance), or hyperdiverse. A basic thrust of current work in biodiversity conservation evaluation is to devise strategies that will give us the ability to make hyperdiverse samples — ideally, the most hyperdiverse samples possible. This applies not only to full representation of all attributes, but also to increasing equability in multiple representations (e.g. Williams *et al.*, in press a).

312

12.2.2 Measurements

Scrabble is simple, life is not. We have little idea how many genes there are (new ones probably originate every day, others are lost), how many species exist (millions remain uncatalogued, while many known and unknown taxa are going extinct), or what rules govern the assembly and stability of ecosystems. A primary task is to estimate diversity in a way suitable for conservation evaluation.

The interplay of local species richness and individual abundance was formalized as α-diversity by Whittaker (1965), who proposed measuring it by the α statistic of Fisher *et al.* (1943). Following Whittaker (1972: α-, β- and δ-diversity) and McNaughton (1983: γ-diversity), ecological diversity has been accepted to have a more complex, spatially hierarchical configuration (McNaughton, 1994), and a considerable variety of procedures for its measurement have now been proposed (Magurran, 1988).

Whittaker (1972) also suggested that time, or the evolutionary dimension, and other components in addition to richness and spatial turnover, should be included in a more complete expression of diversity—but did not propose a solution. Only a decade after IUCN (IUCN *et al.*, 1980) recognized the need to factor taxonomic rank into measurements of diversity, as the only rule of thumb available to estimate degree of genetic loss represented by any particular extinction, Vane-Wright *et al.* (1991) proposed the first taxic diversity metric designed to be sensitive to both taxonomic rank and number of species found within an area. Vane-Wright *et al.* (op. cit.) also formalized complementarity, the degree to which specific areas, singly or in combination, represent the taxic diversity of an entire group or set of groups. Complementarity has much in common with β- and δ-diversity but, crucially, instead of just reducing taxon turnover to a numerical value, information on the identity of taxa between areas is retained (see Section 12.3.3).

Since then, in a series of papers mainly by Williams (e.g. Williams *et al.*, 1991, 1993, 1994; Williams, in press), Weitzman (1992) and Faith (e.g. 1992, 1994), the concept of taxic diversity has been refined with respect to measurement procedures—and even more critically with regard to the fundamental question of what is being measured, and why. Williams and Humphries (Chapter 3) conclude that the current goal is to assess *option value*. This concept, originally proposed in economics (Weisbrod, 1964), has been adopted within conservation as 'a means of assigning a value to risk aversion in the face of uncertainty' (McNeely, 1988), or with maximizing human capacity to adapt to changing ecological conditions (Reid, 1994). If this is accepted, we can abandon the intractable problem of trying to assign differential values to species as such (Ehrenfeld, 1988), and shift our attention to the fundamental level where the objects of future utility exist—to the level of expressible and heritable characters (genes, traits, features etc.—Faith, 1992; Williams *et al.*, 1994).

313

In practice it is impossible to measure character differences between taxa directly on a large enough scale, and so the distribution of characters across taxa must be modelled. While agreed that this should be based on the genealogical hierarchy, as expressed more and more completely by taxonomic ranks, cladograms and metric phylogenetic trees, debate continues over which model of character change should be applied (e.g. empirical *versus* anagenetic *versus* cladogenetic), and whether or not differences should then be assessed in terms of simple character richness, or character combination richness (Williams *et al.*, 1994). Despite some fundamental differences in philosophy, differences in results are not so great — partly because, when large numbers of taxa are involved, species richness itself appears to be a good surrogate for character diversity (Williams & Humphries, 1994).

12.2.3 Surrogates

Problems with the measurement of diversity do not end with the question of option value and how to assess it. They also exist at more prosaic levels — what data are relevant, and how should we collect them?

Margules and Austin (1994) have discussed basic questions about data requirements for conservation evaluation. They note that most available data consists simply of records of presence — without adequate records of absence, or abundance. Existing data have generally been collected in an *ad hoc* fashion, constrained by subjective interest and ease of access, with little or no regard for environmental gradients or regionalization. While they review fundamental solutions for the improvement of this situation through properly thought out survey methods and protocols, they also acknowledge that existing data, although flawed, are better than nothing. If mobilized and used appropriately, they can still tell us a great deal about the geographical spread and location of different elements of biodiversity.

In some regions of the world, however, even reasonable quality data for birds and mammals may effectively be lacking. What can we do in this situation, faced with the frequent need to make rapid decisions in the face of far-reaching development proposals, such as the clear-felling of forests or damming of rivers? Humphries *et al.* (1995) review this problem as the scale of surrogacy, from the use of proximate measures (such as species richness), in the sense that they are close to the currency of option value (character diversity), through increasingly remote surrogates, such as higher taxa (Williams & Gaston, 1994), vegetation classes, land classes (Mabbutt, 1968), and measures of environmental diversity (Walker & Faith, 1993). Depending on situation, scale and data availability, one or more surrogate levels may be chosen as appropriate for the given task and available resources (Margules & Austin, 1991).

A final complication involves choosing which subset of the potentially overwhelming quantity of data to attend to. This is most obvious in the debate over indicator groups. On the one hand (e.g. ICBP, 1992; Kremen, 1994; Beccaloni & Gaston, 1995) there are those who consider that individual groups can have indicator qualities (i.e. they reflect or predict the diversity of other, unsampled or unmeasured groups), whereas others (e.g. Prendergast *et al.*, 1993) have questioned whether we can reasonably expect concordance at all. More pragmatic approaches have been proposed regarding the selection and use of focal taxa (e.g. Ryti, 1992) and higher taxa (Williams & Gaston, 1994).

While all these problems, both theoretical and practical, must be addressed, one of the strengths of the current approach to conservation evaluation is to devise methods that are applicable, in principle, at any scale and to any data—so long as complementarity (which depends on the identity of attributes, or some aspect of attribute pattern, across areas) can be satisfied. If this is accepted, then despite the current limitations of data and measurement, we can still formulate a strategy for action.

12.2.4 A simple strategy

If the currency units of ultimate value are characters (including the genes that code them), then we can recognize 'vessels' that contain those units: cells, whole organisms, life-cycles, demes, populations, taxa, ecosystems. Characters (or genes) can only exist, *in situ*, insofar as they are expressed or their potential is encoded within whole organisms, and each whole organism necessarily forms part of and is generated by a sequence of reproducing life-cycles. This genealogical relationship, involving descent with modification, continues into phylogeny, reflected by our taxonomy of species within genera, genera within families and so on—including the belief, based most convincingly on the evidence of the genetic code, that all known life forms are inter-connected in this way. However, these historical relationships are paralleled by functional relationships, linking cellular metabolism to the physiological workings of whole organisms, ecological interactions within populations and local ecosystems, and interrelations between biotic regions and the whole biosphere. Eldredge and Salthe (1985) refer to these twin rankings as the genealogical and ecological hierarchies, respectively. The units of diversity are thus produced by a genealogical hierarchy, but function within an ecological hierarchy.

In terms of both our ability to describe the form of the genealogical hierarchy, or to understand the functional relationships of the ecological hierarchy, the species category holds a special place. Many influential biologists consider species to be 'the fundamental unit' of diversity. As discussed above, if the fundamental currency units of diversity are the expressed or expressible characters by which they differ, we then have a rational basis for choosing amongst species in relation to option value. Although we may also cavil about the uncertain connotations of the word 'species', there is clearly a sense in which species, as evidence of both the being and becoming of evolution (Rieppel, 1988), do play a central role in our understanding of biological diversity. If we can accept this, and faced with our lack of knowledge of the total magnitude of character diversity and its distribution over the Earth, then we can adopt a simple strategy for identifying *in situ* priority areas: survey the geographic ranges of a wide variety of species (either directly, or through the use of higher surrogates for character diversity) as indicators, analyse these data to identify an efficient network of ecosystems capable of supporting as diverse a set of the indicators as possible (and other taxa likely to occur in the same ecosystems), and then manage this network sustainably (Vane-Wright, 1994).

12.3 Priorities in principle

12.3.1 On priorities

Priorities are about choices at different scales, alternative sequences, and are goal and context dependent. A dictionary definition of priority is 'interest having prior claim to consideration'. Thus interest must be defined – for example, a national agency endeavouring to preserve biodiversity in accordance with requirements of the International Convention, or a local community dependent on the sea for food. National and local interests (the interests of ecosystem people *versus* ecological refugees *versus* biosphere people, or 'omnivores': Gadgil, 1991) often conflict, or major issues at one level may be of no concern at another. If we are to recognize effective priorities (i.e. those prior claims for consideration that can be turned into acceptable and sustainable recipes for action), it will be necessary to ensure that interests at different levels are harmonized. One role for biologists is to provide the scientific information and analyses which will help decision-makers to resolve such conflicts.

A problem with identifying priorities, and in particular priority areas for conservation, is that it can lead to the false conclusion that other land has no significance for biodiversity – is of 'no priority'. On the contrary, the ideal is to manage all areas for their contribution to ecosystem function (e.g. Cousins, 1994). A second problem relates to a widespread misunderstanding of the nature of priorities.

Consider the concept of *triage* (Myers, 1979), best known in relation to crisis management in military hospitals. Faced with overwhelming injuries, from superficial to fatal, the surgeon has to make terrible decisions about who to assist. The injured must be divided rapidly into three classes: the 'walking wounded' who will survive without surgery, those so badly injured that they are unlikely to survive whatever is done, and those who will probably survive if attended to urgently. Given the goal of maximizing the total number of survivors, the surgeon should concentrate on the third group.

A dictionary definition of triage gives two meanings that are not mutually exclusive: 'the act of sorting according to quality', and 'the assignment of degrees of urgency to decide the order of treatment of wounds, illnesses etc.' Unspoken in these definitions is the idea of a clear goal – in the case of the surgeon, to minimize the loss of life. In what follows, I will explore sorting according both to quality and degrees of urgency.

12.3.2 On principles

In planning the use of inevitably limited resources for the conservation of biological diversity, we can recognize a number of 'principles' (e.g. complementarity and flexibility: Vane-Wright *et al.*, 1991; Pressey *et al.*, 1993; Vane-Wright, 1994; Humphries *et al.*, in prep.). Underhill (1994) has queried whether or not complementarity can be considered a principle. Complementarity, like parsimony, offers a criterion for choice, and is basic to general procedures for priority setting in relation to explicit goals. Currently, I recognize four such principles (efficiency, flexibility, vulnerability and viability), all of which can interact to guide the focus of conservation effort. Because they all necessarily affect the outcome of the priority assessment process, the order in which they are dealt with is arbitrary. However, I start with efficiency, as this sets a baseline for the whole procedure.

12.3.3 Complementarity and efficiency

Given a measure of biodiversity, and agreement on a representation goal, then efficiency (Pressey & Nicholls, 1989) is a primary guiding principle for translating the goal into priorities. Every action should contribute towards the goal, and the collective set of all actions should achieve the goal with the minimum use of resources. To take any other course is to err in the direction of *ad hoc* selection which, as Pressey (1990, 1994; Pressey & Tully, 1994) has shown, runs the severe risk of failing to reach the target even when all available resources have been exhausted. A second reason for preferring efficient solutions is that, with respect to politically managed conservation efforts, normal accounting procedures demand cost-effectiveness. If we accept that all areas have at least some unique attributes or value for biodiversity, then efficiency can be seen even as a moral virtue or necessity. The better we select and husband our limited and precious lands, seas, funds and human energies, the more biodiversity we can protect for the benefit of ourselves, future generations, and nature itself.

In addition to the general notion, efficiency has also been used to indicate some minimum subset necessary to achieve a specified goal. Pressey and Nicholls (1989) defined efficiency, E, as $1 - (X/T)$, where X is the minimum number (or total area, or cost) of sites needed, and T is the total number (or area, or cost) of sites sampled (see Humphries *et al.*, in prep. for other

317

measures). Efficiency is achieved by maximizing *complementarity* (Vane-Wright *et al.*, 1991), defined as the degree to which a single area or subset of areas represents the total number of attributes found in the whole system, or adds unrepresented attributes to one or more specified areas. Complementarity can only be applied to measures of biodiversity that retain information on the identity of the component attributes of the surrogate in question (e.g. a focal taxon: Ryti, 1992), or some other aspect of the attribute pattern (Faith, 1994). In this context, simple numerical measures such as species richness are worthless, because they do not permit the attributes represented by two or more areas to be aggregated. As demonstrated by Williams and Humphries (1994) and Humphries *et al.* (1995), complementarity is the most powerful procedure we have for measuring biodiversity in the context of conservation evaluation. This is undoubtedly because it flows from, and represents a direct implementation of the efficiency principle.

Thus an idealized task for *in situ* biodiversity conservation involves identifying the most efficient (and effective) network of areas for protection or specific management that will satisfy the agreed representation goal (Rebelo & Siegfried, 1992). In reality, it has to be recognized that it may not be possible, due to lack of resources, or a range of other political, social and other non-biological factors, to create an entire network at one stroke. Given this uncertainty, we may often have to be content with progressing towards the goal, one or a few steps at a time (Sætersdal *et al.*, 1993). Complementarity is of major importance in this context also (see Sections 12.3.6 and 12.3.7).

12.3.4 Vulnerability and triage

Resources and opportunities are always limited, and so action must be directed where it will do most good. Efficiency is dedicated to this end. However, we must also acknowledge that certain species and areas are more threatened than others. Does this mean that priority areas analysis should concern only the most threatened or vulnerable? If we apply triage (Section 12.3.1), some of the most threatened areas or attributes are often beyond hope—these lost causes are not sensible objects of our attention (cf. Weitzman, 1992). Others (like the walking wounded) will probably survive with little more than minimal attention. There is a risk that the conservation lobby will continue to be 'fobbed off' with areas of low commercial value little threatened by development or other forms of transformation (Pressey, 1994). Investing in soft options is, in effect, avoiding the issue.

Estimates of relative vulnerability can be based on areas or attributes. There is much literature on rarity and vulnerability of different species (especially vertebrates), representing the traditional concern of much of the conservation movement—preservation of the rare and curious. Faced with the extraordinary pressures that the growth of urban society and rural poor are placing on ecosystems world-wide, it is now evident that fragmentation represents an even greater threat to many supposedly 'common and widespread' species previously seen as of little concern for conservation (e.g. Kareiva & Wennergren, 1995). As many of these probably form key elements of ecosystems more important to humanity than the mountain fastness of a rare eagle, or the desert home of some curious plant, their active conservation through appropriate landscape management may be of greater general significance for sustainability and human welfare.

Our ability to understand and integrate the collective threats to individual species over whole ecosystems or landscapes seems limited. A practical way to incorporate vulnerability within systematic evaluation may simply be in terms of areas—it should be relatively easy to work out which ecosystems, based on soils and microclimate for example, and which locations, based on physical geography and human demography, are most likely to be the target of transformation for agricultural or other development (R.L. Pressey, pers. comm., 1994). Focus on these areas will keep the conservation movement away from soft options—but will bring greater need for conflict resolution in the face of political reality. Biologically tractable areas may be rendered lost causes because of political pressures. This type of problem is discussed under the term *contingent areas* (Section 12.4.4).

12.3.5 Viability and the ecological imperative

Viability concerns a boundary between systematic evaluation and management. Preliminary systematic analysis may identify an area as important because it contains a vulnerable species, or other attribute, needed to fulfil a representation goal. Ecological assessment of the area may indicate, however, that the attribute is not viable there—because, for example, the region is a population sink, or it has passed some fragmentation threshold—both with inevitable consequences (but cf. Murcia, 1995). Investing in areas that have no prospect of sustaining the biodiversity for which they were chosen is clearly a bad decision—literally throwing good money after bad. It is imperative that sound ecological principles (both biological and political) are applied not only to management of priority areas, but also to a screening procedure before investment is made. An efficient solution on paper may comprise a hopelessly unrealistic set of areas in terms of viability.

There is a considerable literature devoted to the spatial design of both individual reserves and networks to meet viability goals. Much of this is based, sometimes uncritically, on equilibrium biogeography (see Shafer, 1990), including such factors as size, shape, buffer zones, proximity, interconnection, population size, density and genetics, extinction vulnerability, and multiple representation (see e.g. Margules *et al.*, 1988; Cowling & Bond, 1991; Burgman *et al.*, 1993). Given the widely varying situations encountered in different regions of the world, and the singular properties of different ecosystems and individual taxa (e.g. Margules *et al.*, 1994a), taking full account of all these factors is an almost irreducibly complex process. A major way in which systematic conservation evaluation can help is through exploration of flexibility in area selection.

12.3.6 Flexibility and goal-satisfying networks

Worldwide, almost every river, forest, sea basin or handful of soil could contain some unique element of biodiversity—be it endemic fish, plant assemblage, obscure worm or bacterial gene. Total representation would require everything. In practice we have incomplete knowledge, and will be forced to base our assessments on one or more surrogates (e.g. indicator taxa, plant assemblages, or environmental classes). As we focus at finer scales, we generally find that the component elements of the surrogate are distributed over the landscape in a complex mosaic, such that most occur at many discrete locations. The extent to which these can be considered to represent independent viable units varies enormously (as does, almost certainly, the degree to which they indicate comparable patterns in the distribution of other, unmeasured biodiversity), but at the outset it is generally assumed, with the exception of large vertebrates and migratory organisms, that this is potentially the case. If we make this assumption, then it is usually possible at national or local scales to choose a large number of alternative networks capable (subject to viability screening) of achieving the representation goal. Both the desire and ability to explore these alternatives constitute the principle of flexibility.

Defined narrowly, flexibility is the degree to which different configurations of the network can be found, for a given number of sites or total area, that will satisfy the conservation goal. If the goal is set at representation of all attributes that make up the surrogate (species, families, land classes, etc.), then flexibility can be seen to be an expression of complementarity. In an attempt to transform the idea of flexibility into a measure applicable to every individual area under consideration, Pressey *et al.* (1994) introduced the notion of irreplaceability.

The irreplaceability value of an area can be defined simply as the proportion of all possible fully representative combinatorial networks in which the area occurs. These are enumerated or estimated according to the series of fully representative sets $(M, M + 1, M + 2, \ldots T)$, where M represents all fully representative subsets found at the size of the minimum-set, plus all fully representative subsets at size $M + 1$, $M + 2$, etc., up to the whole set, T. Thus, if an area contains only common and widespread species, found at many other areas under consideration, the particular site will only occur in a small proportion of all fully representative networks, and thus have a low irreplaceability value. On the other hand, any area which has an attribute found nowhere else in the entire data set (globally irreplaceable) will necessarily form part of every fully representative set – and so have an irreplaceability value of 100%. Areas including attributes which occur in only a few other areas will have intermediate values.

Globally irreplaceable areas, if threatened, inevitably represent high priorities for conservation action because, if their biodiversity is lost, the representation goal is permanently compromised through extinction. Similarly, areas with high irreplaceability values, even if not wholly irreplaceable by one or more other areas in terms of the attributes they include, are also likely to be prime targets for conservation action. Faced with urgent decisions about parcels of land in isolation, or without a comprehensive plan, irreplaceability value can give some indication as to the degree to which development proposals can be received with relative equanimity or should be met with determined resistance. Goal irreplaceable areas (those only replaceable by two or more other areas) will often be of high conservation value because their loss will decrease efficiency.

There has been some debate as to whether or not irreplaceability values have practical utility (but cf. Section 12.4.3). However, the most important thing about flexibility, whether construed as a hyperspace of network alternatives, or a set of fixed site-by-site values, is that we have the ability to explore it, to compare alternatives, and make intelligent trade-offs between areal efficiency, network design, vulnerability, viability, resource cost and human impact. The most effective way devised yet for the exploration of this flexibility is through use of iterative algorithms.

12.3.7 Interactive iterative analyses

Following Kirkpatrick (1983), most proponents of systematic conservation evaluation have used iterative procedures. Underhill (1994), however, has made an attack on iterative procedures that employ a greedy algorithm (constrained to choose the greatest increment at each step), claiming that the results will sometimes be 'grossly suboptimal' in relation to the problem of identifying the most efficient fully representative network. To arrive at this minimum-set, Underhill proposes that we follow the lead of Cocks and Baird (1989) to develop rigorous mathematical linear or integer programming methods. Quite apart from some severe practical limitations of mathematical programming (Pressey & Possingham, in press), Underhill has failed to appreciate that stepwise, *interactive* iterative analyses give us control over flexibility. Thus, in a program such as Williams' WORLDMAP, *any* choice can be explored at any step, including rapid comparison of a range of different measures, not just that dictated by the greedy algorithm. In addition, backward checks can be made for redundant choices.

In practice, because of problems posed by viability, politics and lack of resources, we will rarely be able to implement a minimum-set solution

(Pressey *et al.*, in press a). More typically, we will be trying to optimize a subset (Sætersdal *et al.*, 1993), or explore a set larger than the minimum to allow for contiguity, cheaper land options, and so on. In all of these applications, the exploration of flexibility based on complementarity offers the best means to proceed. Furthermore, the selection of priority areas should not be coupled with disregard for the rest, but with differential application of resources across the whole landscape (see Section 12.4.3). While it is important to be rigorous wherever possible, to substitute mathematical elegance for powerful heuristics would be a mistake. Interactive analyses (Margules *et al.*, 1994b; Pressey *et al.*, in press b; Williams *et al.*, in press b) give us control over flexibility — in the widest sense of the word.

12.4 Priorities in practice

If setting priorities, or taking any form of action to conserve biological diversity, is to be effective, it is essential that a 'bottom up' approach is included (Mittermeier *et al.*, 1994). Only if the people who live in an area appreciate the need to protect its biodiversity, wish and have the power to do so, is there any realistic chance of success (Gadgil, 1991, 1992). When actions are imposed from above, 'top down', they are resented and resisted, and fail to benefit from vital local knowledge (Gadgil *et al.*, 1993). Imposition represents an infringement of self-respect and even human rights. If the conservation of biological diversity is seen, as it is within the Convention, in the context of human welfare and sustainable development, then those who are most directly affected by conservation action must be willing partners in the process.

In other words, if local people do not 'own' the management proposals, they will find ways to get round them. A supposed system of benefit that is insensitive to legitimate rights and interests of individual people is doomed, because it is so clearly open to partiality, hypocrisy, corruption and subversion. However, a purely bottom-up, individual-case-by-individual-case approach will never satisfy the needs of biodiversity conservation. Setting priorities involves goals that are context- and area-dependent. A top-down element is necessary to deal with the transboundary phenomena that affect almost all aspects of natural ecology — as Forman (1987) put it, 'no boundary in a landscape is impermeable and no action is isolated'. Thus only by taking an overview can the needs of representation, urgency and efficiency be appreciated and acted upon. If local councils, national governments and regional blocs are legitimate and are charged with making decisions on behalf of the people they represent, then it is essential that they not only take appropriate overviews, but that they also have some power, or some mechanism to put decisions into effect. To admit otherwise is to deny those who form governments their corporate responsibilities.

Thus the challenge becomes, how are different and often conflicting legitimate interests, from individual through to supra-national, to be integrated into an acceptable plan for conservation action, which can take into account, inter alia, *biological* priorities? Can compromises be achieved that do not vitiate all that has gone before? The way forward is to replace the idea of an action plan (static) with a planning system (dynamic), within the context of a continuing bottom-up, top-down review process. This would have to include open debate, wide consultation, local knowledge, individual needs and respect for all legitimate levels of responsibility, together with the requirement for a continuing overview, if effective action is to be taken. The experience reported by Mittermeier *et al.* (1994) of participatory 'workshops' to examine priorities at regional levels in Latin America gives insight into what might be possible, and offers a valuable model for further development.

12.4.1 Levels of responsibility and a hierarchical framework

As already discussed, all areas of the Earth should be managed for biodiversity (Chapter 13; Westman, 1990; Gore, 1992). Each area of land is usually occupied, owned or controlled by an individual, a small group, or an institution. I will assume a widely held ideal, that every individual should own, or be in some way responsible for a more or less extensive land area ('home'), and that such parcels of land collectively constitute most of terra firma (the high seas present problems over common rights, reflected for example in fishing disputes, that are not addressed here). Common lands, state lands, and lands owned by corporations, insofar as they are held in trust on behalf of groups of individuals, can be included for most purposes of the argument (where common resources are open to any user, they are typically degraded – Gadgil *et al.*, 1993).

Such a model is in accord with the 20th century position on individual human rights, including the rights of mutual respect and self-determination. But with individual rights must come responsibilities and obligations – otherwise the cult of individualism (including corporate autonomy) will rob humanity of its most precious qualities. If each individual or group has some control over a piece of land, then they also have responsibilities towards that land because their actions have consequences. Pouring chemical effluent from metal extraction into a stream is vandalism, whether carried out by a state-owned industry, private mining corporation, or an individual prospector.

Obligations extend to fauna and flora and the resources they represent – the landowner or user has a responsibility to minimize impact on local diversity, and might be expected to account for or justify any changes. Like any sensible home owner, he or she should make some investment in environmental maintenance. Each of us, ideally, should create 'Individual Environmental Funds' (IEFs), in cash or kind, to include biodiversity maintenance costs. These could be seen as the personal or corporate equivalent of marginal opportunity costs (Pearce, 1987) — the net cost of choosing not to do something that would otherwise generate personal gain or profit, if the environmental cost or impact is too high.

In addition to individual organization, we can imagine a minimum of three layers of socio-political organization — local, national and global. Each of these would, ideally, be managed by individuals democratically delegated to fill the positions of responsibility. In environmental terms, these layers should give perspective — an integrated system of checks and balances, with ability to sanction infringement of local, state or international laws, and to promote more widely beneficial actions. The orderly hierarchical organization of local governments within national governments subject to federal control is appropriate for this task, so long as each layer recognizes that it too has obligations as well as rights — and these obligations include respect for human rights and the legitimate responsibilities of those operating elsewhere within the hierarchy.

By creating, at successive levels, through agreed processes of taxation and/ or donation, local environment funds (LEFs), national environment funds (NEFs) and global environment funds (GEFs), the social hierarchy can provide both framework and resources to improve environmental protection from wider and wider perspectives. But, as already noted, different perspectives lead almost inevitably to different views about what should or should not be done. How can conflicts be resolved without infringing the rights of those 'lower' in the hierarchy, or rendering well-informed proposals from those 'higher' in the socio-political structure ineffective? Continuing debate over the role of the World Bank's Global Environment Facility in relation to the Financial Mechanism of the Convention on Biological Diversity illustrates how conflicts and frustrations can arise at any level. Amongst many criticisms, low value attached to conservation achieved through local sustainable use (cf. Pearce & Moran, 1994), failure to make environmental impact assessments, and lack of public participation (Groombridge, 1992: Chapter 32) are of particular concern in the present context.

In a simple framework model for levels of responsibility (Fig. 12.1), a key to rectifying some of these undesirable properties may lie in the idea that no 'lower' level should have imposed upon it an environmental action which it does not recognize, at some level, as a priority or acceptable option of its own. Environmental decisions taken in New York should not force indigenous people from tribal lands in Africa against their will, however long-sighted, or however important for some paper conservation plan. In other words, a national government should not impose on a local government, or a local government on an individual, an action that, subject to availability of resources, a responsible local government or individual would not agree to in their own right (even though, in the real world, this may often seem to go against the interests of nature protection — cf. Oates, 1995).

324

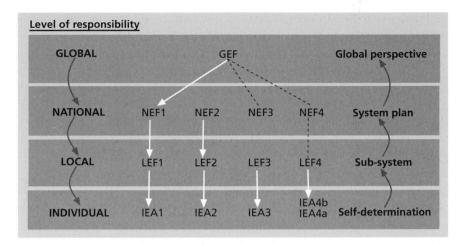

Fig. 12.1 Schematic hierarchical framework for managing and implementing priorities for the conservation of biological diversity. Data, action plans, taxes and some delegation of responsibility pass up from each lower level to the level above. Information, counter proposals, subsidies and empowerment flow down from each higher level to the level below. Priorities set at one level are primarily funded at that level, and can only receive additional resources directly from one level above. See also text.

What higher levels should do, in the case of conflict, is to inform local groups why certain proposals are being made (including the local and wider consequences of ignoring them), and endeavour to alter priorities or priority rankings by offering incremental costs, opportunity costs or even total project costs (Glowka *et al.*, 1994: box 19). This presupposes that the principles outlined above (Section 12.3) are being applied at every level, so that efficient use of available resources is being proposed to secure maximum option value. This would be vital to counter avarice or indolence (Glowka *et al.*, loc. cit.) — those negotiating too high a price on conservation action would run the risk of gaining nothing, in the knowledge that resources could be switched to alternative (flexible) areas offering better biodiversity returns on the investment. Those, on the other hand, who refuse to take action where action is seen to be needed, also forgo environmental subsidies. This is the overwhelming reason why, in my opinion, despite philosophical and other objections to measurement and valuation of biodiversity (such as those voiced by Norton, 1994, and Ehrenfeld, 1993), in practice it is necessary to do so. Without explicit and transparent arguments, including comparative measures, there can be no effective appeal to external arbitration, no baseline, no proxies (Glowka *et al.*, loc. cit.) — in other words, no sanctions or subsidies made legitimate by appeal to reason.

The hierarchical framework for responsibility has wider implications for information, power and resource flow. What passes up the system at each successive level is data, proposals for action, taxes (direct, indirect, donations) and delegation; what should flow down is value-added information, counter proposals, subsidies and empowerment. Action should only be taken with local agreement – be the proposal endogenous to the individuals concerned and merely rubber-stamped locally, or modified following argumentation and the application of subsidies from above, to serve a greater good. In this scheme, individuals can only get subsidies from LEFs; LEFs have their own sources which can be supplemented by NEFs but not GEFs directly; GEFs can only fund NEFs in relation to proposals already agreed at local level, and so on. If, as already noted, short-cuts are applied (e.g. GEF or NEF resources applied directly at individual level), then democracy and equality are threatened. If it is to work, every level of the system must be satisfied that its level of responsibility has not been undermined. Of course, such systems are notoriously cumbersome and can be administratively inefficient, but if the added costs help prevent despotism, corruption and other forms of self-interest, then they are necessary.

12.4.2 Resource distribution profiles

Based on the hierarchical model, priorities can also be considered as relative distribution of resources rather than specific actions. Resources could be distributed to reflect the degree to which each component area potentially contributes to the overall goal. This approach would not replace the need to identify specific priority actions (indeed, it would not make sense without them), but would act as a guide to assess how resources are being mobilized in relation to the whole. This idea leads to what could be termed *resource distribution profiles*. Idealized profiles could be derived from priority areas analysis.

Consider the example in Appendix 12.1 (see Fig. A12.1; analysis in Tables A12.1, A12.2). It has seven local administrative areas. Suppose that each local area generates one biodiversity resource unit for its own LEF, which it distributes according to its own plan, that the national government calls on each local area to contribute an additional 4/7ths of a unit to the NEF (as a biodiversity conservation levy), and that it has also negotiated a three-unit GEF subsidy in support of its own national action plan. Thus the governmental NEF has a total seven resource units to distribute amongst the seven areas. For the purposes of arriving at the idealized profile, assume that all areas have arrived at equally reasonable, democratic and appropriate local action plans. How should the government divide up the NEF biodiversity cake? What options are there?

One possibility would be to distribute the seven units, with respect to biological factors, in an *ad hoc* manner, on the basis of whim or political expedience. This is modelled by a random allocation of whole units in Fig. 12.2a – note that the nationally distributed funds are *additional* to the local funds, and so even this extreme does not result in total disregard of any areas. Alternatively, the government could distribute the NEF funds equally,

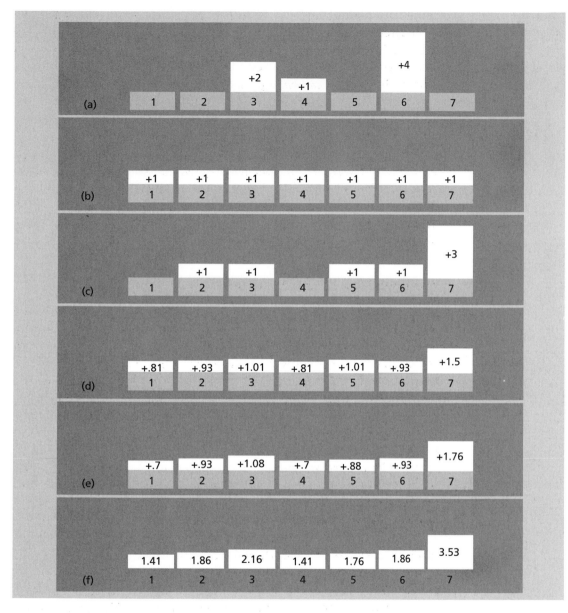

Fig. 12.2 (a–f). Resource distribution profiles for seven administrative areas, 1–7 (Fig. A12.1). In the first five scenarios (a–e), each area generates and independently applies one unit of local (LEF) funding (lower portion of each column). In addition, a further seven national (NEF) units have been distributed (upper portion of each column), in accordance with various national schemes and/or formulae: (a) ad hoc; (b) uncritical; (c) minimum set of flexibility; (d) shades of irreplaceability; (e) irreplaceability x taxic diversity. In scenario (f), the seven local and seven national units have first been aggregated, and then distributed according to a centrally applied index (in this notional example, irreplaceability x taxic diversity). Compared with an idealised profile (f), a resource profile such as (d) or (e) represents a socio-politically preferable solution, having been produced by a combination of independent local and national allocations (see also text).

without any regard to expediency or biological priorities (Fig. 12.2b). Both approaches represent a failure of government to discharge its obligations properly – and could be sanctioned from above (GEF funds might be withdrawn) or below (through the ballot box).

A more defensible option could be based on a narrow interpretation of priority areas analysis, such as dividing the funds among areas of the minimum-set (Fig. 12.2c; see Appendix for details of minimum-sets). This would still leave a problem of how to apportion the available resources amongst that subset – all sorts of schemes could be proposed, all more or less *ad hoc*, and thus difficult to justify. A more general approach needs to be based on some procedure that compares all areas. In Fig. 12.2d, the seven units have been allocated in strict proportion to the irreplaceability values derived in Table A12.2. Irreplaceability is based on complementarity, but does not take direct account of option value as measured by taxic diversity, nor does it deal with threat or viability. So a more sophisticated approach would involve calculation of a complex coefficient for each area (such as the product of irreplaceability and taxic diversity), and then using this to apportion resources (e.g. Fig. 12.2e – based on area values for irreplaceability × character richness, as given in Tables A12.2 and A12.1, respectively; with suitable data on vulnerability, etc., further adjustments could be made). While it would be tempting to add the national and local resources together, to give a total of 14 units, and apportion these according to the national formula (as in Fig. 12.2f), this would violate the principle of hierarchical area management, by effectively eliminating the local level of decision making – and is therefore rejected.

Such an approach to resource allocation could easily be dismissed as pseudo-science. Attention to decimal places is not intended. What I am trying to suggest is that, if profiling indicates that an increment of five units would justifiably be added at the next level below, then if what is actually being applied or proposed is 10 units, or none, then the planning system may be out of balance in relation to the biological needs. Of course, in reality the resources needed to take effective action, and the ability of communities inhabiting different regions to pay, will all be highly disparate, and these factors would have to be taken into account in a more realistic evaluation. However, such complications do not invalidate the idea of resource distribution profiles as a planning tool. If it is not enough just to manage special areas, then we have to find a logical way to allocate resources differentially. Some attempt to distribute them so as to reflect the relative potential contribution that each piece of land can make to the maintenance of biodiversity as a whole seems a reasonable way to proceed. If an acceptable way of computing 'resource profiles' can be agreed, then this procedure is clearly amenable to representation as funding or value maps, geographic information systems (GIS) overlays, etc.

12.4.3 National biodiversity agencies

Gámez and Gauld (1993) have characterized the work of a national biodiversity institute. For them, its role is to facilitate 'three sequential and somewhat overlapping steps: saving biodiversity, knowing it and putting it to use.' They claim that in Costa Rica 'the first has largely been completed', with the result that the National Biodiversity Institute of Costa Rica (INBio) can focus on 'these vital second and third steps'. According to Gámez and Gauld, if you cannot know nature through Latinized binomens, you cannot put it to good use. The second and third stages are concerned with information flow through the 'taxasphere' (Janzen, 1993), to ensure that those who are willing to pay for biodiversity products (e.g. drug companies for pharmaceutical screening, the agricultural industry for potential biological control agents, etc.) get to know what is out there, and how to find it. This vision of the biodiversity institute is intensely technical.

Such institutes will no doubt play an increasing role within developing, and even developed countries. However, the vision does not go far enough. To focus biodiversity on commercial value is a contentious issue in itself (Porritt, 1993; Adams, 1994; but cf. Pearce & Moran, 1994), but my main reservation is the implication that 'saving biodiversity' is a one-stage process. On the contrary, conservation is a continual war in which every battle has to be fought again and again, and no victory is secure. Through effective lobbying and education, conservationists hope to do better in future, but that future is still far distant. If biodiversity institutes are to be purely technical affairs, we need an additional type of organization, national biodiversity agencies, to prosecute the war, organize the saving of biodiversity, and go on saving it. Obviously the two types of organization should be closely linked. If we assume national biodiversity institutes will take care of taxonomists, parataxonomists, the national taxonomic inventory, identification, and getting bio-technical information to end-users, what should NBAs set out to do?

The most basic function for an NBA would be to act as a national integrating centre for all activities relating to the Convention. In particular, it should foster the 'coalition for biodiversity'—a network linking ordinary citizens, professionals and entrepreneurs on the one hand, and communities, businesses, non-government and government agencies on the other, to satisfy their needs for power to conserve biological diversity, and the provision of information about its values and uses (Vane-Wright, 1994). Such a role could not be passive. The NBA would need to promote dialogue for the planning and decision cycle, and so build consensus within the coalition (cf. Mittermeier *et al.*, 1994). If the fixed action plan is to be replaced by a dynamic planning system, involving differential analysis to identify the potential contribution of each land unit and integration to establish practical system and subsystem goals, then implementation will depend on setting and managing priorities within the hierarchical socio-political framework.

To achieve this, the NBA will have to promote agreed goals, gathering suitable data, and application of appropriate measures. Setting practical goals will depend on agreements about surrogates for 'wholesale biodiversity'. This will involve, in addition to managing information on the existence and distribution of crop varieties, species and higher taxa, gathering data on vegetation types and other land classifications, geology, and climatological variables. Most of this data will need to be mapped, subjected to priority areas analysis (Humphries *et al.*, in prep.), and integrated within GIS or comparable technology (e.g. Scott *et al.*, 1993; Pressey *et al.*, in press b). Except for some sensitive information (e.g. occurrence of rare and persecuted species), such data would ideally be available in the public domain, including the Internet, with a policy of openness adopted wherever possible. Richardson (1994) warns of the technical and sociological problems associated with data management alone, but the function of the NBA would go well beyond this. It would have to become part of the fabric of the decision-making process affecting every level of environmental and biological resource control. The challenge is a formidable one. Central to this task would be the *action register*.

12.4.4 An action register

Many of us keep a diary, not only to record past events, but also to keep track of important future events. Personal organizers offer a more sophisticated approach than scraps of paper or old-fashioned appointment diaries — but all are attempts to solve the same problem: how not to forget an array of upcoming tasks, and when to do them. In effect, they provide a register of activities, things that we must do and/or want to do. For the conservation of biological diversity, each NBA will need something similar.

If the ideal of managing all areas for biodiversity is to be achieved, then priority areas analysis applied at local, national, regional or global levels will result, at the coarsest cut, in a form of triage. Areas will be grouped into three major classes: *action areas* (where supportive action is needed and can be taken); *conditional areas* (where intervention is not considered necessary, due to low threat and/or value in relation to the overall goal and current priority areas); and *contingent areas* (where action is desirable but entirely dependent on additional resources and/or a change in political reality).

Generally, contingent areas will include endemic species or other irreplaceable attributes, and be afflicted by extreme political or environmental factors (such as war, oppression, inundation, etc.), such that they are judged beyond help or hope at the present time. Some of the (non-unique) attributes associated with such high quality sites may have to be represented elsewhere, with an expected loss of efficiency due to increasing the number of action areas (or *ex situ* projects) likely to be required. If contingent areas retain their endemic attributes despite prevailing conditions, their transfer to the action list at the first opportunity is desirable. So long as the mountain gorillas of Rwanda survive, their plight should not be forgotten.

All action areas will have some clearly identified function (e.g. essential complement) in relation to the achievable conservation goal. Implementation must be followed by monitoring, to give not only regular appraisal of each action area, but all areas, including those on the conditional as well as the contingent lists. Loss of biodiversity at an action site may elevate the significance of a flexible area currently listed as conditional. On the other hand, conditional areas may also include many unique or otherwise desirable attributes that can move rapidly from unthreatened to threatened status. Changes in political and biological reality as well as resource availability must not go unrecognized if system planning is to work. While a balance in management has to be struck between inflexibility and inconstancy, priorities will inevitably change in time. A system geared to updating information and reviewing priorities on a regular basis can provide both early warning and advanced planning. If this is not done, conservationists face being forever reactive, rather than helping to mould the political-ecology agenda. An action register capable not only of keeping track of data and progress, but also able to underpin proactive policy and decision making to match current resources and changing priorities, would perform a key role in an effective biodiversity agency.

Jenkins (1988) outlines the work of the natural heritage data network, organized by the United States Nature Conservancy. The Conservancy has, as a stated objective, the conservation of biodiversity through establishment of reserves, 'selected and designed to protect examples of the widest possible spectrum of native ecosystems and species habitats' (Jenkins, 1988). The Conservancy long ago realized that, to implement this representation goal effectively, gathering and organization of scientific information was a primary task. Jenkins relates how early attempts failed but, since the first State Natural Heritage Inventory was established in 1974, a system has emerged in which state data-centres record information in a standardized format, and are linked through national centres—information flow is both bottom-up, and top-down.

Most interesting, in the present context, is Jenkins' (1988) list of applications for the extensive data managed by the network: *facilitating continuing inventory; conservation priority setting; preserve selection; preserve design; land conservation administration; element status monitoring; element management; site management; development planning; environmental impact analysis; access to additional information, predictive modelling;* and *research.* It is evident that the United States heritage data network provides a model system for most, if not all, of the requirements for an effective action register. What may be lacking is a comparable local subsystem (e.g. a state or subregional agency linked to a network of local biodiversity agencies), and the idea that *all* attributes and

331

areas need, somehow, to be included. Unless based on extensive voluntary work, such elaborations might appear far too costly in human resources. The conservation of biodiversity as a whole represents a vast undertaking, requiring the mobilization of existing data, huge amounts of new information, and monitoring wildlife on a scale hitherto undreamt of (Vane-Wright, 1994). Advances in information technology, growing public awareness, and the experience of the Nature Conservancy all suggest that this seemingly impossible enterprise may be within our grasp. If it is, then we will only be able to make sense of the avalanche of information if the emergent networks and agencies act in an open and democratic fashion, promote the coalition, and establish an interlinked action register to identify and keep track of priorities at every level of the hierarchical framework (Fig. 12.1).

12.5 The limits of biology and the power of reason

Approaches to setting conservation priorities have moved on from individual species and areas of outstanding wilderness, to an increasing concern for the whole of biological diversity. This has been coupled with formalization of the concept of representation. Technical means of setting priorities have also advanced from initial scoring approaches, as it became apparent that such procedures failed to guarantee representation and were more or less grossly inefficient (Pressey & Nicholls, 1989). Priorities are now discussed in terms of a representation goal based on maximizing option value, and the establishment of networks of conservation or biodiversity management areas capable of satisfying such a goal (Humphries *et al.*, in prep.). May's (1990) call for a *calculus of biodiversity* is being answered.

Given the pressing need for efficiency if limited resources are not to be misdirected, the importance of complementarity has been recognized. The degree of threat (vulnerability) must also be taken into account, and viability of the target biota of chosen areas if action is to be effective. A further need is awareness of alternative network configurations that will satisfy the representation goal, so that intelligent trading-off can be undertaken. The goal, a reasonable one in itself, is to maximize the amount of character (expressible genetic) diversity supported by a network of managed *in situ* functional ecosystems that is possible for any given level of investment of material and human resources. By this means it is hoped to serve the widest range of human interests through the preservation of biological diversity, be those interests economic or cultural, present or future, material or ethical.

This is the point we, as biologists, seem to have reached. Some, such as Norton (1994), believe that no one measurement system can serve the whole of biodiversity, while others (such as Ehrenfeld, 1993) suggest that any economic valuation beyond respect for the sheer existence of biological variety is bound to undermine its continuation. Such views do not accept even the notion of option value, adopted by practical conservationists desperately seeking a way forward within *realpolitik* (e.g. McNeely, 1988). But what is most lacking, with some notable exceptions (e.g. Chapter 13; Gadgil, 1991, 1992) from the purely biological constructions of priority, is a human context. Unless we can see how to place biologically determined priorities within an appropriate socio-political framework, further refinements will be futile.

Our ability to live sustainably, in some form of harmony with nature, will depend on the exercise of responsibility at all levels of the socio-political hierarchy. If the need to manage the whole of nature is accepted, responsibility held at the level of the individual will be crucial. Thus the main threat to biological diversity lies in denial of individual responsibility – be that denial self-imposed, circumstantial, or the result of inequity or repression. From this it follows that the threat posed by the almost overwhelming increase in human population is not a function of numbers *per se*, but the growth of poverty – because poverty robs people of dignity, choice and the ability to take control of their environment. Greed and over-consumption are also root-causes of our environmental problems, reflecting individual and collective inability to regulate our actions when we do hold power. If the leagues of nations and institutions of international trade are unwilling (Middleton *et al.*, 1993) or unable to solve these difficulties, conservation will continue as a rearguard and downhill battle – death by a thousand cuts.

Thus, as often observed, conservation is primarily a political, not a biological problem. Does this mean that all the efforts of conservation biologists to develop appropriate management theories and skills stand for nothing? Are priorities meaningless self-indulgence, academics' playthings? Are we simply bewitched by models of our own making (Ehrenfeld, 1989)?

Current ideas about conservation evaluation are based on explicit goals, defined measures, available data, logical principles and increasingly efficient analytical procedures. I have found that presenting these ideas to the public, to students – even to sceptical professionals – can be a stimulating, even cathartic experience. Cynically, it may be that the quasi-objective style of the systematic approach simply appeals to those who seek order in chaos, and find the doubt and uncertainty of so many areas of conservation biology difficult to accept. More generously, the development of a structured approach to the evaluation of priorities gives a sense of purpose and understanding. With open access to the data, *and simple procedures to extract meaning from it*, people may be more willing to accept, or act upon such notional priorities, simply because they appeal to reason. In the end, transparency and a degree of logic may do more for conservation than the narrow goals of conservation planning itself. In terms of setting priorities, the biology of numbers and difference must translate to the politics of engagement, transparency, self-determination and equality.

Acknowledgments

This chapter represents an amalgam of ideas derived from many different scientific, literary and other sources. While the views, and particularly the political opinions expressed, are entirely my own responsibility, I am very happy to acknowledge my colleagues at the NHM Biogeography and Conservation Laboratory, Chris Humphries, Ian Kitching, Campbell Smith and Paul Williams, for discussion and debate. I am similarly grateful to fellow biologists involved with the Wissenschaftskolleg biodiversity project, and other colleagues working in or visiting Berlin during 1993–94, who have influenced my ideas in various ways —in particular, Katrina Brown, Ashok Desai, Dan Faith, Kathleen MacKinnon, Wolf Lepenies, Chris Margules, Jeff McNeely, Sandra Mitchell, Norman Myers, Joachim Nettelbeck, Bob Pressey, Eduardo Rabossi, Gustaf Ranis, Tony Rebelo, Josef Settele, Wolfgang Streeck, and Rüdiger Wehner. I would also like to thank Clas Naumann, Frank Rittner and Martin Uppenbrinck. Neil Chalmers, Kevin Gaston, Christoph Häuser, Chris Humphries, Chris Margules, Paul Williams, and two anonymous referees who read the first draft made many helpful suggestions for clarification and improvement. Finally, I must also acknowledge Kevin Gaston again —for his unfailing patience.

References

Adams J.G.U. (1994) Environmentalists with the Midas Touch. *Global Ecol. Biogeog. Lett.* **4**, 63–64. [12.4.3]

Austin M.P. & Margules C.R. (1986) Assessing representativeness. In: *Wildlife Conservation Evaluation* (ed. M.B. Usher), pp. 45–67. Chapman & Hall, London. [12.2] [12.2.1]

Beccaloni G.W. & Gaston K.J. (1995) Predicting the species richness of Neotropical forest butterflies: Ithomiinae (Lepidoptera: Nymphalidae) as indicators. *Biol. Conserv.* **71**, 77–86. [12.2.3]

Burgman M.A., Ferson S. & Akçakaya H.R. (1993) *Risk Assessment in Conservation Biology*. Chapman & Hall, London. [12.3.5]

Cocks K.D. & Baird I.A. (1989) Using mathematical programming to address the multiple reserve selection problem: an example from the Eyre Peninsula, South Australia. *Biol. Conserv.* **49**, 113–130. [12.3.7]

Cousins S.H. (1994) Taxonomy and functional biotic measurement, or, will the Ark work? In: *Systematics and Conservation Evaluation* (eds. P.L. Forey, C.J. Humphries & R.I. Vane-Wright), pp. 397–419. Oxford University Press. [12.1] [12.3.1] [Appendix 12.1]

Cowling R.M. & Bond W.J. (1991) How small can reserves be? An empirical approach in Cape Fynbos, South Africa. *Biol. Conserv.* **58**, 243–256. [12.3.5]

Ehrenfeld D. (1988) Why put value on biodiversity? In: *Biodiversity* (eds E.O. Wilson & F.M. Peter), pp. 212–216. National Academy Press, Washington DC. [12.2.2]

Ehrenfeld D. (1989) Hard times for diversity. In: *Conservation for the Twenty-first Century* (eds D. Western & M. Pearl), pp. 247–250. Oxford University Press, New York. [12.5]

Ehrenfeld D. (1993) *Beginning Again. People and nature in the new millennium.* Oxford University Press, New York. [12.1] [12.4.1] [12.5]

Eldredge N. & Salthe S.N. (1985) Hierarchy and evolution. *Oxford Surveys in Evolutionary Biology* **1**, 184–208. [12.2.4]

Erwin T.L. (1991) An evolutionary basis for conservation strategies. *Science* **253**, 750–752. [Appendix 12.1]

Faith D.P. (1992) Conservation evaluation and phylogenetic diversity. *Biol. Conserv.* **61**, 1–10. [12.2.2]

Faith D.P. (1994) Phylogenetic pattern and the quantification of organismal biodiversity. *Phil. Trans. Roy. Soc., Lond. B* **345**, 45–58. [12.2.2] [12.3.3]

Fisher R.A., Corbet A.S. & Williams C.B. (1943) The relation between the number of species and the number of individuals in a random sample of an animal population. *J. Anim. Ecol.* **12**, 42–58. [12.2.2]

Fjeldså J. (1994) Geographical patterns for relict and young species of birds in Africa and South America and implications for conservation priorities. *Biodiv. Conserv.* **3**, 207–226. [Appendix 12.1]

Forey P.L., Humphries C.J. & Vane-Wright R.I. (eds) (1994). *Systematics and Conservation Evaluation.* Oxford University Press. [12.2]

Forman R.T.T. (1987) Emerging directions in landscape ecology and applications in natural resource management. In: *Conference on Science in National Parks: the Fourth Triennial Conference on Research in the National Parks and Equivalent Reserves* (eds R. Herrmann & T.B. Craig), pp. 59–88. The George Wright Society and US National Park Service. [12.4]

Gadgil M. (1991) Conserving India's biodiversity: the societal context. *Evol. Trends Plants* **5**, 3–8. [12.3.1] [12.4] [12.5]

Gadgil M. (1992) Conserving biodiversity as if people matter: a case study from India. *Ambio* **21**, 266–270. [12.4] [12.5]

Gadgil M., Berkes F. & Folke C. (1993) Indigenous knowledge for biodiversity conservation. *Ambio* **22**, 151–156. [12.2] [12.4] [12.4.1]

Gámez R. & Gauld I.D. (1993) Costa Rica: an innovative approach to the study of tropical biodiversity. In: *Hymenoptera and Biodiversity* (eds J. LaSalle & I.D. Gauld), pp. 329–336. CAB International, Wallingford. [12.4.3]

Glowka L., Burhenne-Guilmin F. & Synge H. (1994) *A Guide to the Convention on Biological Diversity.* IUCN, Gland, Switzerland. [12.2] [12.4.1]

Gore A. (1992) *Earth in the Balance: ecology and the human spirit.* Houghton Mifflin, Boston. [12.4.1]

Groombridge B. (ed.) (1992) *Global Biodiversity: status of the Earth's living resources.* Chapman & Hall, London. [12.4.1]

Humphries C.J., Williams P.H. & Vane-Wright R.I. (1995) Measuring biodiversity value for conservation. *Annu. Rev. Ecol. Syst.* **26**, 93–111. [12.2.3] [12.3.2] [12.3.3] [12.5] [Appendix 12.1]

Humphries C.J., Margules C.R., Pressey R.L. & Vane-Wright R.I. (eds) (in prep.) *Priority Areas Analysis: Systematic Methods for Conserving Biodiversity.* Oxford University Press. [12.2] [12.3.2] [12.3.3] [12.4.3] [12.5] [Appendix 12.1]

ICBP (1992) *Putting Biodiversity on the Map: priority areas for global conservation.* ICBP (BirdLife International), Cambridge. [12.2.3]

IUCN, UNEP, WWF (1980) *World Conservation Strategy: living resource conservation for sustainable development.* IUCN, Gland, Switzerland. [12.2.2]

Janzen D.H. (1993) Taxonomy: universal and essential infrastructure for development and management of tropical wildland biodiversity. In: *Proceedings of the Norway/UNEP Expert Conference on Biodiversity* (eds O.T. Sandlund & P.J. Schei), pp. 100–113. NINA, Trondheim. [12.4.3]

Jenkins R.E. (1988) Information management for the conservation of biodiversity. In: *Biodiversity* (eds E.O. Wilson & F.M. Peter), pp. 231–239. National Academy Press, Washington DC. [12.4.4] [Appendix 12.1]

Kareiva P. & Wennergren U. (1995) Connecting landscape patterns to ecosystem and population processes. *Nature* **373**, 299–302. [12.1] [12.3.4]

Kershaw M., Williams P.H. & Mace G.M. (1994) Conservation of Afrotropical antelopes:

consequences and efficiency of using different site selection methods and diversity criteria. *Biodiv. Conserv.* **3**, 354–372. [Appendix A12.1]

Kirkpatrick J.B. (1983) An iterative method for establishing priorities for the selection of nature reserves: an example from Tasmania. *Biol. Conserv.* **25**, 127–134. [12.3.7] [Appendix 12.1]

Kremen C. (1994) Biological inventory using target taxa: a case study of the butterflies of Madagascar. *Ecol. Appl.* **4**, 407–422. [12.2.3]

Mabbutt J.A. (1968) Review of concepts of land classification. In: *Land Evaluation* (ed. G.A. Stewart), pp. 11–28. MacMillan, Melbourne. [12.2.3]

Magurran A.E. (1988) *Ecological Diversity and its Measurement*. Croom Helm, London. [12.2.2]

Margules C.R. & Austin M.P. (eds) (1991) *Nature Conservation: cost effective biological surveys and data analysis*. CSIRO, Australia. [12.2] [12.2.3]

Margules C.R. & Austin M.P. (1994) Biological models for monitoring species decline: the construction and use of data bases. *Phil. Trans. Roy. Soc., Lond. B* **344**, 69–75. [12.2.3]

Margules C.R. & Usher M.B. (1981) Criteria used in assessing wildlife conservation potential: a review. *Biol. Conserv.* **21**, 79–109. [12.2.1]

Margules C.R., Nicholls A.O. & Pressey R.L. (1988) Selecting networks of reserves to maximise biological diversity. *Biol. Conserv.* **43**, 63–76. [12.3.5]

Margules C.R., Milkovits G.A. & Smith G.T. (1994a) Contrasting effects of habitat fragmentation on the scorpion *Cercophonius squama* and an amphipod. *Ecology* **75**, 2033–2042. [12.3.5]

Margules C.R., Cresswell I.D. & Nicholls A.O. (1994b) A scientific basis for establishing networks of protected areas. In: *Systematics and Conservation Evaluation* (eds P.L. Forey, C.J. Humphries & R.I. Vane-Wright), pp. 327–350. Oxford University Press. [12.3.7]

May R.M. (1990) Taxonomy as destiny. *Nature* **347**, 129–130. [12.5]

McNaughton S.J. (1983) Serengeti grassland ecology: the role of composite environmental factors and contingency in community organization. *Ecol. Monogr.* **53**, 291–320. [12.2.2]

McNaughton S.J. (1994) Conservation goals and the configuration of biodiversity. In: *Systematics and Conservation Evaluation* (eds P.L. Forey, C.J. Humphries & R.I. Vane-Wright), pp. 41–62, 1 plate. Oxford University Press. [12.2.2]

McNeely J.A. (1988) *Economics and Biological Diversity: developing and using economic incentives to conserve biological resources*. IUCN, Gland, Switzerland. [12.2.2] [12.5]

Middleton N., O'Keefe P. & Moyo S. (1993) *The Tears of the Crocodile: from Rio to reality in the developing world*. Pluto Press, London. [12.5]

Mittermeier R.A., Bowles I.A., Cavalcanti R.B., Olivieri S. & da Fonseca G.A.B. (1994) *A Participatory Approach to Biodiversity Conservation: the regional priority setting workshop*. Conservation International, Washington DC. [12.4] [12.4.3]

Murcia C. (1995) Edge effects in fragmented forests: implications for conservation. *Trends Ecol. Evol.* **10**, 58–62. [12.3.5]

Myers N. (1979) *The Sinking Ark. a new look at the problem of disappearing species*. Pergamon Press, Oxford. [12.2] [12.3.1]

Norton B.G. (1994) On what we should save: the role of culture in determining conservation targets. In: *Systematics and Conservation Evaluation* (eds P.L. Forey, C.J. Humphries & R.I. Vane-Wright), pp. 23–29. Oxford University Press. [12.4.1] [12.5]

Oates J.F. (1995) The dangers of conservation by rural development—a case-study from the forests of Nigeria. *Oryx* **29**, 115–122. [12.4.1]

Pearce D. & Moran D. (1994) *The Economic Value of Biodiversity*. Earthscan/IUCN, London. [12.4.1] [12.4.3]

Pearce D.W. (1987) Marginal opportunity cost as a planning concept in natural resource management. *Univ. College Lond. Disc. Pap. Econ.* **87(6)**, 1–21. [12.4.1]

Porritt J. (1993) Translating ecological science into practical policy. In: *Large-scale Ecology and Conservation Biology* (eds P.J. Edwards, R.M. May & N.R. Webb), pp. 345–353. Blackwell Scientific, Oxford. [12.4.3]

Prendergast J.R., Quinn R.M., Lawton J.H., Eversham B.C. & Gibbons D.W. (1993) Rare species, the coincidence of diversity hotspots and conservation strategies. *Nature* **365**, 335–337. [12.2.3]

Pressey R.L. (1990) Reserve selection in New South Wales: where to from here? *Aust. Zool.* **26**, 70–75. [12.3.3]

Pressey R.L. (1994) *Ad hoc* reservations; forward or backward steps in developing representative reserve systems? *Conserv. Biol.* **8**, 662–668. [12.3.3] [12.3.4]

Pressey R.L. & Nicholls A.O. (1989) Efficiency in conservation: scoring versus iterative approaches. *Biol. Conserv.* **50**, 199–218. [12.3.3] [Appendix 12.1]

Pressey R.L. & Possingham H.P. (1995) Effectiveness of alternate heuristic algorithms for identifying indicative minimum requirements for conservation purposes. *Biol. Conserv.* in press. [12.3.7] [Appendix 12.1]

Pressey R.L. & Tully S.L. (1994) The cost of ad hoc reservation: a case study in western New South Wales. *Aust. J. Ecol.* **19**, 375–384. [12.3.3] [12.3.4]

Pressey R.L., Humphries C.J., Margules C.R., Vane-Wright R.I. & Williams P.H. (1993) Beyond opportunism: key principles for systematic reserve selection. *Trends Ecol. Evol.* **8**, 124–128. [12.2.3]

Pressey R.L., Johnson I.R. & Wilson P.D. (1994) Shades of irreplaceability: towards a measure of the contribution of sites to a reservation goal. *Biodiv. Conserv.* **3**, 242–262. [12.3.3] [12.3.6] [Appendix 12.1]

Pressey R.L., Posssingham H.P. & Margules C.R. Optimality in reserve selection algorithms: when does it matter and how much? *Biol. Conserv.*, in press a [12.3.7] [Appendix 12.1]

Pressey R.L., Ferrier S., Hutchinson C.D., Siversten D.P. & Manion G. Planning for negotiation: using an interactive geographic information system to explore alternative protected area networks. In: *Nature Conservation: the role of networks* (eds D. Saunders *et al.*), Surrey Beatty, Sydney, in press b. [12.3.7] [12.4.3]

Rebelo A.G. & Siegfried W.R. (1992) Where should nature reserves be located in the Cape Floristic Region, South Africa? Models for the spatial configuration of a reserve network aimed at maximising the protection of floral diversity. *Conserv.Biol.* **6**, 243–252. [12.3.3]

Reid W.V. (1994) Setting objectives for conservation evaluation. In: *Systematics and Conservation Evaluation* (eds P.L. Forey, C.J. Humphries & R.I. Vane-Wright), pp. 1–13. Oxford University Press. [12.1] [12.2.2]

Richardson B.J. (1994) The industrialization of scientific information. In: *Systematics and Conservation Evaluation* (eds P.L. Forey, C.J. Humphries & R.I. Vane-Wright), pp. 123–131. Oxford University Press. [12.4.3]

Rieppel O. (1988) *Fundamentals of Comparative Biology*. Birkhäuser, Basel. [12.2.4]

Ryti R.T. (1992) Effect of the focal taxon on the selection of nature reserves. *Ecol. Appl.* **2**, 404–410. [12.2.3] [12.3.3]

Sætersdal M., Line J.M. & Birks H.J.B. (1993) How to maximise biological diversity in nature reserve selection: vascular plants and breeding birds in deciduous woodlands, western Norway. *Biol. Conserv.* **66**, 131–138. [12.3.3] [12.3.7]

Scott J.M., Davis F., Csuti B., Noss R., Butterfield B., Groves C., Anderson H., Caicco S., D'Erchia F., Edwards Jr T. C., Ulliman J. & Wright R.G. (1993) Gap analysis: a geographic approach to protection of biological diversity. *Wildlife Monogr.* **57**, 1–41. [12.4.3]

Shafer C.L. (1990) *Nature Reserves: island theory and conservation practice*. Smithsonian Institution Press, Washington. [12.3.5]

Spellerberg I.F. (1992) *Evaluation and Assessment for Conservation: ecological guidelines for determining priorities for nature conservation*. Chapman & Hall, London. [12.1] [12.2]

Underhill L.G. (1994) Optimal and suboptimal reserve selection algorithms. *Biol. Conserv.* **70**, 85–87. [12.3.2] [12.3.7] [Appendix 12.1]

Vane-Wright R.I. (1994) Systematics and the conservation of biodiversity: global, national and local perspectives. In: *Perspectives on Insect Conservation* (eds K.J. Gaston, T.R. New & M.J. Samways), pp. 197–211. Intercept, Andover. [12.2.2] [12.2.4] [12.3.2] [12.3.3] [12.4.3] [12.4.4]

Vane-Wright R.I., Humphries C.J. & Williams P.H. (1991) What to protect?–systematics and the agony of choice. *Biol. Conserv.* **55**, 235–254. [12.3.2] [Appendix 12.1]

Walker P.A. & Faith D.P. (1993) *Diversity: a software package for sampling phylogenetic and environmental diversity* (version 1.0). CSIRO Division of Wildlife and Ecology, Canberra. [12.2.3]

Weisbrod B.A. (1964) Collective-consumption services of individual-consumption goods. *Quart. J. Econ.* **78**, 471–477. [12.2.2]

Weitzman M.L. (1992) On diversity. *Quart. J. Econ.* **108**, 157–183. [12.2.2] [12.2.3]

Westman W.E. (1990) Managing for biodiversity: unresolved science and policy questions. *BioScience* **40**, 26–33. [12.1] [12.4.1]

Whittaker R.H. (1965) Dominance and diversity in land plant communities. *Science* **147**, 250–260. [12.2.2]

Whittaker R.H. (1972) Evolution and measurement of species diversity. *Taxon* **21**, 213–251. [12.2.2]

Williams P.H. (1993) Choosing conservation areas: using taxonomy to measure more of biodiversity. In: *International Symposium on Biodiversity and Conservation* (ed. T-Y. Moon), pp. 194–227. Korean Entomological Institute, Seoul. [Appendix 12.1]

Williams P.H. (1994) *Using WORLDMAP. Priority areas for biodiversity, version 3.1*. London (distributed by the author). [Appendix 12.1]

Williams P.H. Biodiversity value and taxonomic relatedness. In: *The Genesis and Maintenance of Biological Diversity* (eds M.E. Hochberg, J. Clobert & R. Barbault). Oxford University Press, Oxford, in press [12.2.2]

Williams P.H. & Gaston K.J. (1994) Measuring more of biodiversity: can higher-taxon richness predict wholesale species richness. *Biol. Conserv.* **67**, 211–217. [12.2.3]

Williams P.H. & Humphries C.J. (1994) Biodiversity, taxonomic relatedness, and endemism in conservation. In: *Systematics and Conservation Evaluation* (eds P.L. Forey, C.J. Humphries & R.I. Vane-Wright), pp. 269–287. Oxford University Press. [12.2.2] [12.3.3] [Appendix 12.1]

Williams P.H., Humphries C.J. & Vane-Wright R.I. (1991) Measuring biodiversity: taxonomic relatedness for conservation priorities. *Aust. Syst. Bot.* **4**, 665–679. [12.2.2] [Appendix 12.1]

Williams P.H., Vane-Wright R.I. & Humphries C.J. (1993) Measuring biodiversity for choosing conservation areas. In: *Hymenoptera and Biodiversity* (eds J. LaSalle & I.D. Gauld), pp. 309–328. CAB International, Wallingford. [12.2.2]

Williams P.H., Gaston K.J. & Humphries C.J. (1994) Do conservationists and molecular biologists value differences between organisms in the same way? *Biodiv. Lett.* **2**, 67–78. [12.2.2]

Williams P.H., Gibbons D., Margules C.R., Rebelo A.G., Humphries C.J. & Pressey R.L. Hotspots, rarity areas and complementary areas: a comparison of three area selection methods for conserving diversity using British breeding birds. *Conserv. Biol*, in press a. [12.2.1]

Williams P.H., Prance G.T., Humphries C.J. & Edwards K.S. Promise and problems in applying quantitative complementary areas for representing the diversity of some neotropical plants (families Proteaceae, Dichapetalaceae, Lecythidaceae, Caryocaraceae and Chrysobalanceae). *Biol. J. Linn. Soc.*, in press b. [12.3.7]

Appendix 12.1

A hypothetical example to illustrate some principles of systematic area selection

Despite their simple and almost self-evident nature, the measurements and guiding principles outlined in this paper are relatively new. Although applied for the first time by Kirkpatrick (1983), it has taken over a decade for the notion of complementarity to have much impact on mainstream conservation literature. A formal measure of efficiency did not appear until 1989 (Pressey & Nicholls), formal taxic measures intended to distinguish different ways of assessing character diversity did not appear until 1991 (e.g. Williams *et al.*, 1991), and a scalar measure of irreplaceability was not proposed until very recently (Pressey *et al.*, 1994). As a result, there have as yet been few, if any, practical analyses using the whole range of existing measures and principles. This section, based on an artificial data set, is intended to give some idea of how, once goals and measures have been agreed, and appropriate data collected, the available information can be explored—in preparation for action.

Underhill (1994) proposed a model data set of eight species distributed amongst five areas to demonstrate how, if the goal was to identify a maximally efficient network of reserves to represent all species at least once, a single pass by a 'greedy' iterative algorithm using complementarity at each step could lead to a suboptimal solution. In his carefully constructed example, the greedy algorithm requires three areas to reach 100% representation (efficiency, E = 40%, using Pressey & Nicholls' 1989 formula), whereas an integer-programming solution finds that only two sites are needed (E = 60%). The step-wise algorithm fails in this context because site 1 (five species) is more species-rich than either site 2 or 3 (four species each), all eight species occur at sites 2 and 3 taken together (these two sites show maximum complementarity in combination), but all eight cannot be represented by any combination of site 1 with one other area.

Differences between optimizations for step-wise and set-wise analyses (both important in different contexts), and the additional computational challenge posed by the setwise approach, have been recognized for some time (e.g. Vane-Wright *et al.*, 1991, p. 250). In order to explore a variety of problems affecting priority areas analysis, I have placed Underhill's matrix within a slightly expanded version, to create a data-set for nine species distributed across seven administrative areas—presented as a matrix of nine columns (each with a separate letter, a–i) and seven rows (areas 1–7) in Fig. A12.1. The first eight columns for rows 1–5 correspond exactly to Underhill's array for species 1–8 and sites 1–5. The sixth area (row 6) is added to duplicate site 2 (to help illustrate flexibility), and a seventh area to provide for a ninth, narrowly endemic species (i). This last area is used to give an example of global irreplaceability and, through inclusion of three other species, also to help illustrate the impact of taxic diversity measurement (based on the classification hierarchy given in Fig. A12.1).

Possible conservation priorities for this system have been explored using three measures of taxic diversity (species richness, character richness and character combination richness), using the iterative procedures within Williams' WORLDMAP 3.1 – which includes a backtracking process to eliminate areas unnecessarily introduced by stepwise optimization when looking for minimum-set solutions (Williams, 1994). Although such heuristic procedures do not match the formal elegance of integer programming preferred by Underhill, and do not guarantee truly optimal results, in speed they currently outperform mathematical programming solutions in real applications by several orders of magnitude (Pressey & Possingham, 1995; Pressey *et al.*, in press a). The results of various analyses are presented in Tables A12.1 and A12.2.

Assuming, to begin with, that threats, costs and viability are the same for all seven administrative areas, the best choice for a single area (Table A12.1a) is site 1 (based on species richness), site 7 (character richness), or site 3 (character combinations), illustrating how choice of biodiversity measure can affect the results (cf. Williams *et al.*, 1993; Kershaw *et al.*, 1994). For the 21

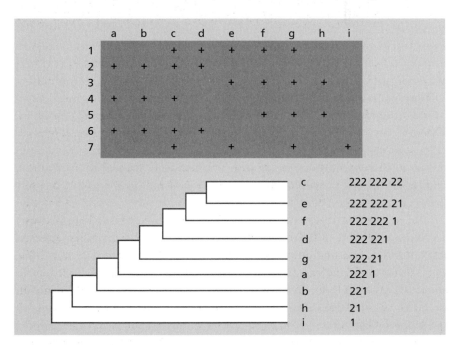

Fig. A12.1 Matrix for distribution of nine species (letters/columns) in seven administrative areas (numbers/rows) (adapted from Underhill, 1994 – for full explanation, see text). The tree specifies a hierarchical classification linking the nine species together; the set of binary codes specifies these same interrelationships, and was used to explore taxic diversity within WORLDMAP (Williams, 1994).

Table A12.1 Diversity scores for seven administrative areas (see Fig. A12.1, and text), singly and in combination, obtained by application of three taxic measures implemented in WORLDMAP (Williams, 1994). (b) (lower part of table), scores for each of the seven areas separately, as percentage species richness (left-hand entry in each column), percentage character richness (centre entry in each column), and percentage character combination richness (right-hand entry in each column); species richness is at a maximum in area 1, character richness is at a maximum in area 7 and character combination richness is at a maximum in area 3 (area scores can only be compared using the same measure – numerical comparisons between the different measures have no meaning). (a) (upper part of table), scores, measured as in (b), for all 21 pairwise combinations of the seven areas, after application of complementarity. See text for full explanation.

CHAPTER 12
Identifying priorities for conservation

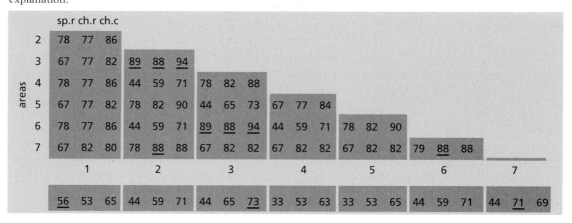

possible combinations of two areas (none of which is fully representative), the best choice is either sites 2 + 3 or 3 + 6, which score a maximum for all three measures, with combinations 3 + 7 and 6 + 7 also scoring the maximum for character richness (Table A12.1b). (As noted by Williams & Humphries, 1994, in the absence of branch length information, cladistic estimates of character richness are relatively insensitive.)

The 'near-minimum-set' heuristic of WORLDMAP shows that four of the 35 possible three-area configurations (2 + 3 + 7, 2 + 5 + 7, 3 + 6 + 7, 5 + 6 + 7) are fully representative at the minimum level of one representation per species, giving a maximum efficiency of 57%. As in Underhill (1994), because of area 1, the simple iterative algorithm using species richness as its measure of biodiversity takes an additional step, requiring four areas to find a fully representative (FR) solution. For four areas (E = 43%), a total of 14 of the 35 combinations are FR, for five areas (E = 29%) 14 out of 21 are FR, while at six areas (E = 14.3%) six out of seven are FR (Table A12.2). Levels of irreplaceability range from 54% for areas 1 and 4, to 100% for area 7.

Thus, if the biodiversity of area 3 or area 5 was obliterated, given that both have an irreplaceability value of 67%, then 26 of the 39 options for constructing a minimal FR network would be lost (none of the other 88 possible combinations is FR). If we move to a goal of two representations (possible for all species except the endemic, i), then five areas (all but 1 and 4) are

Table A12.2 Irreplaceability values for seven administrative areas (Fig. A12.1 —see also text), based on enumeration of all fully representative sets (4 at three areas, 14 at four areas, 14 at five areas, 6 at six areas, 1 for all seven). Area 7 appears in all 39 FR combinations and is globally irreplaceable; areas 1 and 4 both only appear in 21 of the FR combinations, and have the lowest irreplaceability value (54%). No single area, or pairwise combination of two areas (see Table A12.1), contains all nine species.

1	2	3	4	5	6	7
	2	3				7
	2			5		7
		3			6	7
				5	6	7
1	2	3				7
1	2			5		7
1		3	4			7
1		3			6	7
1			4	5		7
1				5	6	7
	2	3	4			7
	2	3		5		7
	2	3			6	7
	2		4	5		7
	2			5	6	7
		3	4		6	7
		3		5	6	7
			4	5	6	7
1	2	3	4			7
1	2	3		5		7
1	2	3			6	7
1	2		4	5		7
1	2			5	6	7
1		3	4	5		7
1		3	4		6	7
1		3		5	6	7
1			4	5	6	7
	2	3	4	5		7
	2	3	4		6	7
	2	3		5	6	7
	2		4	5	6	7
		3	4	5	6	7
1	2	3	4	5		7
1	2	3	4		6	7
1	2	3		5	6	7
1	2		4	5	6	7
1		3	4	5	6	7
	2	3	4	5	6	7
1	2	3	4	5	6	7
21	24	26	21	26	24	39 (irreplaceability score)
54	62	67	54	67	62	100%

required, with no flexibility. Each of the three species found at site 4 occurs in areas 2 and 6, and so site 4 can be seen as perhaps the most negotiable of all—and like site 1, it has the lowest irreplaceability value (54%).

Is site 1 as negotiable as site 4? If the taxonomic hierarchy given in Fig. A12.1 is interpreted as a phylogenetic tree, then area 1 could be seen as a 'species dynamo' zone. According to the arguments of some biologists (e.g. Erwin, 1991; Fjeldså, 1994), these are the very places we ought to protect because they are where development of new species is most likely in future. While my colleagues and I (e.g. Williams, 1993) would not agree with the assumptions about future evolution that lie behind such a conclusion, or the implications for conservation practice, this underlines that priorities are dependent on values (e.g. evolutionary potential, species, characters, genes, ecosystem function), as well as how we measure those values (e.g. character richness versus character combinations) and our representation goal (partial representation, sectoral agenda, minimal full representation, multiple full representation). When these interact with viability and threat, then almost any amongst all practical possibilities might be the preferred course of action.

Thus, if it turns out only species (a), (b) and (c) are vulnerable, and that site 4 currently offers biologically the most viable conditions for their persistence, then site 4, instead of being seen as the most negotiable, might be seen as the only area needing or worthy of conservation effort. If, on the other hand, the economic cost of conservation at site 4 far exceeded available resources, then improving conditions for (a, b, c) at flexible sites 2 or 6 might be a better option. If the inhabitants of site 2 democratically opposed conservation, whereas those of site 6 welcomed it, then site 6 might become the final choice.

Entertaining the vast array of possibilities that confront us need not be cause for despair—on the contrary, full exploration of all possibilities (flexibility), with concern for efficiency (based on complementarity) evaluated in relation to threat and viability, all in relation to well-defined goals and some agreed comparative valuation system for biodiversity, gives the only hope of progress in the real world. Only when all the relevant factors affecting such decisions are made explicit is there any real prospect that biological differences will be taken note of, or used as some form of external value or arbiter between different political claims and interests.

One of the advantages of an explicit and systematic approach to setting priorities is that, in relation to the representation goal, it is possible to identify some primary function of each chosen area in relation to that objective. Within an efficient network, all areas make some unique or particular contribution to the goal—but this contribution will be determined (at least in part) by choice. In the example, if we examine (2, 3, 7), one of four minimum-set solutions for single representation, we find that irreplaceable area 7 contributes species (i) uniquely—it is the only region that can do so. Area 2, however, *within this specified minimum-set*, even though all species occur elsewhere, gives the only representation for three species (a, b, d). Similarly, area 3 is the only site within the set that can contribute species (f) and (h). For the three remaining species, each of them is found in two of the three areas – (e) and (g) in 3 and 7, (c) in 2 and 7. We distinguish the species that are uniquely contributed by each area as the *essential complement* of that area (Humphries *et al.*, in prep.) – and in a true minimum-set for just one representation, this means that every area will have a distinct essential complement, composed of one or more species (attributes) not found anywhere else in the chosen set. The essential complement represents a special function for each area because, if one or more of the attributes that specify it are lost, then the conservation goal for the entire system is compromised immediately. In a sense, the essential complement is, to a chosen network, what global irreplaceability is to the entire system of which the network forms a part.

In the example, because areas 2 and 6 are identical in their attributes, the essential complements for alternative configuration (3, 6, 7) are the same as for (2, 3, 7 – with area 6 replacing 2). Alternative configurations (2, 5, 7) and (5, 6, 7) are, for the same reason, identical to each other, but they differ from those for the (2, 3, 7) and (3, 6, 7) configurations. In this pair, the essential complement for 2/6 is again (a, b, d), but that for 7 includes (g) in addition to (i), while that for 5 is (f, h). In this latter pair of configurations, only two species (c, e) are duplicated, while in the former pair, three are (c, e, g). This difference might help give preference to (2, 3, 7) or (3, 6, 7) over (2, 5, 7) or (5, 6, 7), especially if species (g) is considered more vulnerable and/or less viable than the others. In general, within a set of alternative minimum-set configurations, the lower the total number of attributes accorded essential complement status the better – other factors being equal – because this minimizes the number of attributes represented only once. This again emphasizes the need for efficiency because it also helps to limit the decreasing margins for error inherent in conserving biodiversity in less and less space, and under greater and greater pressure (Jenkins, 1988; cf. Cousins, 1994).

If the areas considered as priorities have been limited to those that are vulnerable in some way, this means that the threatening factors will already have been identified, and can thus better be taken care of. Viability should also have been assessed, particularly with respect to the essential complements, and this will direct attention to specific monitoring needs. Flexibility, and degree of area irreplaceability, alerts us to the extent to which an area can be substituted, and this can be monitored and updated on a regular basis. The very process of assigning explicit priorities means that, in effect, we have appraised the most critical goal-oriented requirements to be contributed or maintained by each chosen area. Thus a primary expectation of conservation management will be to monitor relevant factors and, so far as possible, ensure that those attributes responsible for priority status are maintained as effectively as possible – and warn if they start to deteriorate.

13: Managing biodiversity

MADHAV GADGIL

13.1 Introduction

Managing the Earth's heritage of biodiversity is an immense challenge. It is a twofold challenge; the magnitude of biological diversity is staggering, with tens of millions of species, each with many variants of thousands to millions of genes, constituting an intricate mosaic of biological communities distributed from the depths of the seas to the tops of the mountains. Equally staggering are present day human impacts on the biosphere, with mobilization of perhaps 40% of all primary production towards human ends, and with man-made entities ranging from molecules of dichlorodiphenyltrichloroethane (DDT) and chlorofluorocarbons (CFCs) to buildings and transport vehicles affecting all parts of the atmosphere, hydrosphere, and surface of the lithosphere (Vitousek *et al.*, 1986; Gadgil, 1993a). Humans are, therefore, bringing about ever more drastic changes in the magnitude and distribution of biological diversity on the Earth, on many different space and time scales. Yet we remain largely ignorant of the magnitude and distribution of biological diversity, as also of the driving forces behind and the precise course of impacts of human activities on the stock of biodiversity.

Confronted with this challenge, people have sought to focus biodiversity management action by looking for conservation priorities. Taxa at the level of species have tended to provide the point of departure for arriving at these. Thus species which are taxonomically distinctive, species that have a restricted geographical range, and species that are endangered have been assigned high priorities. So have species with economic potential such as wild relatives of cultivated plants. Management plans have then been drawn up for such high priority taxa as the spotted owl *Strix occidentalis* and its habitat of old growth coniferous forest in the northwestern United States. Developing the management plans for just this one species has entailed enormous efforts, while winning social acceptance for the implementation of these plans has run into serious difficulties. Evidently, extending such a species-based approach to managing all of Earth's biological diversity is not a practical proposition. A somewhat different approach has led to the identification of biomes, which are at once rich in species and under severe pressures of

345

depletion of biodiversity, such as the 18 hotspots identified by Myers (1988, 1990). Focusing attention more directly on the broader issues of biological diversity, this approach forces us to confront the social and economic realities that must be addressed if the biological diversity of Amazonia, Madagascar or the Eastern Himalayas is to be managed.

13.2 Of people and resources

There are few answers available as to how one may set about this task, and the purpose of this chapter is to explore this challenge. A key consideration in such an approach has to be people, and how they relate to biological resources. Dasmann's (1988) classification of humanity into ecosystem people and biosphere people provides an appropriate starting point (Fig. 13.1). The category of ecosystem people refers to many forest dwellers, peasants, herders, and fishers of the non-industrial world who primarily depend on biological resources of a circumscribed ecosystem to meet the bulk of their material requirements—through gathering, grazing, or low input agriculture. These are people with limited capabilities of purchasing from the market substitutes for what they acquire with their own labour. Their well-being is closely tied to that of the ecosystems of which they themselves

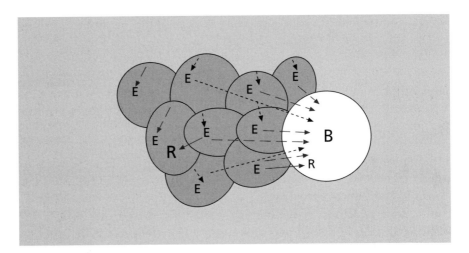

Fig. 13.1 People may be classified on the basis of their patterns of resource use as ecosystem people (E), biosphere people (B) and ecological refugees (R). Ecosystem people derive resources from a relatively limited catchment through fluxes that are either strong (— — —▶) or weak (-------▶). Biosphere people derive resources from much larger catchments through strong fluxes (— — —▶) from nearby catchments of ecosystem people, or weaker fluxes (-------▶) from further afield. Ecosystem people who can derive only very low levels of resources from their own catchments may then move to other areas as ecological refugees (————▶). These movements may bring ecological refugees to areas from which biosphere people are deriving only weak fluxes as, for instance, encroachers on forests. Alternatively, ecological refugees may join biosphere people in urban shanty towns.

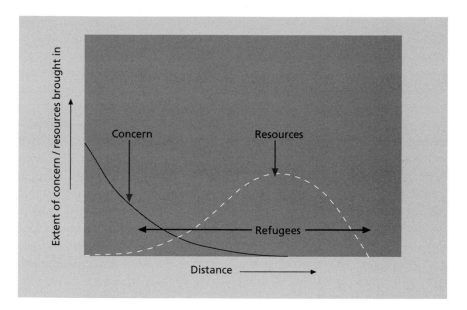

Fig. 13.2 Biosphere people's concern for maintenance of a healthy environment declines with distance from their own habitation. They therefore prefer to bring in resources from areas at some distance. Since the costs of obtaining resources increase with distance, the pressure of resource exploitation is heaviest on localities at some intermediate distance from the habitation of biosphere people. The ecosystem people living in the localities subject to heavy resource exploitation are converted to ecological refugees who may move either further away to localities less subject to resource exploitation pressures of biosphere people, or to areas of concentrations of biosphere people.

constitute a part. The biosphere people, in contrast, have access to the resources of much of the biosphere—resources that are brought to them through increasingly integrated global markets. They are largely city dwellers involved in the organized industries-services sector, or practise high input industrialized agriculture or animal husbandry. Most citizens of the developed world belong to this category, as do the elite of the non-industrial world. The biosphere people do not depend on the biological resources of any particular locality; if the resources of a given locality are exhausted, they always have the option of drawing resources from others. The biosphere people also have options of substituting exhausted resources with alternatives through technological innovation. They, therefore, have little stake in the health of the biological resource base of any given locality for their own material well-being. However, they do have an interest in maintaining a healthy, pleasing ambience in their immediate vicinity, and therefore prefer to shift the pressure of their resource demands, or generation of pollutants, to geographically distant localities (Fig. 13.2). The biosphere people therefore tend to generate pressures for exhaustion of biological resources in distant localities, principally in the non-industrial world countryside, in which are concentrated the ecosystem people. Given the far greater per capita resource demands of the biosphere people, these demands play a significant role in determining the fate of biological resources of the non-industrial world countryside, often exceeding in magnitude the role of localized resource demands of ecosystem people. Such resource exhaustion through combined action of the growing resource appetite of the biosphere people and growing populations of the ecosystem people creates a third category of humanity, that of ecological refugees (Gadgil, 1995). These are ecosystem people deprived of traditional access to biological resources in their immediate vicinity, who are forced to migrate in search of a fresh resource base. Such are the poor peasants who move into the Amazonian forest to clear it and eke out a living for a few years before they move on to another patch of forest. Such are the basket weavers of India deprived of access to bamboo resources exhausted by the pressures of the paper industry, who flock to shantytowns of cities like Bombay in search of livelihood. Footloose as they are, the ecological refugees too have little stake in sustainable use of biological resources of the localities to which they are forced to migrate (Gadgil & Guha, 1992).

The level of motivation for maintaining a healthy biological environment in their immediate vicinity is then rather strong for the biosphere people. The biosphere people are also in a position to accomplish this, since they have fairly firm control over not only their own resource base, but access to resources from elsewhere. This is how Japanese are able to maintain 60% of their country under forest cover, while importing timber from southeast Asia, Brazil, or the northwestern United States. In a way, therefore, problems of managing biodiversity in the industrially developed countries are simpler. But the biodiversity-rich countries tend to be largely developing countries, who have large populations of ecosystem people and ecological refugees. Of these, the ecosystem people do have a stake in the health of the biological resources of their immediate vicinity, they also have rich traditions of conservation of biodiversity (Gadgil *et al.*, 1993). Today they have little control over their biological resource base, and therefore little motivation or capability of prudent use of their environment. The ecological refugees of the non-industrial world are in the worst plight of all, and are today significant agents of erosion of biodiversity. The problem of management of biodiversity is therefore above all a human problem of how to involve the ecosystem people of the world in a positive fashion in managing biodiversity and how to help ecological refugees to settle down to a reasonable subsistence and thereby also come to play a positive role in this endeavour. Of course, this could only be brought about through the active cooperation and support of the biosphere people (Gadgil, 1995).

13.3 Small-scale societies

In seeking answers to this question, it may be useful to review how human societies in different parts of the world have approached the problem of managing biodiversity at various periods in their historical development (Gadgil, 1987). An interest in conservation of biodiversity is by no means a post-Rio Summit phenomenon. Indeed, the small-scale, egalitarian societies of hunter-gatherers, shifting cultivators and horticulturalists which encompass much of human history exhibit a number of practices of restraint in the use of biological resources that promote conservation of biodiversity (Gadgil & Berkes, 1991). Such societies view the world around them as a community of beings. To them rocks and rivers, trees and birds are fellow beings, often viewed as kin or as benefactors. These may then be accorded respect and protection in several ways. The restraints so motivated include regulation of two main kinds of processes that could erode biodiversity, namely harvests of biological populations and transformation of habitats. Regulation of harvests may range all the way from occasional lowering of hunting pressure to complete protection (McNeely & Pitt, 1985). Thus, several Pacific islanders give up fishing from particular lagoons if the catches from these lagoons decline. Some New Guinea tribes stop hunting birds-of-paradise when their populations have declined. In modern terminology these are instances of adaptive management (Walters, 1986). In southern India, storks and pelicans breeding at a heronry may be given full protection, although they may be hunted at other times of the year. Also in southern India, the large fruit bat *Pteropus giganteus* may be protected at the day time roost, although it will be hunted at night away from the roost. Other plants and animals may be totally protected at all times, for instance as kin accorded totemic status. Thus a farming community called the Bishnois of Rajasthan in northwestern India will never cut a Khejari *Prosopis cinerarea* tree even if it grows in the midst of a field. They also give total protection to all antelope species and peafowl around their villages. Amongst the more notable of such instances of total protection is that accorded to many tree species of the genus *Ficus* in many parts of Africa and Asia. Because of this, *Ficus* trees are often left standing when forests are clear-cut in India, and huge trees of *Ficus religiosa* dot India's thickly settled rural countryside, and persist even in city centres. *Ficus* is considered a keystone resource of tropical forests since it often fruits in months when none of the other plants are in fruit and therefore promotes the persistence of frugivorous birds and primates (Terborgh, 1986). Today many forest dwellers of India appear to be aware of this role of *Ficus* trees, and it is quite plausible that the widespread protection to *Ficus* came to be accorded in the interest of maintaining populations of favoured prey species of humans such as fruit-eating pigeons.

Small-scale societies regulate habitat transformation by protecting samples of natural communities on sacred sites (e.g. sacred groves, sacred ponds). These sacred sites may be associated with nature spirits resident in trees or pools of water or with more formalized worship as with Buddhist temple groves of Thailand or groves associated with Shinto shrines in Japan. Sacred groves, ponds and lagoons persist to this day in many parts of Asia, Africa and some of the Pacific islands, a sacred cacao grove has also been described from Mexico (Gomez-Pompa *et al.*, 1990). Brandis, the first Inspector General of Forests in India, noted over a century ago that a network of sacred groves once covered much of the subcontinent, but that it had dwindled considerably under the forest management system introduced by the British. Indeed in 1801, just after the British conquest of southern India, Francis Buchanan (1870), a surgeon in the employ of the British East India Company, wrote of a sacred grove near the town of Karwar:

> The forests are property of the gods of the villages in which they are situated, and the trees ought not to be cut without having obtained leave from the . . . headman of the village, whose office is hereditary, and who here also is priest to . . . the village god. The idol receives nothing for granting this permission; but the neglect of this ceremony of asking his leave brings his vengeance on the guilty person. This seems, therefore, merely a contrivance to prevent the government from claiming the property.

This hostile attitude towards sacred groves has continued in independent India with the forest department taking over and clear-cutting groves as extensive as 400 ha to raise eucalyptus plantations in the 1970s (Subash Chandran & Gadgil, 1993).

It is, however, possible to reconstruct the system of sacred sites that might have once prevailed amongst the small-scale gatherer–horticulturalist societies in many parts of the world. We have attempted one such reconstruction for a 5 km × 5 km area in Siddapur taluk of Western Ghats of southern India, based on the local names of the landscape elements. This exercise suggests that as much as 9.3% of the land was originally maintained under sacred groves, which occurred in 10 patches ranging in size from 1 ha to 134 ha (Fig. 13.3). Gangtes, a group of shifting cultivators in the Churchandapur district of the state of Manipur in northeastern India, report that between 10 and 30% of the land was left completely inviolate as sacred groves during their shifting cultivation cycles prior to their conversion to Christianity in the 1950s (N.S. Hemam & M. Gadgil, pers. comm., 1993). Following religious conversion many of the groves were clear-cut. This led to a series of difficulties, in particular the spread to their hamlets of fires set during the slash and burn cycle, which led to the burning down of houses. In response many Christian shifting cultivator groups have revived protection to some of the erstwhile sacred groves, in particular those forming a ring around the villages. Of course, these groves are no longer considered abodes of deities, instead they are termed 'safety forests' (Malhotra, 1990). Nevertheless, the social sanctions ensuring protection to the groves are of the same form as those prevalent in pre-Christian times. Such protection to sacred groves is paralleled by protection to sacred ponds, or pools along rivers. Indeed, the only surviving population of a freshwater turtle *Trionyx nigricans*

350

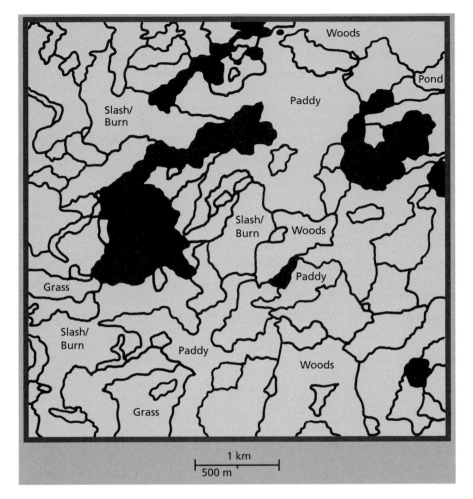

Fig. 13.3 Presumed patterns of land-use prior to British occupation in the early 19th century in a 5 km × 5 km area of Siddapur taluk in Karnataka Western Ghats (14°16′N lat. and 74°54′E long.) The land-use was reconstructed on the basis of local names of individual landscape elements. The black areas were sacred groves, remnants of which persist to this day. The identity of a few other representative elements has also been indicated.

now occurs in a sacred pond dedicated to a Moslem saint in Bangladesh (Daniel, 1983). Many coastal and island communities also afford protection to mangrove swamps, lagoons and other coastal ecosystems (Ruddle & Johannes, 1985).

These systems of regulation of harvests and of habitat transformations were most probably motivated by the interest of small-scale societies in ensuring the availability of a diversity of biological materials from within a limited area to which they were largely confined. The systems of shifting cultivation that many of these societies followed create a mosaic of different successional stages of vegetation. When coupled to maintenance of patches of climax vegetation as sacred groves, and populations of keystone resources—

such as *Ficus* protected as sacred trees—these management systems would have created high levels of diversity at the ecosystem and landscape scales. While such a system may fail to conserve populations of a small proportion of species requiring very large areas of primaeval habitats, the dispersed system of plentiful, patchy refugia may still have promoted the maintenance of the bulk of diversity at the species level as well (Fig. 13.4).

Remnants of many of these regulatory practices of ecosystem people persist to this day in Papua New Guinea, India, Ghana, Mexico, on several Polynesian islands and elsewhere in the world. It is then possible that such management practices did in an earlier era perform a valuable function of conservation of biodiversity; indeed a function that they continue to perform to this date, albeit in an attenuated form (Gadgil *et al.*, 1993). This view may appear to contradict much evidence marshalled by Diamond (1994) and others that the first colonizers of the Americas, Madagascar, New Guinea, and many Polynesian islands were probably responsible for large scale extinctions. It would, however, appear that on first colonizing new environments people are unlikely to be concerned with sustainable use of seemingly unlimited resources. Equally, they would have little understanding of the

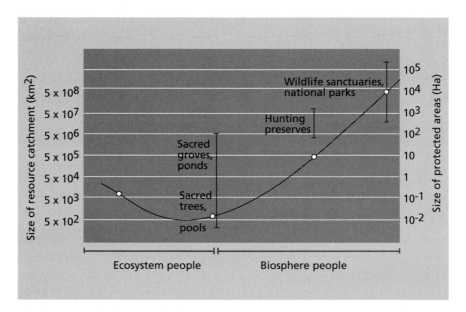

Fig. 13.4 Relation between sizes of resource catchments and sizes of protected areas. Pure hunter-gatherer societies have catchments of the order of $5 \times 10^3 \, \text{km}^2$. It is not certain if they maintain any protected areas. Slash and burn and low input settled cultivators have somewhat smaller resource catchments around $10^3 \, \text{km}^2$ and maintain protected areas ranging in size from around 0.01 to 100 ha. More advanced agrarian societies have larger resource catchments of the order of $10^5 \, \text{km}^2$, and they maintain hunting preserves of the order of 100 to 1000 ha in size. The modern industrial societies have resource catchments spanning the whole biosphere and concentrate on protected areas of a thousand to a million hectares.

ongoing ecological processes on which to devise effective regulatory measures. It is only after human populations have settled in a locality for a length of time, perhaps some generations, that they may become motivated to work towards sustainable use of the biological resources. On becoming so rooted, the societies may also acquire sufficient understanding of the ongoing ecological processes to be able to devise simple rules of thumb for regulation of harvests and habitat transformations that could promote sustainable use of biological resources coupled to conservation, indeed local enhancement of diversity on species, ecosystem, and landscape scales (Gadgil & Berkes, 1991).

While in the long-term interests of the group, such restraint is often likely to be against the immediate, short-term interest of individual group members. Small-scale societies of ecosystem people achieve acceptance of such restraint largely through attribution of sacred qualities to the individual plants, animals or whole biological communities to be protected, coupled to social sanctions against violation of the regulations. Thus, in southern India many sacred groves are dedicated to cobras. People believe that they will be safe from death due to snake bite so long as they respect protection of the serpent grove, but will incur the wrath of cobras, or some associated deities, if they violate it. However, the protection of the sacred groves need not always be absolute; in the case of a special calamity such as a fire consuming houses in the village, the deity, through the agency of the priest, may permit selective harvesting of the trees (Gadgil & Vartak, 1976). Of course, the sanctions may become entirely social sanctions divorced from religious context as in the case of the revival of sacred groves as safety forests in Manipur and Mizoram (Malhotra, 1990). In all cases, however, regulation is enforced primarily on the basis of fear of undesirable repercussions visited either by supernatural forces, or one's own social group.

13.4 Agrarian societies

As small-scale gatherer-horticulturalist societies gave way to larger agrarian societies, regimes of regulation of the use of the biological resource base underwent a transformation. The agrarian societies are characterized by far greater levels of social stratification, with an elite supported by the surplus generated by the peasantry (Lenski & Lenski, 1978). The elite has access to surplus produced from many different localities, it therefore has little motivation to promote sustainable resource use in any particular locality. They in fact belong to Dasmann's (1988) category of biosphere people. While the ecosystem people may hold on to some of their traditions of sustainable resource use and conservation of biodiversity, many such traditions may be weakened. The elite of agrarian societies, however, have their own new conservation techniques; particularly the hunting preserves where game is strictly protected for hunting by the aristocracy. Such hunting preserves played an important role in biodiversity conservation in medieval Europe and Asia. The Mughal emperors of India, for instance, maintained large areas of several hundreds of square kilometres each as hunting preserves in northern India; in fact, these hunting preserves have served as the nuclei of India's modern system of wildlife sanctuaries and national parks (Gadgil, 1991). Again it was fear of punishment that maintained the protection to these hunting preserves; fear of punishment at the hand of the aristocracy and their armed forces, rather than the wrath of deities or sanctions of one's own social group.

Such stratified agrarian societies gave rise to several organized religions. Those of the east, such as Buddhism, absorbed some elements of the conservation traditions of the ecosystem people, as witness the temple groves of Thailand. On the other hand, religions of the Middle East, Christianity and Islam, rejected the attribution of any sacred quality to nature. As Christianity spread to Europe, churches were built by cutting down sacred groves of oak trees. Indeed medieval Europe saw rejection of the so-called pagan traditions of conservation and rather wholesale liquidation of the region's natural stock of biodiversity, except in the game parks of the aristocracy (Whyte, 1967).

13.5 European expansion

Much of the European expansion beginning in the fifteenth century carried this spirit of rejection of any attempts at restraints on the harvest of biological resources or transformation of natural habitats. It was this ethic which promoted large-scale felling of forests in the northeastern United States and the massacre of hundreds of thousands of buffaloes *Bison bison* of the prairies. It led to the fur trade by the Hudson Bay Company dramatically increasing the hunting pressure on Canadian wildlife (Crosby, 1986). Berkes (1989) has documented the response of Amerindians drawn into this trade as commercial suppliers of pelt; this involvement led to a breakdown in their traditional hunting practices which rejected killing except for consumption as food or pelt for their own use. They did, however, develop new sets of yardsticks coupled to the demarcation of territories that promoted more sustainable use even under pressures of commercial harvests.

The European impacts also related to large-scale introduction of exotic plants and animals on the many continents and islands newly colonized by them. Domesticated animals such as cattle and sheep, and rabbits running wild, were responsible for a significant decline in Australia's indigenous marsupial fauna (Wilson, 1990). Interestingly enough, one of the few placental mammals that had reached Australia ahead of Europeans, the dog dingo *Canis familiaris* became a major pest for the European ranchers. Dingo then attracted bounty, a reward paid on presentation of its tail as evidence of having killed one. The indigenous people of Australia took to this as a welcome source of income, but while hunting dingos they ensured that they

354

would not kill lactating females or disturb the dens, so that the dingo population would continue to thrive! The Australian aborigines had thus arrived at a simple yardstick for ensuring sustainable harvests of a biological resource that supported them (Meggitt, 1965).

Notions of conservation arose amongst Europeans only in the second half of the nineteenth century as frontiers of new lands to be colonized seemed to be closing and deforestation led to serious negative consequences in parts of Europe such as Switzerland. These led to protection of forests for their watershed function in colonial possessions, and the revival of forest cover under communal management in Switzerland (Grove, 1992). This also led to the National Park movement in the United States with the setting aside of large scenic areas such as Yellowstone National Park (Koppes, 1988). Notably enough, these areas were thought of as primeval nature untouched by man. In reality, they were humanized landscapes where nature was lightly trodden upon and much of its diversity protected over thousands of years of use by indigenous people. So in the National Park system the emphasis was on regulating human access by deploying a bureaucratic machinery. For the first half of the twentieth century there was little interest in the conservation of biodiversity for its own sake in Europe, in neoEuropes like the United States or Australia, or in the state-sponsored efforts which came to be influenced by the European world view in all countries of the world.

13.6 Of guns and guards

The nature conservation effort that developed in India following independence in 1947 illustrates well the strengths and weaknesses of this approach. Prior to independence India had its aristocratic hunting preserves, such as the wetland of Keoldeo Ghana at Bharatpur visited by hundreds of thousands of wintering waterfowl, or the Gir lion hunting preserve of the Nawab of Junagarh. It also had equivalent shooting preserves for the European planters, such as those of the Nilgiri Game Association in the upper Nilgiris with its Nilgiri tahr *Hemitragus hylocrius*. Broader, state-sponsored conservation efforts started only in the 1950s, led principally by erstwhile princes, with the Maharaja of Mysore being the first president of the Indian Board for Wild Life and the Maharaja of Baroda the first president of the Indian branch of the World Wildlife Fund. The whole nature conservation approach has tended to focus on protection of a few flagship species such as the Indian elephant *Elephas maximus*, rhinoceros *Rhinoceros unicornis* or lion *Panthera leo*, with the firm belief that such protection can be achieved primarily through elimination of the subsistence demands of India's ecosystem people by the force of guns and guards.

Consider as an example Gir National Park, dedicated to the conservation of the lion. At the time of British conquest, the lion was distributed over much of the northern peninsula. It was a prime hunting trophy and was

355

eliminated through most of its range during the nineteenth century. In 1920 the Nawab of Junagarh had to pretend that it had become extinct in his hunting preserve of Gir in order to resist the pressure of organizing a lion hunt for the Governor General of India. In fact, just 22 lions still survived in Gir at that time, and by the 1950s their population had grown to over 200 (Seshadri, 1969). When Gir was constituted a National Park, an important measure immediately introduced was removal of Maldharis, buffalo keepers who had coexisted with the lion in the Gir forest for centuries. Maldharis traditionally accepted occasional kills of their livestock by lions as inevitable; their animals were an important source of food, especially for the lazier male lions. Being removed from the Gir forest has meant serious hardships and a decline in living standards for Maldharis. They are currently living on the periphery of the National Park, denied access to the rich grazing within the park. But the lions are addicted to buffaloes and come out at night to feed on them. Now when a buffalo is killed, the impoverished Maldharis are unwilling to tolerate the loss and poison the carcass to kill the lion when it comes to feed on it a second time. So the displacement of Maldharis has served little positive purpose from the perspective of the flagship species in whose interest the National Park is being managed (Gadgil, 1991).

There has been little interest in the broader objective of conserving biodiversity on the part of the state machinery. Indeed the National Commission on Agriculture recommended in 1976 that all mixed species forests of India should be replaced by more productive monoculture plantations (National Commission of Agriculture, 1976). In consequence, the state forest departments have resorted to clear-cutting sacred groves, considering them, as one Chief Conservator of Forests once put it to me, as merely stands of over-mature timber.

The inevitable consequence has been the ever-growing conflict between the state machinery striving to protect nature reserves by force of arms and the local communities all over the country. Given this hostility, there is little political support for the ongoing conservation effort in India, outside of a narrow circle belonging to part of the urban middle classes. The political leaders are in fact now in the process of dismantling the protected area network of the country (covering some 4% of the land surface) by denotifying some of the wildlife sanctuaries (Nambiar, 1993). There have been similar experiences in other parts of the non-industrial world as well, with management focusing on flagship species, often in the interests of ecotourism, and concentrating on excluding local communities, with little participation or benefit sharing on the part of the latter (Gomez-Pompa & Kaus, 1992). Only in a few exceptional cases, such as Papua New Guinea, has there been some recognition of the traditions and rights of the local communities and attempts to involve them in the conservation of biodiversity.

13.7 The American experience

The 1960s witnessed the beginnings of an interest in biodiversity issues on the part of the developed countries. This interest began with a realization that pesticides persisting in the environment were reaching excessive levels of concentrations in the bodies of animals high up in the food chain, with serious consequences, such as reproductive failure in the case of the peregrine falcon *Falco peregrinus*. The discovery of DDT in the bodies of penguins from Antarctica also highlighted the ubiquity of these substances. The result was the passage of the Endangered Species Act of 1973 in the United States (Primack, 1993). This act committed the United States Federal Government to protect critical habitats of endangered species, thereby promoting conservation of a diversity of natural habitats outside the National Park system. The 1970s also saw the coming on the scene of Nature Conservancy, a Washington-based non-governmental organization (NGO) with considerable influence. The Nature Conservancy initiated the process of systematizing the assignment of conservation priorities to various elements: individual taxa, or populations of particular taxa in a given region, as well as land and water elements supporting populations of one or more species of high conservation priority. The habitats so identified may be of very limited size, of a few hectares or less. The Nature Conservancy has followed up on the identification of such habitats by either their outright purchase or by organizing agreements with the owners, compensating them in some way for giving up the option of developing the property in ways incompatible with the conservation objectives. These developments have brought onto the biodiversity management scene the practice of protecting large numbers of patchy, widely dispersed elements of the landscape (Grove, 1988).

The Nature Conservancy approach also shifted attention from the protection of flagship species and spectacular landscapes to habitats rich in taxa of high conservation value. This habitat focus also obtains in programmes in other countries, such as protection of Sites of Special Scientific Interest (SSSIs) in Great Britain. These programmes have also brought in the new element of paying individual landowners for behaving in ways conducive to the conservation of biodiversity (United Nations Environment Programme, in press).

The 1970s saw the beginning of serious attention being paid not only to protecting specific areas, but also to regulating processes that have an impact on biodiversity levels. Such regulation pertains to the production of harmful substances such as pesticides and a variety of industrial effluents that directly affect living organisms, as well as molecules such as CFCs that have an indirect effect through destruction of ozone and consequent increase in levels of ultraviolet radiation. The regulatory process has also broadened to assessment of environmental impacts of a range of developmental activities, such as the damming of rivers or extraction of timber. There is, however, as yet little explicit consideration of broader biodiversity issues in the process of environmental impact assessment, with attention remaining focused on endangered species. The list of these endangered species does include many flowering plants and smaller vertebrates as in the famous case of the snail darter—but there is still very inadequate attention paid to invertebrates or microbes.

The efforts at management of biodiversity in the developed countries have thus begun to incorporate a number of new, significant elements. These include: (i) interest in a broader range of living organisms; (ii) focus on a wide range of habitats, often small in extent and widely dispersed; (iii) positive rewards to private landowners for adopting biodiversity-friendly practices; and (iv) overall regulation of processes affecting biodiversity.

13.8 Biosphere reserves

Another major positive initiative to have emerged in the 1970s is the worldwide programme of Biosphere Reserves. The Biosphere Reserves do emphasize the need to manage overall biodiversity, to base such an effort on a scientific understanding of the underlying processes, to recognize people as an integral part of the ecological world, and to involve them positively in the conservation effort. While an impressive worldwide network of biosphere reserves comprising 300 reserves in 77 countries has come into being, most of these are merely old national parks dressed up in new jargon without any fundamental change in the management approach (Groombridge, 1992). An important exception includes some of the biosphere reserves in Mexico, where serious attempts have indeed been launched to involve local people. But in most places such attempts remain restricted to programmes of creating awareness and educating local communities, and of extending state patronage through schemes such as creation of village woodlots or supply of fuel efficient cooking stoves. Nowhere do local communities actually play a role in shaping the programmes. This ignores the important fact that local communities may indeed be more knowledgable about local biodiversity resources and may in fact have a greater stake in their conservation, if only permitted genuinely to participate in the effort.

13.9 Imperatives of biotechnology

By the 1970s the remarkable developments that followed the elucidation of the molecular basis of heredity began to be applied to the manipulation of life-forms to human ends. These, in combination with a number of other technological advances, have led to the developments that go by the generic name of biotechnology. Biotechnological industry is already an important player on the world economic scene; it is expected to play a much bigger role in the coming decades. Its bag of tricks includes the capability to move genes from one organism to another; from a mouse to a bacterium, from a virus to a tobacco plant. This implies that organisms once believed to be insignificant could turn out to be of considerable commercial value. Industrial concerns are therefore now greatly interested in access to the world's resources of biological diversity. They would also benefit from information on traditional uses of such biodiversity, as drugs, dyes, cosmetics and so on. Much of this biodiversity is outside industrially developed countries, much knowledge of the uses of plants and animals is with the ecosystem people of the non-industrial world. Organizing access to this biodiversity and this information is therefore now a priority of biotechnology industry, and of the industrial nations (Fig. 13.5). These players are also interested in establishing monopolistic control over these resources and this information to the extent possible (Reid *et al.*, 1993). In the meantime, at least until the biotechnological applications are in place and monopolistic control established over biodiversity resources – perhaps held in *ex situ* collections – as well as information on their uses, the industry and industrial nations are

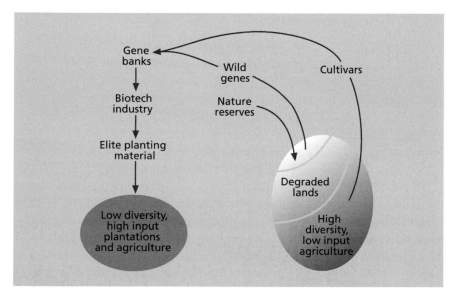

Fig. 13.5 The current strategy of conservation and utilization of biological diversity of biosphere people. This strategy derives genetic diversity of cultivars from low-input–high-diversity subsistence agriculture, and wild biological diversity from nature reserves in developing countries. With little economic rewards flowing into these areas, pressures of poverty are converting extensive tracts of these countries into degraded lands. In consequence, nature reserves are becoming difficult to maintain enclaves of diversity surrounded by degradation from all sides. These genetic resources are being used as raw material for the production of new high yielding plant material, as well as new drugs and other industrial products. The value-added genetic resources support high-input–low-diversity plant production, and other industrial production that brings in handsome economic returns to those with access to advanced technologies.

interested in the conservation of natural biodiversity and knowledge of its uses. It is likely that this interest will be evanescent; to be given up once monopoly over biodiversity resources brought into *ex situ* collection is established (Gadgil, 1993b).

Nevertheless, for the present, there is considerable worldwide concern with managing biodiversity, concern that has led to the signing of the international convention on biological diversity by over one hundred countries. This convention asserts national sovereign rights over the biological diversity of the countries of origin of these resources; encourages all countries to document and conserve biological diversity and to make it available to other countries on the basis of prior, informed consent, and assures countries of origin of some share in the commercial profits that may flow from utilization of biodiversity resources. Additionally, the convention urges nations to recognize the role of local communities in the conservation of biodiversity and to share with them benefits from the utilization of biodiversity. The convention has constructed a radically different framework for the management of biodiversity than what has prevailed so far (Reid *et al.*, 1993).

13.10 Bioregional approach

This is the setting for the current global effort at managing biodiversity. The objective of this management ought to be to conserve, as far as possible, the entire spectrum of genetic diversity in multiple populations of each of the millions of species, in the great variety of ecosystems, both natural and man-made, for the benefit of all of humanity. While potential economic utilization will undoubtedly remain an important motivation for such conservation, it should equally be motivated by the rights of all life to persist and to continue to evolve. In fact, the philosopher Bryan Norton (1987) concludes that the most important reason for conserving biodiversity is its transformative value, its influence in moulding human values to be more friendly towards the natural world. If transformative value is a significant reason for conserving biodiversity, then we must surely strive to ensure that as large a cross-section of humanity as possible has the opportunity to experience a biodiversity-rich world. The inescapable conclusion is that the primary objective of managing the earth's biodiversity should be to maintain a biodiversity-rich milieu over the entire earthscape, and organize people having easy access to this heritage of biodiversity. This is a very different programme from that of guarding a few biodiversity-rich enclaves from which people are totally excluded.

Such a programme is being widely accepted today under what has been termed the bioregional approach (World Resources Institute *et al.*, 1992). This visualizes the earthscape being divided into a number of relatively homogeneous bioregions, each with their characteristic life forms and human cultures; with people feeling an affinity for the well-being of the environment of their own region. The management of the biodiversity resources of the region would be based on ecologically prudent use of all elements of the landscape, which may of course include total protection of certain representative ecosystems. This regime of ecologically prudent practices would be sought to be arrived at through a democratic, participatory process involving all people. It would attempt to motivate people primarily through positive rewards to behave in a biodiversity-friendly fashion. While the focus of the effort would be a bioregion, it would also consider, and try to appropriately regulate, the impacts of people outside of their specific bioregion.

Such an approach will have to be adapted to the very different human settings of the different parts of the world. These may vary from a region like southern Sweden, with low densities of biosphere people, to one like Papua New Guinea, with low densities of ecosystem people and a region like the African Sahel with large numbers of ecological refugees. Obviously, one can only sketch some broad principles of how a bioregional approach could be implemented under such widely divergent conditions.

13.11 The knowledge base and building motivation

Two questions need to be answered in order to work out a biodiversity management strategy: (i) how is the necessary knowledge base created? and (ii) how is the required motivation generated?

13.11.1 The knowledge base

The knowledge presently available to us on the distribution and dynamics of biological diversity is extremely inadequate. Such knowledge is required on many different spatial and temporal scales. We need to understand global levels of diversity of, say, freshwater fishes; we need information on countrywide distribution of such diversity (especially now that the principle of national sovereignty over genetic resources has been accepted), we need to understand fish diversity levels in particular river basins, and we need knowledge of the fish faunas of particular streams. We also need to understand how such biodiversity levels have been changing over time, on scales of days, for instance, in relation to the discharge of effluents in rivers, over seasons in response to annual rainfall, migration or breeding cycles; over years as river beds may have been gradually silted up; over decades as dams and other human activities may have markedly affected river flow regimes; and over centuries and millenia in response to geological changes. Much of this information is simply not available; some of it is available with professional scientists and fishery managers; much else only with local fisherfolk. Thus in the Nilgiri Biosphere Reserve in south India the fish fauna in most river systems has changed drastically over the last few decades due to the construction of a series of dams and intensification of agriculture in the upper catchment. There is no formal record of these changes, but local fishermen who have been fishing the rivers for decades can provide a detailed account of the changes, as well as their interpretation of the causes behind them. Today there is little recognition of this knowledge. But such practical ecological knowledge of the ecosystem people would obviously be of considerable value in management of biodiversity at the ground level. In fact, through most of the biodiversity-rich developing countries, levels of formalized scientific information are low. At the same time in these countries, the levels of practical ecological knowledge of the masses of people necessarily living close to the earth are high. The challenge therefore is to put this knowledge to practical use in managing biodiversity (Gadgil *et al.*, 1993).

In any event, ecologists must acknowledge that they are as yet not in a position to offer many general guidelines for managing biodiversity that would be of practical value in the field. What is required is to try out various options, monitor the consequences and make corrections as we go along, applying the so-called adaptive approach (Walters, 1986). As Slobodkin (1988) puts it, ecologists at their best remain to some degree naturalists, aided by modern technology and computational devices, but for most practical purposes relying on accumulated experience. Now much accumulated experience of relevance is with the ecosystem people of the non-industrial world, and it is imperative that we develop a new mutually beneficial synthesis of their knowledge with formally organized knowledge especially relevant on larger spatial scales to effectively craft an approach of adaptive management of biodiversity.

Such adaptive management calls for continual monitoring of levels of biodiversity. Again, the ecosystem people of the non-industrial world are engaged in such monitoring day in and day out as a result of their manifold subsistence activities. This level of detailed monitoring can never be organized as a formal scientific effort, in spite of all our advances in remote-sensing, instrumentation and informatics. What is then needed is to organize a system of utilizing the information being thus continually gathered by the ecosystem people in the task of adaptive management of biodiversity.

13.11.2 Building motivation

The second important question is that of motivating people to maintain as high a level of distinctive elements of biodiversity in their own bioregion, or in their own localities as possible. This would not only involve protection to specific species, and maintenance of some protected areas, but also management of the entire landscape in a biodiversity-friendly fashion. The last is the more difficult but also the most significant of the tasks from the perspective of the long-term future of global biodiversity. It could best be accomplished through a continual programme of monitoring of biodiversity levels in the different landscape elements making up the bioregion. The task of identifying the landscape elements and following their fate can be greatly facilitated by the use of remote-sensing techniques and geographical information systems. But this must be complemented by monitoring of selected taxa in a representative sample of the various landscape element types on the ground. Such a programme could be organized through local educational institutions and largely manned by practical ecologists amongst the ecosystem people, be they graziers, fisherfolk or women gathering fuelwood for subsistence use. This would create a network of people continually aware of the impact of the various developments taking place with regard to biodiversity in their bioregion. Mechanisms should then be developed for translating this understanding into inputs for the region's development strategy. In the spirit of adaptive management, the implications of any changes made in the development strategy should be continually monitored from the perspective of biodiversity, and the development plans continually modified in the light of the observations. Similar programmes for monitoring of biodiversity and adaptive management at more intensive levels should be organized for protected areas. The monitoring programmes should have a special focus on protected species (Gadgil, 1994).

The ultimate motivation for people to conserve high levels of distinctive elements of biodiversity must in modern times come not from the age-old negative, but rather from appropriately devised positive incentives. In part, such incentives may derive from people gaining better access to a diversity of biological resources that they value for subsistence as well as a source of income. This is the experience of the joint forest management programmes of West Bengal. In these programmes local communities are assigned the responsibility of protecting patches of regenerating *Shorea robusta* forests with support from the state forest department. They then have much better access to enhanced levels of a great variety of non-wood forest produce such as leaves, fruit, honey and mushrooms (Deb & Malhotra, 1993). The extractive reserves of Amazonia are similar in spirit.

However, such improved access to biological resources is unlikely to be an adequate incentive to maintain high levels of biodiversity everywhere. This could be far better ensured by people being paid service charges for maintenance of biodiversity each in their own bioregion, or some more restricted territory for which they are assigned responsibility. The amount of these service charges would have to be linked firmly to the value of biodiversity maintained within their territory. This valuation would have to be based on some global system of assessment of conservation priorities such as that developed by the Nature Conservancy in the United States, coupled to periodic monitoring of biodiversity levels within each territory or bioregion. The service charges should not be a one-time payment, but paid annually on the basis of performance, so that any failure on the part of the local community to deliver goods would be reflected in a reduction in, or cancellation of service charges (Gadgil & Rao, 1994). Such a system could easily be made a part of programmes such as the tripartite alliances of the tropical forest action plan of Mexico (SARH, 1994a,b). To be put in practice on a wider scale would of course require the elaboration of a whole series of mechanisms for co-ordination and conflict resolution amongst neighbouring territories within bioregions, amongst neighbouring bioregions within states and so on. Such a system, firmly linking service charges to conservation performance, could be the most effective means of involving local communities in modern times when traditional belief systems are inevitably in decline. Such a system may also be highly cost effective, for alternate systems relying on bureaucracies have proven to be thoroughly inefficient (Gadgil & Rao, 1994).

The system sketched above would be particularly appropriate for parts of the world predominantly inhabited by ecosystem people (i.e. people rooted in a given locality in intimate contact with it). It would not be appropriate to a region predominantly inhabited by biosphere people (i.e. much more mobile people dependent on markets to bring them resources from many different parts of the world). For such regions the more conventional systems of formally protected areas (including many smaller ones), protection of en-dangered species, control of pollution, assessment of environmental impacts (all largely implemented by professionals), are much more appropriate. But even in these cases, elements of the bioregional approach sketched above may be of value. For instance, residents of northwestern United States paid some special service charges based on the extent of old growth coniferous forests in their locality may feel far more inclined to support the preservation of the spotted owl than they do today.

Acknowledgments

This work has been supported by the Ministry of Environment and Forests, Government of India.

References

Buchanan F.D. (1870) *A Journey from Madras Through the Countries of Mysore, Canara and Malabar, vol. 2*. Higginbothams and Co., Madras. [13.3]

Berkes F. (1989) Cooperation from the perspective of human ecology. In: *Common Property Resources: ecology and community based sustainable development* (ed. F. Berkes), pp. 70–88. Belhaven Press, London. [13.5]

Crosby A.W. (1986) *Ecological Imperialism. The biological expansion of Europe*, 900–1900. Cambridge University Press. [13.5]

Daniel J.C. (1983) *The Book of Indian Reptiles*. Bombay Natural History Society, Bombay. [13.3]

Dasmann R.F. (1988) Toward a biosphere consciousness. In: *The Ends of the Earth* (ed. D. Worster), pp. 277–288. Cambridge University Press. [13.2] [13.3]

Deb D. & Malhotra K.C. (1993) People's participation: the evolution of joint forest manage-ment in south-west Bengal. In: *People of India: biocultural dimensions* (eds S.B. Roy & A.K. Ghosh), pp. 329–342. Inter India Publications, New Delhi. [13.11.2]

Diamond J. (1994) Ecological collapses of ancient civilizations: the golden age that never was. *Bull. Am. Acad. Arts Sci.* **47**, 37–59. [13.3]

Gadgil M. (1987) Diversity: cultural and biological. *Trends Ecol. Evol.* **2**, 369–373. [13.3]

Gadgil M. (1991) Conserving India's biodiversity: the societal context. *Evol. Trends Plants* **5**, 3–8. [13.4]

Gadgil M. (1993a) Of life and artifacts. In: *The Biophilia Hypothesis* (eds S. Kellert & E.O. Wilson), pp. 365–377. Island Press, Washington DC. [13.1]

Gadgil M. (1993b) Tropical forestry and conservation of biodiversity. In: *Norway/UNEP Experts Conference on Biodiversity, Trondheim, Norway, 24 May 1993.* [13.9]

Gadgil, M. (1994) Inventorying, monitoring and conserving India's biological diversity. *Current Science* **66**, 401–406. [13.11.2]

Gadgil M. (1995) Prudence and profligacy: a human ecological perspective. In: *The Economics and Ecology of Biodiversity Decline* (ed. T. Swanson). Cambridge University Press, in press. [13.2]

Gadgil M. & Berkes F. (1991) Traditional resource management systems. *Res. Manage. Optim.* **18**, 127–141. [13.3] [13.4] [13.6]

Gadgil M. & Guha R (1992) *This Fissured Land: an ecological history of India.* Oxford University Press, New Delhi and University of California Press, Berkeley. [13.2]

Gadgil M. & Rao P.R.S. (1994) A system of positive incentives to conserve biodiversity. *Economic & Political Weekly* **Aug. 6**, 2103–2107. [13.11.2]

Gadgil M. & Vartak V.D. (1976) Sacred groves of Western Ghats of India. *Econ. Bot.* **30**, 152–160. [13.3]

Gadgil M., Berkes F. & Folke C. (1993) Indigenous knowledge for biodiversity conservation. *Ambio* **22**, 151–156. [13.2] [13.3] [13.11.1]

Gomez-Pompa A. & Kaus A. (1992) Taming the wilderness myth. *BioScience* **42**, 271–279. [13.6]

Gomez-Pompa A., Flores J.S. & Fernandez M.A. (1990) The sacred cacao groves of the Maya. *Latin American Antiquity* **1**, 247–257. [13.3]

Groombridge, B. (ed.) (1992) *Global Biodiversity: status of the Earth's living resources.* Chapman & Hall, London. [13.8]

Grove N. (1988) Quietly conserving nature. *National Geographic* **174** (January), 818–844. [13.7]

Grove R.H. (1992) Origins of western environmentalism. *Sci. Am.* July, 22–27. [13.5]

Koppes C.R. (1988) Efficiency, equity, aesthetics: shifting themes in American conservation. In: *The Ends of the Earth* (ed. D. Worster), pp. 230–251. Cambridge University Press. [13.5]

Lenski G. & Lenski J. (1978) *Human Societies: an introduction to macrosociology.* McGraw-Hill, New York. [13.4]

Malhotra K.C. (1990) Village supply and safety forest in Mizoram: a traditional practice of protecting ecosystem. In: *Abstracts of the Plenary, Symposium Papers and Posters Presented at the V International Congress of Ecology, 23–30 Aug. Development of Ecological Perspectives for the 21st Century*, p. 439. INTECOL, Yokohama, Japan. [13.3]

McNeely J.A. & Pitt D (eds) (1985) *Culture and Conservation: the human dimension in environmental planning.* Croom Helm, London. [13.3]

Meggitt M.J. (1965) The association between Australian aborigines and dingoes. In: *Man, Culture and Animals* (eds A. Leeds & P. Vayda), pp. 7–26. American Association for the Advancement of Science, Washington DC. [13.5]

Myers N. (1988) Threatened biotas: 'hot spots' in tropical forests. *Environmentalist* **8**, 187–208. [13.1]

Myers N. (1990) The biodiversity challenge: expanded hot spots analysis. *Environmentalist* **10**, 243–256. [13.1]

Nambiar P. (1993) A change for the worse. *Down to Earth.* Nov. 15, p. 45. [13.6]

National Commission of Agriculture (1976) *Report of the National Commission of Agriculture, Volume IX.* Ministry of Agriculture, Delhi. [13.6]

Norton B.G. (1987) *Why Preserve Natural Variety?* Princeton University Press, New Jersey. [13.10]

Primack R.B. (1993) *Essentials of Conservation Biology.* Sinauer Associates, Sunderland, Massachusetts. [13.7]

Reid W.V., Laird S.A., Meyer C.A., Gamez R., Sittenfeld A., Janzen D., Gollin M.A. & Juma C. (1993) *Biodiversity Prospecting: using genetic resources for sustainable development.* World Resources Institute, Washington DC. [13.9]

Ruddle K. & Johannes R.E. (eds) (1985) *The Traditional Knowledge and Management of Coastal Systems in Asia and the Pacific.* UNESCO, Jakarta Pusat, Indonesia. [13.3]

SARH (1994a) *Tropical Forest Action Plan of Mexico, vol I: synthetic document of the Paft-Mexico and lines of action.* Secretary of Agriculture and Hydraulic Resources, Mexico. [13.11.2]

SARH (1994b) *Tropical Forest Action Plan of Mexico, Volume II: basic document.* Secretary of Agriculture and Hydraulic Resources, Mexico. [13.11.2]

Seshadri B. (1969) *The Twilight of India's Wildlife.* John Baker, London. [13.4]

Slobodkin L.B. (1988) Intellectual problems of applied ecology. *BioScience* **38**, 337–342. [13.11.1]

Subash Chandran M.D. & Gadgil M. (1993) Kans – safety forests of Uttara Kannada. *Proceedings at IUFRO Forest History Group Meeting on Peasant Forestry, Freiburg, Germany* **40**, 49–57. [13.3]

Terborgh J. (1986) Keystone plant resources in the tropical forest. In: *Conservation Biology: science of scarcity and diversity* (ed. M.E. Soulé), pp. 300–344. Sinauer Associates, Sunderland, Massachusetts. [13.3]

United Nations Environment Programme (1995) *Global Biodiversity Assessment, Section 10: measures for conservation of biodiversity and sustainable use of its components.* UNEP, Nairobi, Cambridge University Press. [13.7]

Vitousek P.M., Ehrlich P.R., Ehrlich A.H. & Matson P.A. (1986) Human appropriation of the products of photosynthesis. *BioScience* **36**, 368–373. [13.1]

Walters C. (1986) *Adaptive Management of Renewable Resources.* Macmillan, New York. [13.3] [13.11.1]

Whyte L. (1967) The historical roots of our ecological crisis. *Science* **155**, 1203–1207. [13.4]

Wilson A.D. (1990) The effect of grazing on Australian ecosystems. In: *Australian Ecosystems: 200 years of utilization, degradation and reconstruction* (eds D.A. Saunders, A.J.M. Hopkins & R.A. How), pp. 16. Surrey Beatty, Chipping Norton, New South Wales. [13.5]

World Resources Institute, IUCN, UNEP, FAO & UNESCO (1992) *Global Biodiversity Strategy.* [13.10]

14: Biodiversity and global change

RIK LEEMANS

14.1 Introduction

The concern that changing atmospheric composition will have significant impact on climate, vegetation and human society, has increased over the last decades. Initially, most of this concern was focused on local and regional consequences of air pollution, such as acidification, but recently also major global consequences, such as climate change, have drawn attention. Currently, one of the major issues is the increase in the concentration of so-called greenhouse gases (GHGs: H_2O, CO_2, CH_4, CO, CFCs, N_2O, tropospheric O_3 and its precursors) that influence the Earth's radiative balance. Such increases could well lead to large changes in regional and global climate (Houghton *et al.*, 1992). The whole complex of causes, processes and impacts of these changes is generally summarized with the term *'global environmental change'* or *'global change'*.

The sources of GHGs are diverse. The major source is the worldwide combustion of fossil fuels. Other important sources are, for example, deforestation, the burning of biomass and agricultural practices. Consequently, atmospheric concentrations of GHGs have been rising continuously during the last decades. Although most sources and sinks of GHGs are relatively well understood globally (Watson *et al.*, 1992), it is still difficult to obtain a complete understanding of all GHG-fluxes and their trends. Such understanding requires knowledge of many atmospheric (both chemical and climatic; e.g. Prinn, 1994), oceanic (e.g. Siegentaler & Sarmiento, 1993), and terrestrial (e.g. Leemans, 1995) processes together with complex interactions and feedbacks. In particular, the role of terrestrial ecosystems (through their photosynthesis and respiration, their role in global hydrology, vegetation dynamics and subtle changes in land-use and land-cover) can only be approximately quantified globally, because ecosystem processes and properties are strongly differentiated through local conditions. The significance of ecosystem processes and changes therein globally operates through their cumulative effects (Turner *et al.*, 1990a; Leemans & Zuidema, 1995). This makes a reliable estimation of trends in GHGs that involve the biosphere extremely difficult. Many atmospheric, oceanic and climatic processes possess strong systemic properties globally, which are easier to comprehend and thus more unambiguously implementable in advanced assessment models.

This is one of the main reasons why the global change debate has been controlled by atmospheric scientists, physicists and oceanographers and less by biologists, ecologists and geographers. Fortunately, this domination is changing rapidly. Through large international research programmes, such as the International Geosphere Biosphere Programme (IGBP, 1994a) and the Human Dimension Programme (HDP, 1994), the importance of the structure and functioning of ecosystems has already been more strongly emphasized and has now obtained a recognized position in global change research (cf. Fig. 14.1).

Another event, the UNCED (United Nations Conference on Environment and Development) further highlighted the importance of terrestrial ecosystems for global environmental change. The resulting climate treaty addresses the increased concern over how profound the impacts of global change might be. The Treaty has now been ratified by most countries and its main objective (Article 2) is:

> To achieve stabilization of greenhouse gas concentrations in the atmosphere at a level that would prevent dangerous anthropogenic interference with the climate system. Such a level should be achieved within a time-frame sufficient to allow ecosystems to adapt naturally to climate change, to ensure that food production is not threatened and to enable economic development to proceed in a sustainable manner.

From the language used it is clear that impacts on ecosystems, human resources and human activities are the most important concerns. The treaty implicitly requires the identification of vulnerable eco- and agrosystems and requests the determination of thresholds and rates of climate change under which adaptations are possible and/or acceptable. This creates a direct link with the biodiversity convention (Anonymous, 1992), which was also adopted at UNCED.

In this chapter I will review the different approaches that have been used to analyse the consequences and impacts of global environmental change on ecosystems and try to emphasize the links with biodiversity. I will start by focusing on the global distribution of biomes, then continue with landscapes and communities and finally discuss the more specific ecosystem-level processes. Although I will limit the presentation mainly to vegetation and plant species, I recognize that the other biota will also be affected by global environmental change. However, the impacts on them are much more difficult to estimate and have not yet been addressed systematically, because the actual impacts mainly result from indirect mechanisms following primarily those on vegetation. This 'impact' review will strongly focus on modelling, because this is one of the few ways to adequately determine the future behaviour of complex systems.

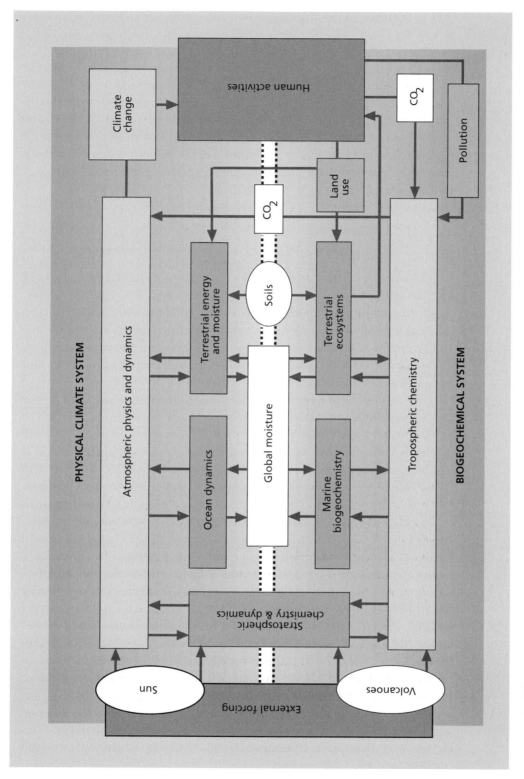

Fig. 14.1 The different components of the Earth system as defined by the core projects of the International Geosphere Biosphere Programme (derived from IGBP, 1994b).

I further discuss the consequences of different aspects of global environmental change for biodiversity at all organizational levels (world, landscape and ecosystem or community). This discussion will strongly focus on the different levels of scale at which both environmental issues (global change and biodiversity) are addressed. Global change, in general terms, involves ecosystem processes and large regions, while biodiversity issues more strongly emphasize local phenomena, such as habitat loss and species-richness. To illustrate this, I first present the results of a literature search on biodiversity, in which I will try to evaluate all the different levels from genes, through species, to biomes. This allows me to highlight more easily the gaps between global change and biodiversity assessments which still need to be addressed and quantified.

14.2 Literature on global change and biodiversity

Biodiversity or biological diversity refers to the variety within the living world and is as such a synonym for 'Life on Earth' (Groombridge, 1992). However, such broad definition is impossible to quantify and does not allow us to assess the impact of global change on the world's biodiversity.

The problems of such assessment do not only lie with the definitions used, but also with the conceptual domains of both biodiversity and (global) environmental change. Biodiversity is generally limited to living organisms and their interactions, while environmental change focuses more strongly on the physical and biogeochemical aspects of the Earth system (Fig. 14.1). The impacts of environmental change lie on the interface between the two: organisms influence and alter the(ir) environment while the environment sets the constraints within which these organisms survive, develop and evolve. Large or rapid environmental change could restrict continued survival of organisms, if their adaptive capability is inadequate to cope with such change. The dilemma of this dynamic interaction is strongly addressed by Article 2, but it still needs to be supported by a comprehensive theoretical and empirical scientific framework.

It is a common practice to define biodiversity on different levels of organization: genes; species; community or ecosystem; and landscape or biome (Fig. 14.2; Solbrig, 1991; Groombridge, 1992). The first level represents the heritable variation within and between populations of organisms. The second level is the most common use of the term biodiversity and is merely a count of total species numbers in a site or region. The global range of total species numbers lies between 3 and 30 million (May, 1992) and appears to consist mainly of insects and micro-organisms. Much research has focused on the geographical and evolutionary patterns of species diversity (e.g. the collected papers in Wilson & Peter (1988) and Solbrig *et al.* (1992)). The third level includes the diversity of communities, the interactions between species in communities, their functioning and the resulting structuring of ecosystems.

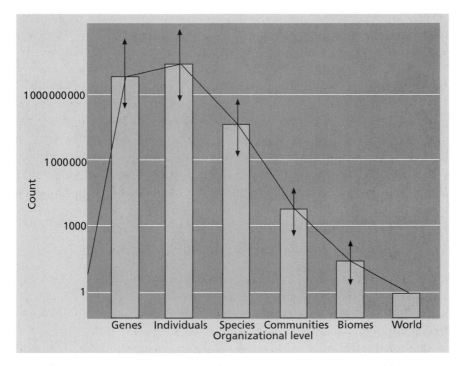

Fig. 14.2 Biodiversity at different organizational levels (calibrated points for this curve: 1 world, *c.* 150 biomes, 15 million species, 1 milliard genes (which are all derived from 2 bp of nucleic acid)). Total biodiversity could well be defined as the integral of such a curve, if the horizontal axis is a continuous one. Unfortunately, the scale and domain of the different levels is difficult to define and there can be considerable overlap.

Many approaches based on relative abundances of species or species-types, have been developed to describe diversity here (e.g. MacArthur, 1972), but no generally accepted or robust quantified approach has yet emerged. The last level includes the diversity that develops in space (patchiness) in relation to environmental and vegetation development (succession, disturbance etc.) and involves, in principle, all other levels. One of the major problems in defining biodiversity on this level is that no generally agreed classifications of the units exist. There are many different landscape and biome classifications which often use a common terminology, but are based on different climatic, soil or vegetation properties or combinations thereof (Leemans *et al.*, in press).

Since the decline in biodiversity during the last century has been caused largely by environmental change, and has resulted from intensified human activities locally, regionally and during the last decades also globally, one would expect that recent papers concerned with biodiversity issues mainly addressed the 'larger' levels. However, although many attempts were made (e.g. Solbrig *et al.*, 1992), most of them still only addressed the 'smaller' levels. After the UNCED conference and its important treaties, I expected a recognizable shift in the literature towards the larger levels. I tested this hypothesis

371

using the papers presented in the literature databases of Current Contents on Disk (ISI, 1991) for two different periods, almost immediately after UNCED and almost 2 years later. I evaluated all publications which listed 'diversity' or 'biodiversity' in either their title or keywords and classified the subjects of the resulting papers into the appropriate level. Each paper was assigned to a single class. To gain additional insight on topics and applications, I classified them also into topical classes. Here, a single paper could be assigned to several classes.

The results of this evaluation (Table 14.1) showed that there was indeed a very slight increase in papers concerned with the larger organizational levels, but the bulk consisted of papers considering only the genetic and species levels. The topical classification resulted in a similar impression. Most papers were only descriptive and did not deal with applications or the relations with environmental change. Local conservation and ecosystem management were among the most common applications of biodiversity research.

I believe that this very skewed distribution is a reflection of our lack of interest in the link between the impacts of global change and biodiversity, despite recently published and timely books on this topic (e.g. Peters & Lovejoy, 1992; Schulze & Mooney, 1993). Below, I will further illustrate this opinion with a review of current approaches. This review is not complete, but could be used as a guide to extract the most important issues at the 'larger' levels.

Table 14.1 The results of a literature search on the term 'biodiversity' (and 'diversity') and classification of the contexts of its usage, using CCOD (ISI, 1991) for the periods January–April 1993 and September–December 1994.

	Spring 1993		Autumn 1994	
	Quantity	%	Quantity	%
Organizational levels				
Genetic	165	44	187	38
Species/population	154	41	215	43
Ecosystem/community	57	15	83	17
Landscape/biome	3	1	12	2
Special topics				
Conservation/management	38	15	61	18
Ecosystem dynamics and structure	7	3	21	6
Local environmental change	17	7	36	11
Regional environmental change	5	2	12	4
Global environmental change	2	1	8	2
Measurement and theory	36	15	43	13
Descriptive	143	58	156	46

14.3 Assessing impacts on biomes

The coarse patterns in the physiognomy and potential species composition of vegetation are primarily determined by climate (Woodward, 1987). Basic climatic parameters, such as average temperature and precipitation, define the major borders between latitudinal zones (e.g. boreal, temperate and tropical), and vegetation types (e.g. deserts, steppes or savannas, and forests). Temperature and precipitation regimes together with soil and land characteristics determine the seasonality of available moisture for plant growth. Climate thus determines the potential appearance of vegetation, influencing the distribution of, for example, deciduous and evergreen trees, shrubs and grasses. However, variation between years, deviation from climatic means and extreme events (e.g. extreme temperatures, storms, or droughts) also influence vegetation dynamics and can lead to changes in vegetation succession and thus appearance (e.g. through frost-damage, windthrow and fire). Such natural events as well as human influences define the actual vegetation in any region and govern resilience to environmental change.

The close correlation between climate and vegetation has long been recognized by natural scientists and has led to the use of vegetation to develop climate maps and vice versa (e.g. Grisebach, 1872; Walter, 1985). Several conceptually simple (i.e. based on only few parameters) correlation schemes have been developed during the last century (for a review see Leemans *et al.*, 1995). All these traditional climate and/or vegetation classifications share the common feature of having specifically defined limits for each climate class, biome, or species-type. All can easily be derived from readily available climate data.

This last feature has resulted in the frequent use of these classifications for assessing the impacts of climate change (e.g. Emanuel *et al.*, 1985; Cramer & Leemans, 1993; Van Campo *et al.*, 1993; Leemans *et al.*, 1995). Climate change scenarios derived from General Circulation Models for the Atmosphere (AGCMs) and climate databases have been used to define patterns for different climatologies (for a discussion of the methodology, see Leemans, 1992a or Carter *et al.*, 1994). All these assessments showed large polewards shifts of the major vegetation zones or biomes (cf. Fig. 14.3). These shifts were most pronounced at high latitudes, because the AGCMs simulated the largest temperature changes here and the vegetation types here are generally characterized by smaller climatic ranges.

Such large shifts in vegetation zones were common throughout the Earth's history. The well-studied patterns and rates of vegetation change in North America and Europe illustrate this (cf. Huntley & Webb, 1988). However, these changes were caused mainly by a gradual warming after the last glaciation. An increase in temperature of *c.* 5°C occurred from the last glacial

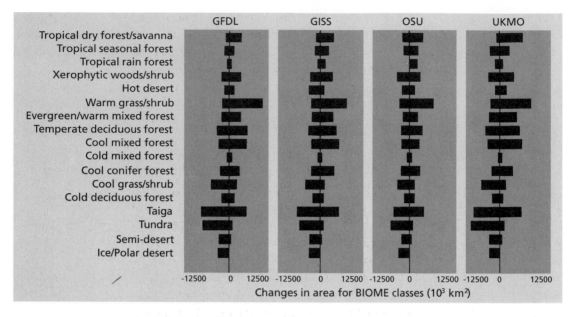

Fig. 14.3 Shift in biome patterns using the BIOME model (Prentice *et al.*, 1992) for scenarios from different climate models (GFDL: General Fluid Dynamics Laboratory in Princeton; GISS: Goddard Institute for Space Studies in New York; OSU: Oregon State University in Corvallis; and UKMO: United Kingdom Meteorological Office in Bracknell, for more detail on the approaches see Leemans, 1992a). The bars to the left illustrate a decline in the original extent, those to the right an expansion.

maximum (18 000 years before present, BP) until the warmer climatic optimum (9000–6000 years BP). Only one relatively short cooler period occurred with more rapid changes in climate, the Younger Dryas (12 000–10 000 years BP), when the ocean currents changed the northwards heat fluxes. The Younger Dryas was especially pronounced in Europe, where a tundra-like vegetation, characterized by *Dryas*, then dominated the landscape again. After 6000 years BP the climate remained relatively constant and the modern vegetation pattern developed rapidly.

Vegetation thus adapts to a changing climate. Such adaptation is governed by the response of individual species and not by displacement of biomes as complete entities. Here lies one of the main disadvantages of impact assessments based on climate classifications, such as the Holdridge Life Zone Classification (Holdridge, 1947). The empirically derived climatic descriptors for biomes bear little relation to the actual climatic determinants of individual species (e.g. frost or drought tolerance). Only the artificial boundaries of biomes are described. Although such models provided an early evaluation of the magnitude of change, the actual simulated pattern of change is not very realistic. Alternative biomes, without a currect analogue, cannot emerge. Modern models, such as BIOME (Prentice *et al.*, 1992) and MAPPS (Neilson, 1993), use a more ecophysiological or mechanistic approach to describe the potential distribution of major plant types (e.g. broad-leaved deciduous trees). In these models, all plant types show an individualistic response to climate, and biomes are simulated as local assemblages of these plant types. The location and ecological characteristics of ecotones between the major biomes are also simulated relatively well. In contrast to the more traditional models (Adams *et al.*, 1990; Prentice & Fung, 1990), simulations with these modern models represent the vegetation changes since 18 000 years BP correctly (Van Campo *et al.*, 1993; Prentice *et al.*, 1994).

These global models seem more appropriate to simulate the response of vegetation to climate change. Unfortunately, severe time-lags of up to several centuries can occur in the response time of the slowest reacting species, such as trees. Davis (1989) analysed maximum migration speeds of several tree species from palaeo-records and found that they ranged from 20 to 40 km per century. These often cited rates lead to the conclusion that vegetation response would be too slow to follow the projected climate change due to the increased concentrations of GHGs (cf. Plate 14.1 (between pp. 214 & 215)). Many impact assessments have consequently emphasized the escalating imbalance between climate and vegetation, probably leading to high extinction rates and consequently negative impacts on species counts (e.g. Markham *et al.*, 1993). Prentice *et al.* (1991) revised this view somewhat by analysing the equilibrium between vegetation and climate through the Holocene. They concluded that on time scales greater than 500 years, no significant mismatches could be recognized. This view is also supported by recent research on vegetation response from vegetation patterns in Europe (Peng, 1994). Relevant vegetation responses to climatic change thus probably lead to a new equilibrium after several centuries, and not millennia.

The delays caused by slow migration do not define the actual vegetation response. Too much emphasis has been directed to mean migration rates which are probably less relevant. The few propagules that disperse and establish over long distances create isolated islands from which this species can spread again. These propagules are much more important. If one assumes that such extreme dispersion distances could be over a thousand kilometres, then the time-lag in vegetation response is strongly determined by the establishment and maturation of these individuals, which is of the order of decades to centuries. This is consistent with actual palaeo-observations (Prentice *et al.*, 1991; Peng, 1994).

However, the above discussion focuses attention on the major weaknesses in the models based on climate–vegetation classifications. No adequate links to any of the 'lower' biodiversity levels have been made. Only shifts of individual biomes are emphasized. These global approaches are appropriate to determine potential future vegetation patterns when new equilibria between vegetation patterns and climate are developed, but not to evaluate short-term responses of vegetation distribution. These approaches are thus too static for such applications. On shorter timescales, dynamic processes such as plant development and ecosystem succession define the response. This response is not only smaller in its temporal domain, but also in its geographical domain. Successional processes are typical for the landscape scale, consisting of a series of patches in different phases of development and response.

14.4 Assessing impacts on landscapes

The most frequently used impact assessments on landscapes use succession models (Smith *et al.*, 1994). Most of these models have been developed for forests, although models for other ecosystems are available (e.g. Lauenroth *et al.*, 1993). I will limit the presentation here to forests, because these ecosystems show the longest time-lags in their response.

Forest succession models have existed since the early seventies (Botkin *et al.*, 1972) and have been applied to many different forest types worldwide (Shugart *et al.*, 1992). They have been strongly improved by incorporating more appropriate routines for environmental processes (e.g. moisture availability, permafrost dynamics and response to temperature) and integrating those with routines for ecophysiological processes of plant growth (photosynthesis, respiration and resource allocation; Smith *et al.*, 1994). The underlying structure of these models, however, has not changed significantly.

The models simulate forest stands as a series of independent patches with shared resources (light, moisture and temperature). Each patch consists of several individual trees, each characterized by its species, diameter, height, leaf area, etc. Light attenuation through the canopy and the subsequent reduction of plant growth under shaded conditions in the lower layers of the canopy is accounted for. Each species has specific environmental conditions under which it can establish, grow and mature. Growth slows down under less favourable conditions and mortality also increases. Growth is thus a combination of the position of any individual tree in a stand and its species' characteristics. The life cycles of individual trees (establishment, growth, maturation and death) are all simulated dynamically with respect to changing environmental and stand conditions. The models are sometimes also called gap-models, because they can simulate some of the processes that follow the creation of a canopy gap by the death of a large tree and the subsequent changes in species composition and growth rates in the lower canopy strata.

The models are suitable for simulating landscapes when large numbers of patches are simulated on environmental gradients (some versions of the models also incorporate disturbances, such as drought, windthrow and fire). Patches are in different stages of development through the outcome of stochastic mortality and establishment processes, and attain different species compositions and stand characteristics along the gradient. In principle, the landscape patterns simulated with these models for a constant climate are consistent with the static approaches (Solomon, 1986), but their strength does not only lie with such long-term dynamics, but especially with the shorter-term transient responses to global change.

Many simulations with forest succession models have been performed to address the transient impact of climate change. The outcomes of these assessments have been very diverse and depend on the initial stage of the simulated forest stands (young, mature or old-growth), the climate change scenario applied, and additional assumptions on species availability. The most common assumption for the last of these is that species are available if the environmental conditions become suitable. This is probably valid for the time sequences modelled (most runs cover *c.* 2000 years), but probably less appropriate for the decadal responses.

The results of the simulations are further dependent on the climate change scenario used. Early simulations changed temperature, without changing precipitation. This led to drier conditions with possible large-scale forest dieback or disturbances (fire). Most AGCM climate scenarios include, besides an increase in temperature, also a (logical) increase in precipitation. The actual available moisture depends on the balance between these two, and is controlled through evapotranspiration. The timing of climate change is also important. Sudden shifts towards a 'doubled CO_2' climate are unrealistic and most climate change scenarios therefore use a century to shift from one equilibrium climate to another. This approach is more realistic, but the outcome of the impact assessment largely depends on which period is used to define the changes. Only the transient period and the early phases for the new steady climate are of immediate interest for response assessments. The conditions of later periods are less plausible, because with extrapolations of current trends in GHG concentrations it is unrealistic to assume a levelling of the atmospheric concentrations at a 'double-CO_2' level (Leggett *et al.*, 1992).

In Fig. 14.4 I illustrate the dynamic response patterns from a simulation of a Scandinavian boreal forest. The model used is derived from FORSKA (Prentice *et al.*, 1993) and applied to an old-growth stand in Central Sweden. At the start of the simulation this stand is characterized by abundant shade-tolerant spruces *Picea abies* in all size classes, fewer large shade-tolerant pines *Pinus sylvestris* and some scattered early successional birches *Betula pubescens* and poplars *Populus tremula*. During initial stages of climate change the spruces thrive and increase in biomass. The other species continue to decline as they would have done if there were no climate change and/or disturbance. After a certain threshold temperature is reached spruce also start to decline. This is not caused by growth limitations of mature individuals, but by an increased limitation on regeneration of the species under warmer conditions. The canopy opens up gradually and after one century the forest stand has opened enough to allow re-establishment of birch, closely followed by more

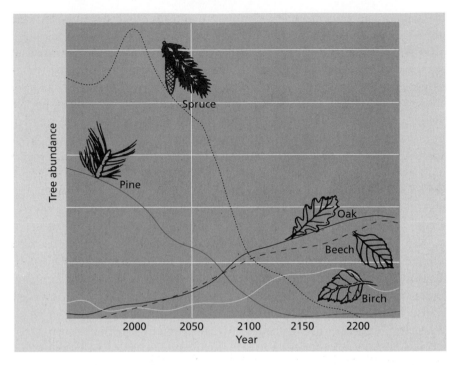

Fig. 14.4 Simulated forest succession under climate change for a southern boreal forest in Sweden using a shift from current climate towards the doubled-CO_2 climate scenario by the GFDL model, during the initial 100 years. Successional time-lags of this shift last several centuries.

temperate species, such as oak *Quercus* species and beech *Fagus sylvatica*. However, it takes several centuries to change the appearance of the forest from one that is boreal coniferous and evergreen to one that is temperate broad-leaved and deciduous.

Discussion of results of these models has focused on the rapidity of responses. The example presented is gradual and a lot of opportunities exist for adaptation. Several researchers using other versions of such models (e.g. Neilson, 1993; Smith & Shugart, 1993) have argued that the response could well be catastrophic with sudden and large-scale forest dieback followed by a slow succession towards better adapted vegetation types. Such catastrophic events could result in large pulses of carbon emissions to the atmosphere, temporally enhancing the greenhouse effect. Although this scenario is possible, it seems less probable, mainly because it assumes initial conditions of mature forests in equilibrium with current climate. In reality the global forests are much more diverse, currently with an abundance of young stands in the temperate and boreal regions (Kauppi *et al.*, 1992; Kolchugina & Vinson, 1993). This heterogeneity will also lead to a much more heterogeneous response and diminishes the likelihood of sudden large-scale continental dieback.

Although these succession models are now well developed and their results clearly emphasize the importance of diverse landscapes and their dynamic responses, they are still limited in scope. The basic philosophy of succession models has also been applied to the succession of grasslands (Lauenroth *et al.*, 1993); these models are not yet capable of incorporating the ecotones between forests, grasslands and scrublands, and other ecosystems with mixed life-forms, and therefore do not provide a broad geographic cover. Only one-third of the terrestrial world is currently forested, and most of these forests contain mixed life forms, not only trees. Although several attempts have been made (e.g. Urban *et al.*, 1992), a complete assessment of the impacts of global change with succession models is still impossible.

The major limitation of these models, however, is the missing linkage with biogeochemical cycling. Soil processes, litter inputs, the flows of nitrogen, carbon and water, and the direct impact of enhanced atmospheric carbon dioxide (CO_2) concentrations are not very well considered (cf. Fig. 14.1). These processes mainly operate at the ecosystem level and influence ecosystem functioning and structure, and are therefore important issues to consider when assessing biodiversity and global change.

14.5 Assessing impacts on ecosystems

Ecosystems are not just collections of individuals or species at a specific site, but consist of interacting organisms that utilize, alter and adapt to, their environment. The primary ecosystem processes, such as photosynthesis, respiration and decomposition, determine the exchange of energy, water and trace gases between the biosphere and atmosphere. Through photosynthesis, carbon is assimilated and accumulated in living biomass. The processes are influenced by changes in temperature, humidity, atmospheric CO_2 concentrations, nutrient availability, etc. (Larcher, 1980), and have received most attention in global change assessments (e.g. Melillo *et al.*, 1990). Only some of the papers in Peters and Lovejoy (1992) and Pernetta *et al.* (1994) address some explicit biodiversity issues on an ecosystem level, and not yet in a very satisfactory manner. The evidence for these impacts is still very general or only anecdotal. The emphasis on ecosystem processes can easily be explained by their importance in determining the fluxes of GHGs. Biogeochemical processes in ecosystems are relevant for global change and have been simplified as comprehensive globally systemic processes, neglecting the diversity of local and regional properties of these ecosystems. These processes have also been proposed to assist in mitigating emissions, and are significant feedbacks in global biogeochemical cycling. Other impacts on ecosystems have never played such dominant roles.

One of the most frequently mentioned impacts of global change influences ecosystem processes directly. The increased atmospheric levels of CO_2 could increase photosynthesis and thus enhance plant growth and increase carbon sequestration. At high CO_2 levels, more CO_2 diffuses into the leaves through the stomata and the photochemical uptake of CO_2 increases. Consequently photosynthesis increases, so-called 'CO_2-fertilization' (e.g. Bazzaz, 1990).

Enhanced photosynthesis occurs mainly in plants with a C_3 photosynthetic pathway. This pathway is most common. The C_4 photosynthetic pathway (mainly in tropical grasses) is seldom CO_2-limited and these plants therefore show little or no reaction. Stomatal conductance largely controls the rate of gas diffusion through stomata and is directly proportional to stomatal pore width. In general, higher CO_2 concentrations reduce stomatal conductance and less water is lost through evapotranspiration (such reduction in stomatal conductance eventually leads to lower growth rates and is often recognized as a growth adaptation to higher CO_2 levels). Lower stomatal conductance improves water use efficiency (WUE) of plants and enhances plant growth under water-limited conditions (e.g. Bazzaz, 1990). This causes a shift of the distribution of plants into more arid regions. Both C_3 and C_4 plants show such enhanced WUE.

Increased atmospheric CO_2 concentrations could affect plants in various other ways as well. Elevated atmospheric CO_2 levels can increase plant growth at low light intensity, and under nutrient, SO_2 pollution or sodium excess stress (Melillo *et al.*, 1990; Mooney *et al.*, 1991). The response could be different for C_3 or C_4 species. The carbon allocation in plants changes as well. In trees, particularly, more carbon is sequestered in roots under these conditions (Melillo *et al.*, 1990; Norby *et al.*, 1992). Moreover, elevated CO_2 can affect tissue density, tissue quality (increasing C:N ratio), phenology and senescence (Melillo *et al.*, 1990). Many of these aspects of plant response to elevated atmospheric CO_2 levels determine competitive abilities of plants.

Other major components of an ecosystem involve the soil, humus and litter layers. Here, decomposition processes determine the rate and magnitude of nutrient, water and CO_2 fluxes. Soil respiration is directly influenced by climate through temperature and available soil moisture. Changes in climate will alter the processes and change the soil characteristics. This is an additional effect that could influence competitive capabilities of species and thus their abundances and functioning.

Increased atmospheric CO_2 levels have thus significant effects on major processes in ecosystems. Many ecosystem models that incorporate such ecophysiological processes only simulate the biogeochemical cycling of the major elements for major biomes (e.g. Melillo *et al.*, 1993). Such biogeochemical models are merely a scaling up of primary ecosystem processes, but do not consider change in species composition or biodiversity. They also often lack landscape processes such as vegetation dynamics and succession.

Little or no understanding of changes in biodiversity on the ecosystem level is generated from such ecophysiological studies. The actual understanding originates at the landscape level, which is more suitable for making ecosystem generalizations. The succession models, for example, are general enough to picture the broad spatial and temporal response pattern, but are incapable of simulating the succession and dynamics of single local stands, not even when the environmental variables are known in detail. These generalized response patterns are often erratically interpreted as being specific for the ecosystem level.

However, several recent studies that address the ecosystem level more clearly recognize the basic ecosystem processes that influence and alter ecosystem structure and functioning. Naeem *et al.* (1994), for example, analysed the consequences of different experimental ecosystems and showed that communities with a simpler structure were more vulnerable to changes. They concluded that less diverse communities could well be less efficient in assimilating resources. Körner (1993) defined different plant types which are likely to gain relative competitive advantages under increased CO_2 concentrations. Annuals, for example, are likely to undergo greater growth increment than deciduous perennials and trees, which in turn do better than evergreen perennials and trees. Young trees would grow best under changed conditions, followed closely by mature trees, but not by old trees. Early successional, large fruited, nitrogen-fixing and mycorrhizal species would gain additional competitive advantages over their counterparts. Ecosystems showing the largest response to CO_2 fertilization could be ranked from steppes and alpine systems to forests, from terrestrial to aquatic, from saline to fresh water, and from those in warm climates compared to cold climates. Although only few experimental studies present changes in competitive ability under environmental change, generalizations can easily be drawn as to which kinds of species are likely to become dominant and which become more vulnerable. These generalizations can, together with advanced experiments on species interactions at the ecosystem level (e.g. Naeem *et al.*, 1994), lead to an improved understanding of the impact of global change on biodiversity.

Solomon and Leemans (1990) extended such reasoning by linking it to assumed patterns of dynamic vegetation response. They evaluated the impact of rapid global change on European ecosystems and concluded that early successional species with broad geographical distributions and adequate seed dispersal and establishment would become dominant. The abundance of late successional, often shade tolerant, tree species would decline. Such decline could well be caused by the greater frequency of disturbances under such rapid change. This would result in a very open forested landscape, with a high abundance of species like birch. Such vegetation development would have large negative consequences on species richness, an important component of biodiversity, because many species confined to more mature forests could slowly disappear.

14.6 Linking global change and biodiversity

The preceding sections all highlight the modelling efforts assessing the impacts of global change on several hierarchical levels. These model approaches have guided us towards the major types of responses, the dynamics of these responses, and the linkages with environmental processes. Although these assessments give a clear impression of the type of impacts, their rates and magnitudes, little has been done to link these impacts directly with biodiversity. Due to the lack of quantification of biodiversity on these levels, the few comprehensive approaches that have been developed only approximate the potential impacts.

One of these approximations was based on using the location and size of large nature reserves. Databases on the status of nature reserves are compiled by the World Conservation Monitoring Centre (e.g. Groombridge, 1992) and about 6% of land can currently be classified as protected. These protected areas are islands where higher levels of natural biodiversity can be maintained. Due to more intensive land-use, construction, ownership and legal or other constraints outside of protected areas, endangered species living here have large difficulties in maintaining viable populations and habitats. These can only survive in the protected areas. The species can generally not adapt to environmental change by moving outside the reserve boundaries. If no dense network of protected areas with an integrating and connecting infrastructure exists for all relevant species (as is generally the case), protected areas have only limited possibilities to cope with global change, especially if the environmental conditions become less favourable for those populations of species and habitats that one wants to preserve and protect.

I have tried to address such threats to biodiversity by utilizing the location of large nature reserves, mainly protected because of their vegetation (Leemans, 1992b). 2500 reserves scattered world-wide over all vegetation zones were considered and a global vegetation model used to locate the reserves that were currently in one vegetation zone, while in another for a doubled CO_2-climate scenario. This simple evaluation showed that up to 50% of the reserves could become 'endangered' (51% for the most severe scenario, used in Plates 14.1, 14.2 (between pp. 214 & 215); Fig. 14.3). In these reserves, one vegetation type should be replaced by other, better adapted vegetation types at equilibrium with the new climatic conditions. The succeeding vegetation will probably not comply with the original conservation purposes of all these reserves. Future composition probably consists of more widespread or common species and could encompass an overall lower species diversity. Although these figures for single climate scenarios seem high, the reserves in regions where all plausible climate scenarios showed change (33%) or a similar change (17%) are still significant. This assessment showed large potential decreases in biodiversity.

Although such assessment highlights the importance of incorporating global change in biodiversity-at-risk assessments, they are still limited in their scope. The above example focuses only on climate change and does not consider changes in land-use and the subsequent changes in land-cover. Land-use change is an important determinant of GHG fluxes between the terrestrial biosphere and the atmosphere. It also strongly influences the biodiversity status of a region. Changes in land-use are governed by socio-economic behaviour (e.g. population pressures, political and legal structures, and technological level), and are influenced by the potential uses of land (crop productivities, tourism, etc.) and the biogeochemical and physical status of a patch of land (including climate, soil, vegetation and land degradation status).

Integrated assessment models, that simulate in a comprehensive way the processes and interactions between society, the biosphere, ocean and the atmosphere, also address land-use change. The most well-developed model hitherto is the Integrated Model to Assess the Greenhouse Effect (IMAGE: Alcamo, 1994). This model includes several modules that simulate changes in global land-cover by reconciling the regional demand for food, fibre and biomass (for energy generation). This demand is satisfied initially on current agricultural land, by evaluating the potential and actual productivity of that land. Regional demand is a function of demography, income and assumed dietary preferences. Potential productivity is a function of climate and soil characteristics, while actual productivity is set by regional technology and management levels. The modules dynamically simulate changes in land-cover patterns by expanding or contracting agricultural areas and these changes are used to determine land-use related emissions of GHGs. Large regional differences occur in land-cover dynamics through time. For example, the intensification of agriculture combined with a low population growth leads to contraction of agricultural land in the developed regions. Land becomes available for other uses, such as forestry (carbon sequestration), biomass production for energy generation (lower dependence on fossil fuels) and nature conservation. Latin America follows a similar pattern after a few decades with a gradual decline in the increase in agricultural extent. In many other developing regions, however, the presumed increase in agricultural productivity is insufficient to cope with the increasing population and changing dietary preferences (there is a tendency to consume more food, and especially more meat when people become more prosperous). In these regions agriculture therefore rapidly expands. Simulations with such a model show the importance of land-use change for GHG emissions from the terrestrial biosphere (Vloedbeld & Leemans, 1993).

Such land-use and land-cover change assessment can be used to update the impact on nature reserves. The earlier assessment showed that most of the 'endangered' nature reserves were located in temperate and boreal zones. However, adding land-use change, increases the threat by including also large parts of the tropical regions. Here, land-use change will strongly impact nature reserves (Plate 14.2). With increasing demands for natural resources, human encroachment of nature reserves will probably be difficult to halt.

14.7 Concluding remarks

Impacts of global environmental change will occur on all organizational levels and will involve individuals, species, communities, ecosystems, landscapes and biomes. Global change assessments should comprehensively integrate the causes, the processes and their feedbacks, and the impacts, without losing sight of the actual cause: human activities (energy- and land-use). All these human activities not only change the habitats of many species, but probably also lead to a decline in overall biodiversity. Such changes could well lead to less possibilities to sequester carbon on land (Naeem *et al.*, 1994), which will additionally accelerate the build-up of the concentrations of GHGs. It is on exactly this land interface where biodiversity and global change research should meet and result in improved understanding of ecosystem processes, their dynamics and directions. This should facilitate a scaling up or down to other organizational levels and, respectively, an integration with biodiversity issues and the more global systemic properties of global change.

The above presentation of issues that determine the impact of global change on biodiversity emphasizes that this interface has not yet been addressed adequately. This is, however, urgently needed, because past and current environmental changes already have had significant impacts on current biodiversity patterns. And, if no appropriate actions are taken, it can safely be assumed that with increased societal pressures, these effects will become more apparent and pronounced in the near future.

Acknowledgments

This paper was funded by the Dutch Ministry of Housing, Planning and the Environment under contract MAP410 to RIVM and the National Research Programme 'Global Air Pollution and Climate Change'. I appreciate useful comments on earlier drafts of the manuscripts from J. Wiertz, B. Walker, J. Latour and two anonymous reviewers.

References

Adams J.M., Faure H., Faure-Denard L., McClade J.M. & Woodward F.I. (1990) Increases in terrestrial carbon storage from the Last Glacial Maximum to the present. *Nature* **348**, 711–714. [14.3]

Alcamo J. (ed.) (1994) IMAGE 2.0: *Integrated Modeling of Global Climate Change*. Kluwer Academic Publishers, Dordrecht. [14.6]

Anonymous (1992) Convention on Biological Diversity. *Biol. Int.* **25**, 22–38. [14.1]

Bazzaz F.A. (1990) The response of natural ecosystems to the rising global CO_2 levels. *Annu. Rev. Ecol. Syst.* **21**, 167–196. [14.5]

Botkin D.B., Janak J.F. & Wallis J.R. (1972) Some ecological consequences of a computer model of forest growth. *J. Ecol.* **60**, 849–873. [14.4]

Carter T.R., Parry M.L., Harasawa H. & Nishioka S. (1994) *IPCC Technical Guidelines for Assessing Impacts of Climate Change.* IPCC Special Report CGER-1015-'94. Intergovernmental Panel on Climate Change, WMO and UNEP, Geneva. [14.3]

Cramer W. & Leemans R. (1993) Assessing impacts of climate change on vegetation using climate classification systems. In: *Vegetation Dynamics Modelling and Global Change* (eds A.M. Solomon & H.H. Shugart), pp. 190–217. Chapman & Hall, New York. [14.3]

Davis M.B. (1989) Lags in vegetation response to greenhouse warming. *Clim. Change* **15**, 75–82. [14.3]

Emanuel W.R., Shugart H.H. & Stevenson M.P. (1985) Climatic change and the broad-scale distribution of terrestrial ecosystems complexes. *Clim. Change* **7**, 29–43. [14.3]

Grisebach A.V.R. (1872) *Die Vegetation der Erde nach ihrer Klimatischen Anordning.* Engelman Verlag, Leipzig. [14.3]

Groombridge B. (ed.) (1992). *Global Biodiversity: status of the Earth's living resources.* Chapman & Hall, London. [14.2] [14.6]

HDP (1994) *Human Dimensions of Global Environmental Change Programme. Workplan 1994–1995.* Occasional Paper Number 6, September 1994. Human Dimensions of Global Environmental Change (ISSC-HDP), Geneva. [14.1]

Holdridge L.R. (1947) Determination of world plant formations from simple climatic data. *Science* **105**, 367–368. [14.3]

Houghton J.T., Callander B.A. & Varney S.K. (eds) (1992) *Climate Change 1992: the supplementary report to the IPCC scientific assessment.* Cambridge University Press. [14.1]

Huntley B. & Webb T. III (eds) (1988) *Vegetation History.* Kluwer, Dordrecht. [14.3]

IGBP (1994a) *IGBP in Action: work plan 1994–1998.* Report No. 28. February 1994. International Geosphere-Biosphere Programme, Stockholm. [14.1]

IGBP (1994b) *IGBP Global Modelling and Data Activities 1994–1998.* Report 30. May 1994. International Geosphere-Biosphere Programme, Stockholm. [14.1]

ISI (1991) *Current Content On Diskette® (CCOD) with abstracts for the Macintosh.* Institute for Scientific Information, Philadelphia. [14.2]

Kauppi P., Mielikäinen K. & Kuusela K. (1992) Biomass and carbon budget of European forests, 1971 to 1990. *Science* **256**, 70–74. [14.4]

Kolchugina T.P. & Vinson T.S. (1993) Carbon sources and sinks in forest biomes of the former Soviet Union. *Global Biogeochem. Cycles* **7**, 291–304. [14.4]

Körner C. (1993) CO_2 fertilization: the great uncertainty in future vegetation development. In: *Vegetation Dynamics and Global Change* (eds A.M. Solomon & H.H. Shugart), pp. 53–70. Chapman & Hall, New York. [14.5]

Larcher W. (1980) *Physiological Plant Ecology.* Springer-Verlag, Berlin. [14.5]

Lauenroth W.K., Urban D.L., Coffin D.P., Parton W.J., Shugart H.H., Kirchner T.B. & Smith T.M. (1993) Modeling vegetation structure-ecosystem process interactions across sites and ecosystems. *Ecol. Modelling* **67**, 49–80. [14.4]

Leemans R. (1992a) Modelling ecological and agricultural impacts of global change on a global scale. *J. Sci. Ind. Res.* **51**, 709–724. [14.3]

Leemans R. (1992b) Ecological and agricultural aspects of global change. In: *The Environmental Implications of Global Change.* (ed. Pernetta J., Leemans R., Elder D. & Humphrey S.), pp. 21–38. IUCN–The World Conservation Union, Gland, Switzerland. [14.6]

Leemans R. (1995) Incorporating land-use change in Earth system models. In: *Proceedings of the First IGBP–GCTE Science Conference* (ed. B. Walker). Cambridge University Press, in press. [14.1]

Leemans R. & Zuidema G. (1995) The importance of changing land use for global environmental change. *Trends Ecol. Evol.* **10**, 76–81. [14.1] [14.2]

Leemans R., Cramer W. & van Minnen J.G. (1995) Prediction of global biome distribution using bioclimatic equilibrium models. In: *Effects of Global Change on Coniferous Forests and Grassland* (eds J.M. Melillo & A. Breymeyer). Wiley, New York, in press. [14.3]

Leggett J., Pepper W.J. & Swart R.J. (1992) Emissions Scenarios for the IPCC: an update. In: *Climate Change 1992: the supplementary report to the IPCC scientific assessment* (eds J.T. Houghton, B.A. Callander & S.K. Varney), pp. 71–95. Cambridge University Press. [14.4]

MacArthur R.H. (1972) *Geographical Ecology: patterns in the distribution of species*. Princeton University Press, New Jersey. [14.2]

Markham A., Dudley N. & Stolton S. (1993) *Some Like it Hot: climate change, biodiversity and the survival of species*. WWF International, Gland, Switzerland. [14.3]

May R.M. (1992) Past efforts and future prospects towards understanding how many species there are. In: *Biodiversity and Global Change* (eds O.T. Solbrig, H.M. van Emden & P.G.W.J. van Oort), pp. 71–81. The International Union of Biological Sciences (IUBS), Paris. [14.2]

Melillo J.M., Callaghan T.V., Woodward F.I. & Salati E. (1990). Effects on ecosystems. In: *Climate Change: the IPCC scientific assessment* (eds J.T. Houghton, G.J. Jenkins & J.J. Ephraums), pp. 283–310. Cambridge University Press. [14.5]

Melillo J.M., McGuire A.D., Kicklighter D.W., Moore III B., Vorosmarty C.J. & Schloss A.L. (1993) Global climate change and terrestrial net primary production. *Nature* **363**, 234–239. [14.5]

Mooney H.A., Drake B.G., Luxmoore R.J., Oechel W.C. & Pitelka L.F. (1991) Predicting ecosystems responses to elevated CO_2 concentrations. *Bioscience* **41**, 96–104. [14.5]

Naeem S., Thompson L.J., Lawler S.P., Lawton J.H. & Woodfin R.M. (1994) Declining biodiversity can alter the performance of ecosystems. *Nature* **368**, 734–737. [14.5] [14.7]

Neilson R.P. (1993) Vegetation redistribution: a possible biosphere source of CO_2 during climatic change. *Water Air Soil Pollut.* **70**, 659–673. [14.3] [14.4]

Norby R.J., Gunderson C.A., Wullschleger S.D., E.G., O.N. & McCracken M.K. (1992) Productivity and compensatory responses of yellow-poplar trees in elevated CO_2. *Nature* **357**, 322–324. [14.5]

Peng C. (1994) '*Reconstruction of the Past Terrestrial Carbon Storage from Pollen Data and Biosphere Models since the Last Glacial Maximum*.' PhD thesis, Universite de Droit, d'Economie et des Sciences d'Aix, Marseille. [14.3]

Pernetta J., Leemans R., Elder D. & Humphrey S. (ed.) (1994) *Impacts of Climate Change on Ecosystems and Species: implications for protected areas*. International Union for Conservation of Nature and Natural Resources (IUCN), Gland, Switzerland. [14.5]

Peters R.L. & Lovejoy T.E. (eds) (1992) *Global Warming and Biological Diversity*. Yale University Press, New Haven. [14.1] [14.2] [14.5]

Prentice I.C., Bartlein P.J. & Webb III T. (1991) Vegetation and climate change in eastern North America since the last glacial maximum. *Ecology* **72**, 2038–2056. [14.3]

Prentice I.C., Cramer W., Harrison S.P., Leemans R., Monserud R.A. & Solomon A.M. (1992) A global biome model based on plant physiology and dominance, soil properties and climate. *J. Biogeogr.* **19**, 117–134. [14.3]

Prentice I.C., Sykes M.T. & Cramer W. (1993) A simulation model for the transient effects of climate change on forest landscapes. *Ecol. Model.* **65**, 51–70. [14.4]

Prentice I.C., Sykes M.T., Lautenschlager M., Harrison S.P., Denissenko O. & Bartlein P.J. (1994) Modelling global vegetation patterns and terrestrial carbon storage at the last glacial maximum. *Global Ecol. Biogeogr. Lett.* **3**, 67–76. [14.3]

Prentice K.C. & Fung I.Y. (1990) Bioclimatic simulations test the sensitivity of terrestrial carbon storage to perturbed climates. *Nature* **346**, 48–51. [14.3]

Prinn R.G. (1994) The interactive atmosphere: global atmospheric-biospheric chemistry. *Ambio.* **23**, 50–61. [14.1]

Schulze E.-D. & Mooney H.A. (eds) (1993) *Biodiversity and Ecosystem Function*. Springer-Verlag, Berlin. [14.2]

Shugart H.H., Leemans R. & Bonan G.B. (eds) (1992) *A Systems Analysis of the Global Boreal Forest*. Cambridge University Press. [14.4]

Siegentaler U. & Sarmiento J.L. (1993) Atmospheric carbon dioxide and the ocean. *Nature* **365**, 119–125. [14.1]

Smith T.M. & Shugart H.H. (1993). The potential response of global terrestrical carbon storage to a climate change. *Water Air Soil. Pollut.* **70**, 629–642. [14.4]

Smith T.M., Leemans R. & Shugart H.H. (eds) (1994). *The Application of Patch Models of Vegetation Dynamics to Global Change Issues*. Kluwer, Dordrecht. [14.4]

Solbrig O.T. (ed.) (1991) *From Genes to Ecosystems: a research agenda for biodiversity*. The International Union of Biological Sciences (IUBS), Paris. [14.2]

Solbrig O.T., van Emden H.M. & van Oort P.G.W.J. (eds) (1992) *Biodiversity and Global Change*. The International Union of Biological Sciences (IUBS), Paris. [14.2]

Solomon A.M. (1986) Transient responses of forests to CO_2-induced climate change: simulation modeling in eastern North America. *Oecologia* **68**, 567–579. [14.4]

Solomon A.M. & Leemans R. (1990) Climatic change and landscape-ecological response: issues and analyses. In: *Landscape Ecological Impact of Climatic Change* (eds M.M. Boer & R.S. de Groot), pp. 293–317. IOS Press, Amsterdam. [14.5]

Turner II B.L., Kasperson R.E., Meyer W.B., Dow K.M., Golding D., Kasperson J.X., Mitchell R.C. & Ratick S.J. (1990) Two types of global environmental change: definitional and spatial-scale issues in their human dimensions. *Global Environ. Change* **1**, 14–22. [14.1]

Urban D.L., Hansen A.J., Wallin D.O. & Halpin P.N. (1992) Life-history attributes and biodiversity. Scaling implications for global change. In: *Biodiversity and Global Change* (eds O.T. Solbrig, H.M. van Emden & P.G.W.J. van Oort), pp. 173–195. The International Union of Biological Sciences (IUBS), Paris. [14.4]

Van Campo E., Guiot J. & Peng C. (1993) A data-based re-appraisal of the terrestrial carbon budget at the last glacial maximum. *Global Planet. Change* **8**, 189–201. [14.3]

Vloedbeld M. & Leemans R. (1993) Quantifying feedback processes in the response of the terrestrial carbon cycle to global change — the modeling approach of image-2. *Water Air Soil. Pollut.* **70**, 615–628. [14.6]

Walter H. (1985) *Vegetation of the Earth and Ecological Systems of the Geo-Biosphere*. Springer-Verlag, Berlin. [14.3]

Watson R.T., Meira Filho L.G., Sanheuza E. & Janetos A. (1992) Greenhouse gases: sources and sinks. In: *Climate Change 1992: the supplementary report to the IPCC scientific assessment* (eds J.T. Houghton, B.A. Callander & S.K. Varney), pp. 27–46. Cambridge University Press. [14.1]

Wilson E.O. & Peter F.M. (eds) (1988) *Biodiversity*. National Academy Press, Washington DC. [14.2]

Woodward F.I. (1987) *Climate and Plant Distribution*. Cambridge University Press. [14.3]

Index